The Origin of Life and Evolutionary Biochemistry

ALEKSANDR IVANOVICH OPARIN

A Volume Commemorating the Fiftieth Anniversary of the Publication of Proiskhozhdenie Zhizni *and the Eightieth Birthday of A. I. Oparin*

The Origin of Life and Evolutionary Biochemistry

Edited by

K. Dose

Institut für Biochemie
Universität Mainz
Mainz, Germany

S. W. Fox

Institute for Molecular and Cellular Evolution
University of Miami
Coral Gables, Florida

G. A. Deborin

A. N. Bakh Institute of Biochemistry
Academy of Sciences of the USSR
Moscow, USSR

and

T. E. Pavlovskaya

A. N. Bakh Institute of Biochemistry
Academy of Sciences of the USSR
Moscow, USSR

PLENUM PRESS • NEW YORK AND LONDON

ITHACA COLLEGE LIBRARY

Library of Congress Cataloging in Publication Data

Main entry under title:

The origin of life and evolutionary biochemistry.

Includes bibliographies.
1. Chemical evolution. 2. Life — Origin. I. Dose, Klaus, ed. [DNLM: 1. Biochemistry. 2. Biogenesis. QH325 067]
QH325.069 577 74-10703
ISBN 0-306-30811-8

© 1974 Plenum Press, New York
A Division of Plenum Publishing Corporation
227 West 17th Street, New York, N.Y. 10011

United Kingdom edition published by Plenum Press, London
A Division of Plenum Publishing Company, Ltd.
4a Lower John Street, London W1R 3PD, England

All rights reserved

No part of this book may be reproduced, stored in a retrieval system, or transmitted, in any form or by any means, electronic, mechanical, photocopying, microfilming, recording, or otherwise, without written permission from the Publisher

Printed in the United States of America

CONTENTS

HISTORICAL INTRODUCTION:

A. I. OPARIN AND THE ORIGIN OF LIFE

A. I. OPARIN AND THE ORIGIN OF LIFE

The authors of papers in this volume have joined in honoring the pioneering contributions of Aleksandr Ivanovich Oparin, whose first written work appeared half-a-century ago, in 1924. In addition, many others who have been prominent in this field of inquiry in recent years enthusiastically endorsed the preparation of a volume honoring the pioneering ideas and the pioneer.

The problem of the origin of life has intrigued man throughout recorded history. Until 1922, however, when the young Russian biochemist, A. I. Oparin, orally presented to a meeting of the Russian Botanical Society the thesis that the origin of life is an event governed by natural laws and is an inalienably integral part of the total process of the evolution of the Universe, this field was generally considered as belonging to religious faith rather than to natural science.

Oparin was the first to emphasize that the evolution of carbonaceous compounds that occurred on the primitive Earth had to lead to the abiogenic synthesis of organic compounds--the components of living systems. This bold yet rational thesis was received with interest by the scientific audience. Nevertheless, at that time people could hardly imagine the whole significance of Oparin's concepts.

The first booklet titled "Proiskhozdenie zhizni" (1924) has outlined a program of scientific research, thus demonstrating that the problem of the origin of life is as accessible to observational and experimental approach as any other field of natural science. This blueprint stimulated many scientists in various fields in different countries to initiate and develop broad researches in biochemistry, paleontology, geology, chemistry, astronomy, etc. and to relate their studies to the new scientific discipline of the origin of life.

To our best knowledge, fourteen years elapsed before experiments expressly testing the concept of the origin of life-related substances were published in the scientific literature. The author of those first experiments is one of the contributors to this volume (p. 143). Substantial public awareness of prospects for experimental progress required yet another fifteen years. In the following twenty years many advances were recorded, and these will be better judged in additional years to come. Today the active serious interest in this area of inquiry is widespread among scientists and the general public both.

After presenting a comprehensive outline of a theory of the origin of life, Oparin has continued to develop these concepts through many books and other publications in the years since 1924. For these refinements, he has drawn from many disciplines and from experiments that have been performed as direct or indirect consequences of his conceptualizations.

In addition to performing the formidable task of reversing the negative view that prevailed since the time of Pasteur, A. I. Oparin has initiated many novel ideas and emphases. He was perhaps the first to state clearly that, "The most important and essential characteristic of organisms is, as is demonstrated by their very name, their organization, their particular form or structure." In support of this emphasis, Oparin has formulated also the concept that at a certain stage of the evolution the appearance of phase-separated systems of molecules was necessary. This was the event that opened the opportunity for further chemical as well as biological evolution of individual systems. Oparin thus clearly recognized that life is characterized as multimolecular systems, to which the component molecules, whichever is considered, can be only a less-than-the-whole component. The emphasis on the crucial structural organization of living systems runs throughout his works.

Nor is it sufficient that life be viewed as systems. Oparin, adapting the ideas of Prigogine and others to the origin-of-life problem, has emphasized the importance of recognizing that cells, including the first, have been thermodynamically open systems; they are not therefore described by the second law of thermodynamics, which applies to closed systems in which energy is not replenished. This view was clearly stated in 1957, and earlier. In a way, it is amazing that continued restatement is necessary.

The recognition that the properties of living units had their roots in other forms of matter was stated by Oparin, in 1924, as, "The specific peculiarity of living organisms is only that in them there have been collected and integrated an extremely complicated combination of a large number of properties and characteristics which are present in isolation in various dead, inorganic bodies. Life is not characterized by any special properties but by a definite, specific combination of these properties." It is in this way that Oparin has viewed the first "living" structure, and its elegant offspring, as a dramatic punctuation mark in the successive transformations of matter.

Oparin was also very early in emphasizing that enzymes are meaningful in the context of cells, rather than in the dilute aqueous solutions from which biochemists have so often unjustifiably extrapolated their data, and which they still use. His comprehensive understanding of this issue is symbolized by the title of his book, "Enzyme Action in the Living Cell," published in 1934.

Another contribution of Oparin was the early recognition that the first organism was a heterotroph. This was a bold resolution of a conceptual difficulty at a time when the "commonsense" view stressing an initial autotroph was popular. It is easy to declare now that aggregation of macromolecules to a protocell was itself an act of heterotrophism, but the basic idea was in the twenties a break with well-entrenched thinking, and a masterful insight into the illogic of a "logical" view.

The genius of Oparin has lain not only in his endless attention to detail (for problems with seemingly endless interface), to his unflagging devotion to scientific concepts that had long been treated otherwise, and to his uncanny ability to see relationships that existed in less ostensible ways in earlier periods in the history of the Earth. His genius has lain also in his openmindedness, as some of us who have discussed opposing views with him, have learned. Oparin has never been a dogmatist in his field; he has exemplified the old observation that he who is most often correct is he who is readiest himself to recognize that he may, on some individual issue, be incorrect. However one analyzes the attitudes, the product is undeniable.

A. I. Oparin is not only an eminent scientist but also an outstanding teacher. He has trained many distinguished biochemists both at Moscow University, where he has taught since 1917, and at the A. N. Bakh Institute of Biochemistry, established by him together with A. N. Bakh in 1935.

The distinguished merits of A. I. Oparin in the formulation of a materialistic theory of the origin of life have been widely recognized. After the formation of the International Society for the Study of the Origin of Life, he became its first president. He is a member and doctor honoris causa of many academies and scientific societies.

In early 1974, his works on the theory of the origin of life have again been honored with the most prestigious national prize of the USSR for scientific and artistic activities--the Lenin Prize.

Today in 1974, the authors and editors of this book are privileged and pleased to honor the fiftieth anniversary of the first publication of A. I. Oparin's theory, an anniversary which coincides with his eightieth birthyear.

S. W. Fox

G. A. Deborin

K. Dose

T. E. Pavlovskaya

CHAPTERS IN HONOR OF

"PROISKHOZHDENIE ZHIZNI"

and

A. I. OPARIN

PROTEIN STRUCTURE AND THE MOLECULAR EVOLUTION OF BIOLOGICAL ENERGY CONVERSION

Herrick Baltscheffsky

Department of Biochemistry, Arrhenius Laboratory

University of Stockholm, S-104 05, Stockholm, Sweden

INTRODUCTION

Already half a century ago, in his original version of Proiskhozhdenie zhizny (The Origin of Life), Oparin (1) was able to bring for the first time the central human problem on how life on Earth originated to the fundamental, molecular level. At that time, in 1924, biochemistry was still very young. Its molecular biology branch was not yet born. No enzyme crystal had yet been obtained. The energy-rich phosphate concept did not yet exist, nor did that of a replicating nucleic acid.

"The specific peculiarity of living organisms is only that in them there have been collected and integrated an extremely complicated combination of a large number of properties and characteristics which are present in isolation in various dead, inorganic bodies. Life is not characterized by any special properties but by a definite, specific combination of these properties." This early statement by Oparin (1) may be added to his detailed arguments about the prebiotic origin of organic carbon compounds, including those containing nitrogen, from simple inorganic molecules and his emphasis on carbohydrates and proteins: "In combination with other and yet more complicated substances they are, as it were, the foundation of life." When taken together, these opinions forecast our more recently acquired knowledge about the molecular origin and essence of life, the interaction between nucleotide and peptide. In our time the problem of the origin and the early evolution of life can be formulated in impressive molecular and atomic detail, based upon the new information about polynucleotide and polypeptide structure and function (2-10).

"However, in the course of this process of change, selection of the better organized bits of gel was always going on. It is true that the less well organized could grow alongside the more efficient but they must have soon stopped growing." Today, 50 years after Oparin (1) made this statement, his extension of the Darwinian principle about the survival of the fittest to the prebiological part of molecular evolution appears as well-founded as ever. Today it is possible to treat and investigate it, both at the prebiological and the biological levels, from a theoretical (5) as well as an experimental (7,9) viewpoint. In Eigen's (5) penetrating study on the self-organization of matter and the evolution of biological macromolecules one finds the conclusion that any competition between different types of replicating systems inevitably leads to the survival of only one system, "the fittest."

The plausible assumption that what we know as life on Earth began with a functioning genetic nucleic acid code does not immediately lead to the corollary that all life on Earth has a monophyletic origin. But it is an attractive hypothesis, shown in Fig. 1a, that such a common origin exists in the first surviving (surviving from generation to generation, during biological evolution of all the known species) nucleotide system performing replication as well as translation, as expressed primarily in amino acid sequence of formed peptide. On the other hand, the possibility that present-day organisms should trace their ancestry back to a situation with parallel origin and early evolution of more than one such system, all based on the same nucleotide-peptide principle, should not be excluded (Fig. 1b), as will be discussed below.

ON SOME RECENT DEVELOPMENTS

Both of the alternatives mentioned above are of immediate interest in connection with some new and very rapid developments relating to the field of molecular evolution in biological energy conversion. Before treating this progress occurring at the level of protein structure it will be necessary to present as a background some other recent developments.

Broda (11,12) has pioneered the more detailed treatment of this research area in its broad and general evolutionary aspects.

With respect to the conservation of energy in energy-rich phosphate compounds, I, some years ago, found biochemical support for a proposal that inorganic pyrophosphate (PPi) is a likely link between prebiological and early biological energy conversion (13,14). This was based upon our discoveries that PPi can be formed at the expense of light energy in bacterial photophosphorylation (15,16) and the subsequent finding that it can be utilized as a donor of biologically useful chemical energy (17,18). Also, high molecular weight in-

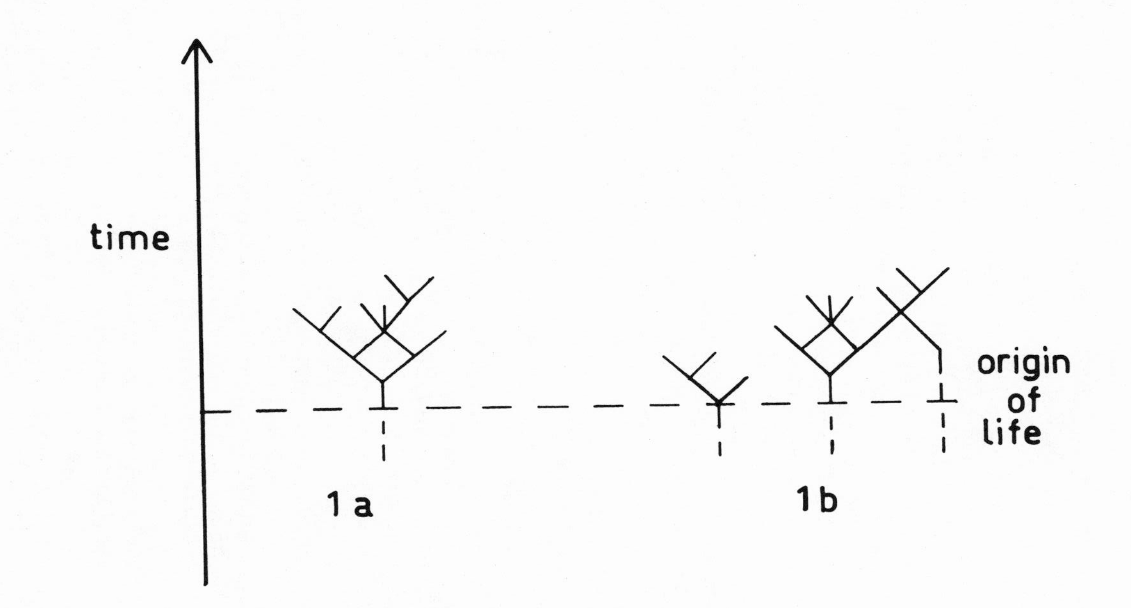

Figure 1. Origin and early evolution of life. (1a) Monophyletic origin; (1b) diphyletic (as an example of polyphyletic) origin.

organic polyphosphates were suggested, by Kulaev (19,20), as such early energy-rich phosphates, based upon his important results with enzyme reactions of bacterial glycolytic systems.

With respect to the electron transport part of biological electron transport phosphorylation systems, I have very recently proposed a hypothesis for the origin and evolution of biological electron transport chains (21,22,23). It was founded on the collected knowledge about the protein parts of the electron carriers in question (iron-sulfur proteins, heme proteins, flavoproteins, and NAD(P)-binding proteins) rather than on coenzymes and prosthetic groups. According to this new hypothesis, the evolution of cellular electron transport has originated with ferredoxin-like molecules at the hydrogen electrode level of the electrochemical potential scale and then proceeded in various advantageous directions, resulting in close evolutionary relationships between various neighbor links in the chains thus formed (22). This possibility was long hidden by the fact that very much information existed about the differences between the coenzymes and prosthetic groups involved, but very little about the corresponding proteins and their possible structural similarities and evolutionary relationships.

GENERAL COMMENTS ON CURRENT DEVELOPMENT

Earlier attention to the molecular evolution of biological energy conversion naturally focussed on available information about the energy transfer at the level of energy-rich phosphates, involving both glycolysis and electron transport phosphorylation reactions. The recent hypothesis on the origin and evolution of biological electron transport was based on the great achievements of many research groups which had been able to determine primary, secondary, and tertiary protein structures of various electron carriers and groups (22). And today, mainly due to the progress in X-ray crystallographic structure determination of proteins, by Rossmann and his group (24, 25), Branden and his group (26), Watson and his group (27), Schulz and his group (28), Blake and Evans (29), and many others, a new molecular pattern of protein structure is beginning to emerge. This would appear to allow soon a deeper understanding of both evolution and mechanism in biological energy conversion. Indeed, it appears to amount to a new integration at the level of polypeptide structure and function over large areas of the entire field of biological energy conversion - to be more specific, an integration with respect to the evolution and function of common secondary structural features of proteins involved in electron transport and of proteins involved in phosphorylation!

In its remaining part, the presentation will try to elucidate this remarkable new perspective on traceable common ancestry in different parts of the pathways for cellular energy transduction. However, restriction to a few selected examples will be necessary. A brief discussion of my hypothesis on the origin and evolution of biological electron transport and some alternative possibilities will be connected with the suggestion of Orgel (8) concerning a possible step in the origin of the genetic code, and related to the "super - secondary structures" of proteins first described by Rao and Rossmann (3). This will lead to the apparent evolutionary link that these structural domains seem to constitute between proteins with different functional roles.

The current picture appears to develop at an increasing pace. It provides at the fundamental level of protein structure the strongest detailed support for the early evolutionary insights of Berzelius (31), Darwin (32), and Oparin (1). The earlier approach with comparisons of the same protein from different organisms is now extended to include different kinds of proteins. This new line of investigation seems to give already increased momentum to the search for evolutionary relationships and molecular mechanisms in the field of biological energy conversion.

ORIGIN AND EVOLUTION OF PROTEINS IN ELECTRON TRANSPORT

The biochemical unity of metabolism is a well-known concept, long realized from investigations of relatively simple organic molecules, such as substrates and products, coenzymes and prosthetic groups, participating in the reactions of living cells. More recent studies, at the level of protein structure, have led to constructions of detailed evolutionary trees for several proteins, including, for example, the iron-containing electron carriers cytochrome c (33) and ferredoxin (34).

Certain bacterial ferredoxins might well relate closely to structures of some of the earliest polypeptides formed by nucleotide-peptide interaction during the first billion years in the history of the Earth, when life originated (35,36). According to Hall et al. (35) "ferredoxins may have been among the simplest proteins, being first formed as a short polypeptide chain of twenty-six amino acids comprising only nine different amino acids which are easily formed in simulated primaeval conditions". The demonstration of biological activity in such "half-ferredoxin" molecules by Orme-Johnson (37) is in line with this assumption. Another remarkable piece of support is, as Hall (35,38) points out, the identity of the most common amino acids in simple bacterial ferredoxins with those found in the Murchison meteorite by Ponnamperuma and his coworkers (39,40), as well as in other celestial bodies.

Hall limited his treatment of evolutionary relationships among biological electron carriers to the iron-sulfur proteins. The recent hypothesis for the evolution of biological electron transport chains

(22) goes one step further. It proposes that common ancestry exists and can be demonstrated between carriers from different groups of proteins, such as (1) iron-sulfur proteins and cytochromes, or (2) flavoproteins and NAD-linked proteins, or even (3) all of these four groups. I consider the experimental evidence in support of (1) and (2) to be rather strong, whereas that in support of (3) is still very weak (22,23).

Rao and Rossman (3) discovered similar super-secondary structures or domains in an FMN-containing flavoprotein, Clostridium MP flavodoxin, and an NAD-binding protein, lactate dehydrogenase (LD) from dog-fish. Such a domain consists of about 60 amino acids forming a three stranded parallel β-pleated sheet with connecting α-helices. This occurs once in flavodoxin, which binds a mononucleotide and twice in LD, which binds a dinucleotide. The fact that a rather similar domain also was found in subtilisin lead to their assumption that this structural similarity was an expression of functional convergence rather than evolutionary divergence. When Branden et al. (26) determined the structure of liver alcohol dehydrogenase (LAD), they observed that it bound the coenzyme NAD to a similar six stranded parallel β-structure as found in LD, and predicted that this substructure may be a general one for binding of nucleotides and, in particular, NAD. The concept and its apparent evolutionary significance have now been further developed with respect to flavodoxin and several NAD-linked dehydrogenases by Buehner et al. (41). This is in excellent principal agreement with my hypothesis concerning evolutionary relationships between electron carriers with different coenzymes or prosthetic groups (22).

Two quite separate items should be considered in connection with the finding that parallel β-sheet structures may be very common in nucleotide-binding proteins. The first is the suggestion by Orgel (8) that the earliest coding polynucleotides contained a high proportion of alternating sequences of purines and pyrimidines, and that these sequences coded for polypeptides in which hydrophobic and hydrophilic amino acids alternated. Such alternating polypeptides may form asymmetric β-sheets with one hydrophobic and one hydrophilic surface (8). Thus, mutually reinforcing relationships may have emerged at a very early stage of an evolving interaction mechanism between nucleotide and peptide; the earliest coding polynucleotides by producing suitable β-sheet structures, may have given the very type of polypeptide secondary structure which has nucleotide binding property! Nucleotide binding may thus have evolved very early, either separately, or in ancestral relationship with the binding of iron-sulfur clusters. In other words (compare Fig. 1), is there a mono- or a di-phyletic molecular origin of biological electron transport? Both these alternatives appear much more attractive and probable today than a third alternative that does not reckon with any evolutionary relationship at all between different groups of proteins. Although the whole spectrum of alternatives should perhaps still be considered, I tend to believe that a common ancestry may eventually be traced for all polypeptide structures involved in biological electron transport, as well as many, and possibly even most, other polypeptide structures of living cells.

Determination of primary structures will also continue to be extremely important in the analysis of polypeptide evolution. However, the first steps of the second item, the remarkable integration at the structural level of proteins with differing functions in energy transduction, resulted from studies on their three-dimensional properties.

SIMILARITIES BETWEEN STRUCTURAL DOMAINS OF PROTEINS IN OXIDATION AND IN PHOSPHORYLATION REACTIONS

As has been vividly expressed by Rossman (25), the last year has been witnessing a breakthrough in our understanding of nucleotide binding in proteins. Not only many oxidation-reduction reactions, but also many phosphorylation reactions, involve nucleotide. To make only a brief comment; myokinase (28), phosphoglycerate kinase (27,29), and possibly also hexokinase (42), seem to contain nearly identical domains for binding the adenosine phosphates, as the nucleotide binding dehydrogenases mentioned above. The evolutionary implications of these findings can only be imagined at the present time. It may suffice here to cite Blake and Evans (29): "The possibility of an evolutionary relationship in the nucleotide binding units of two such important groups of enzymes as the dehydrogenases and the kinases, further suggests that other classes of enzymes may also contain this basic unit. The observed specificity of this unit is apparently for the adenosine group of the co-factors. This group occurs not only in NAD and ATP, but also in FAD, and, as the 3' - phosphate, in co-enzyme A. In view of the present results, it seems possible that the presence of adenosine in these four important co-factors may be a consequence of a mutually conservative evolutionary history with a complementary nucleotide binding protein unit."

AN OUTLOOK

A number of questions that arise about the possibilities in various directions may be inherent in the current patternization at the level of protein structural relationships in biological energy conversion.

The arguments of Horowitz (43) in support of the concepts that all the genes of an operon are descendants of a common ancestral gene and that they originated by tandem duplication, followed by functional differentiation, are of interest also in connection with the metabolic pathways of major cellular energy conversion processes. Whether an operon arrangement exists in the genome for any of the sequences of proteins in electron-transport chains or in pathways of phosphorylation reactions in glycolysis or emerging from "coupling sites" of electron transport chains, I consider the idea of traceable evolution along these reaction sequences as a most useful working hypothesis. At least it may serve as a point of embarkation for experiments along new lines of investigation. It may be appropriate to mention three such lines which could contribute towards the further elucidation of evolutionary relationships, as well as reaction mechanisms, in biological electron transport and phosphorylation. The first concerns electron transport, the second electron transport phosphorylation and the pre-nucleotide level, and the third electron transport phosphorylation of ADP to ATP.

1. Investigation of the structure of solubilized succinate dehydrogenase from bovine heart (44,45) and _Rhodospirillum rubrum_ chromatophores (46). The enzyme contains, on the same polypeptide, binding sites for the dinucleotide FAD and 4 atoms each of iron and "labile sulfur." Details on binding of FAD to polypeptide are still unknown. Also NADH dehydrogenase is interesting in these respects.

2. Investigation of the membrane-bound inorganic pyrophosphatase from photosynthetic bacteria. This enzyme, which has not yet been solubilized, appears to catalyze the final reaction in light-induced phosphorylation of orthophosphate to pyrophosphate (15,47). It may contain some of the common domain in nucleotide-binding proteins. The soluble enzyme from _R. rubrum_ is inhibited by ATP and NADH, apparently allosterically (48,49).

3. Investigation of the "coupling factor" ATPase substructures which are very similar in animals, plants and bacteria (50,51), with respect to primary, secondary, tertiary, and quaternary structure, and to binding of nucleotide.

Success along any of these suggested lines of investigation should, in the last analysis, contribute greatly to our understanding of evolutionary relationships in biological energy conversion.

ADDED IN PROOF

A very recent publication [Carter, C. W., and Kraut, J., _Proc. Nat. Acad. Sci. U.S._ _71_, 283 (1974)] provides new information on β-sheet structures in proteins and their structural complementarity to double-stranded RNA as well as a detailed hypothesis for the earliest events of biological evolution, based upon this complementarity and upon the assumption that both members of the proposed RNA-polypeptide complex can act as primordial "polymerases" for one another.

REFERENCES

1. Oparin, A. I., "Proiskhozhdenie zhizni" ("Origin of Life"), Izd. Moskovskii rabochii, Moskva, 1924.
2. Crick, F.H.C., J. Mol. Biol. 38, 367 (1968).
3. Orgel, L. E., J. Mol. Biol. 38, 381 (1968).
4. Watson, J. D., "The Molecular Biology of the Gene," W. A. Benjamin, Inc., New York, 1970.
5. Eigen, M., Naturwissenschaften 58, 465 (1971).
6. Eigen, M., Quart. Revs. Biophys. 4, 149 (1971).
7. Spiegelman, S., Quart. Revs. Biophys. 4, 213 (1971).
8. Orgel, L. E., Israel J. Chem. 10, 287 (1972).
9. Mills, D. R., Kramer, F. R., and Spiegelman, S., Science 180, 916 (1973).
10. Miller, S. L. and Orgel, L. E., "The Origins of Life on the Earth," Prentice-Hall, Inc., Englewood Cliffs, 1974.
11. Broda, E., Progr. Biophys. Mol. Biol. 21, 143 (1970).
12. Broda, E., "Molecular Evolution," Vol. 2 (Schoffeniels, E., ed.), pp. 224-235, North-Holland, Amsterdam, 1971.
13. Baltscheffsky, H., Acta Chem. Scand. 21, 1973 (1967).
14. Baltscheffsky, H., "Molecular Evolution," Vol. 1 (Buvet, R., and Ponnamperuma, C., eds.), North-Holland, Amsterdam, 1971.
15. Baltscheffsky, H., von Stedingk, L.-V., Heldt, H.-W., and Klingenberg, M., Science 153, 1120 (1966).
16. Baltscheffsky, H., and von Stedingk, L.-V., Biochem. Biophys. Res. Commun. 22, 722 (1966).
17. Baltscheffsky, M., Nature 216, 241 (1967).
18. Baltscheffsky, M., Baltscheffsky, H., and von Stedingk, L.-V., Brookhaven Symp. Biol. 19, 246 (1967).
19. Kulaev, I. S., Szymona, O., and Bobyk, M. A., Biokhimiya 33, 419 (1968).

20. Kulaev, I.S., "Molecular Evolution," Vol. 1 (Buvet, R., and Ponnamperuma, C., eds.), North-Holland, Amsterdam, 1971.

21. Baltscheffsky, H., "VI International Congress of Photobiology," Bochum, Abstracts, no. 368, 1972.

22. Baltscheffsky, H., Origins of Life 5, 1974 (in press).

23. Baltscheffsky, H., "Dynamics of Energy Transducing Membranes," in press, Elsevier, Amsterdam, 1974.

24. Rossmann, M.G., Moras, D., and Olsen, K. W., submitted to Nature.

25. Rossmann, M.G., New Scientist 61, 266 (1974).

26. Branden, C.-I., Eklund, H., Nordström, B., Boiwe, T., Söderlund, G., Zeppezauer, E., Ohlsson, I., and Akerson, A. Proc. Nat. Acad. Sci. U.S. 70, 2439 (1973).

27. Bryant, T.N., Watson, H.C., and Wendell, P.L., Nature 247, 14 (1974).

28. Schulz, G.E., Elzinga, M., Marx, F., and Schirmer, R.H., submitted to Nature.

29. Blake, C.C.F., and Evans, P.R., J. Mol. Biol., in press.

30. Rao, S.T., and Rossman, M.G., J. Mol. Biol. 76, 241 (1973).

31. Berzelius, J.J., "Föreläsningar i Djurkemien," Vol. 1, p. 3, Carl Delen, Stockholm, 1806.

32. Darwin, C., "On the Origin of Species," Cloves and Sons, London, 1860.

33. Fitch, W.M., and Margoliash, E., Science 155, 279 (1967).

34. Dayhoff, M.O., "Atlas of Protein Sequence and Structure," Vol. 5, p. 392, The National Biomedical Research Foundation, Silver Springs,1972.

35. Hall, D.O., Cammack, R., and Rao, K.K., Nature 233, 136 (1971).

36. Hall, D.O., Cammack, R., and Rao, K.K., "Theory and Experiment in Exobiology," Vol. 2 (Schwartz, A.W., ed.), pp. 67-85, Wolters-Noordhoff Publ., Groningen, 1973.

37. Orme-Johnson, W.H., Biochem. Soc. Transact. 1, 30 (1973).

38. Hall, D.O., Origins of Life 5, in press, (1974).

39. Kvenvolden, K., Lawless, J., Pering, K., Peterson, E., Flores, J., Ponnamperuma, C., Kaplan, I., and Moore, C., Nature 228, 923 (1970)

40. Ponnamperuma, C., Quart. Revs. Biophys. 4, 77 (1971).

41. Buehner, M., Ford, G.C., Moras, D., Olsen, K.W., and Rossmann, M.G., Proc. Nat. Acad. Sci. U.S. 70, 3052 (1973).

42. Steitz, T.A., Fletterick, R.J., and Hwang, K.J., J. Mol. Biol. 78, 551 (1973).

43. Horowitz, N.H. "Evolving Genes and Proteins" (Bryson, J., and Vogel, H.J.,eds.), pp. 15-23, Academic Press, New York, 1965.

44. Davis, K.A., and Hatefi, Y., Biochemistry 10, 2509 (1971).

45. Hanstein, W.G., Davis, K.A., Ghalambor, M.A., and Hatefi, Y., Biochemistry 10, 2517 (1971).

46. Hatefi, Y., Davis, K.A., Baltscheffsky, H., Baltscheffsky, M., and Johansson, B.C., Arch. Biochem. Biophys. 152, 613 (1972).

47. Guillory, R.J., and Fisher, R.R., Biochem. J. 129, 471 (1972).

48. Klemme, J.-H., and Gest, H., Proc. Nat. Acad. Sci. U.S. 68, 721 (1971).

49. Klemme, J.-H., and Gest, H., Eur. J. Biochem. 22, 529 (1971).

50. Beechey, R.B., and Cattell, K.J., Curr. Top. Bioenerg. 5, 305 (1973).

51. Baltscheffsky, H., and Baltscheffsky, M., Ann. Rev. Biochem. 43, in press (1974).

CONDENSATION REACTIONS OF LYSINE IN THE PRESENCE OF POLYADENYLIC ACID*

Loring K. Bjornson,** Richard M. Lemmon,
and Melvin Calvin
Laboratory of Chemical Biodynamics
Lawrence Berkeley Laboratory
University of California, Berkeley, California 94720

Since the existence of life depends upon the genetic code, an understanding of the code's origins and evolution is crucial to the understanding of how life began. Recent discussions (1-6) have dealt with amino acid and nucleic acid abiogeneses and how these may have been coupled to give rise to the genetic code. The papers of Crick (6) and Orgel (7) contain detailed theoretical analyses of the problem. One of the conclusions presented was that there may have been prebiological polynucleotides with catalytic properties. There are presently two extreme views on the origin of the genetic code, one that an amino acid's codon resulted from a direct interaction between the amino acid and its codon, and the other that the assignment of a codon to a particular amino acid was purely arbitrary. Since the theory that the codon assignments are "accidental" is not subject to experimental verification, the case would seem to rest on proving or disproving the "direct interaction" hypothesis.

In searching for interactions which could have led to the coupling of the nucleic acid and amino acid systems, our attention was drawn to the ionic interactions between the basic polypeptides and the polynucleotides. These interactions are quite important in biological systems--*e.g.*, in the complexing of histones and DNA (8). They could also play a role in the structure of ribosomes since many of the polypeptides contained in this unit are basic (9). It has been reported that polylysine exhibits a marked selectivity in interactions with adenylic acid and with polynucleotides having a high adenine-thymine content (10,11). It has also been shown that lysyl adenylate incorporates lysine most effectively into microspheres (prepared from basic proteinoid and poly A, poly G, poly C, or poly U) with a high poly A content (12). It therefore seemed to us that the above primarily ionic interaction might be such that the polymeric unit of one system would act as a catalyst in the formation of the other. For example, in the presence of a suitable condensing agent, the polymerization of lysine might be catalyzed by polynucleotides and, likewise, the polymerization of nucleotides might be catalyzed by basic polypeptides. This sort of catalytic behavior could have been important in early stages of chemical evolution when macromolecules were first being formed.

However, before this relationship could be investigated, the reaction of lysine with a condensing agent in aqueous solution had to be studied in detail. From our previous experience, in which a variety of carbodiimide and cyanamide type condensing agents were used, we decided that 1,3-bis-(2-methoxyethyl) carbodiimide (MEC) would be the most suitable reagent for this study. This compound is water-soluble, uncharged (so as not to interfere with lysine-polynucleotide electrostatic interactions), and of a class (carbodiimides) simple enough to have been a product of high-energy processes occurring in the primitive Earth's atmosphere. The new carbodiimide was prepared, both labeled (from ^{14}C-phosgene) and unlabeled, according to the reaction sequence:

$$COCl_2 + 2\ CH_3OCH_2CH_2\text{-}NH_2$$

$$\downarrow$$

$$CH_3OCH_2CH_2\text{-}NH\text{-}\overset{O}{\overset{\|}{C}}\text{-}NH\text{-}CH_2CH_2OCH_3$$

$$\downarrow \ \underline{p}\text{-tosCl}$$

$$CH_3OCH_2CH_2\text{-}N{=}C{=}N\text{-}CH_2CH_2OCH_3$$

The reaction of lysine with MEC is illustrated in Fig. 1. It can be seen that the presence of the second amino group in lysine gives rise to a reaction that cannot occur with monoamino acids. This is the cyclization of lysine through attack of the ε-amino group on the carbonyl carbon to form the seven-membered lactam, 3-aminohomopiperidone-2. This type of cyclization has been observed for the methyl ester of N-α-tosyllysine in ammonia and alcohol, and for the analogs, ornithine and α,β-diaminobutyric acid (13,15). We have found that the cyclization of the lysine N-acylurea in 0.1 N NaOH at 70° is very nearly quantitative, with only 1 or 2 percent of lysine being produced. The lysine N-acylurea is stable to heating at 70° in 1 N HCl for one hour, but is hydrolyzed to lysine in 6 N HCl when heated to 100° for twelve hours.

The lactam is quite stable, and is unaffected by heating in 0.1 N NaOH or 1 N HCl for short periods of time. Heating at 100° in 6 N HCl for twelve hours only partially hydrolyzed this compound to lysine. The hydrolysis was best carried out at 70° in 4 N NaOH (13), and was about 70 percent complete after twelve hours.

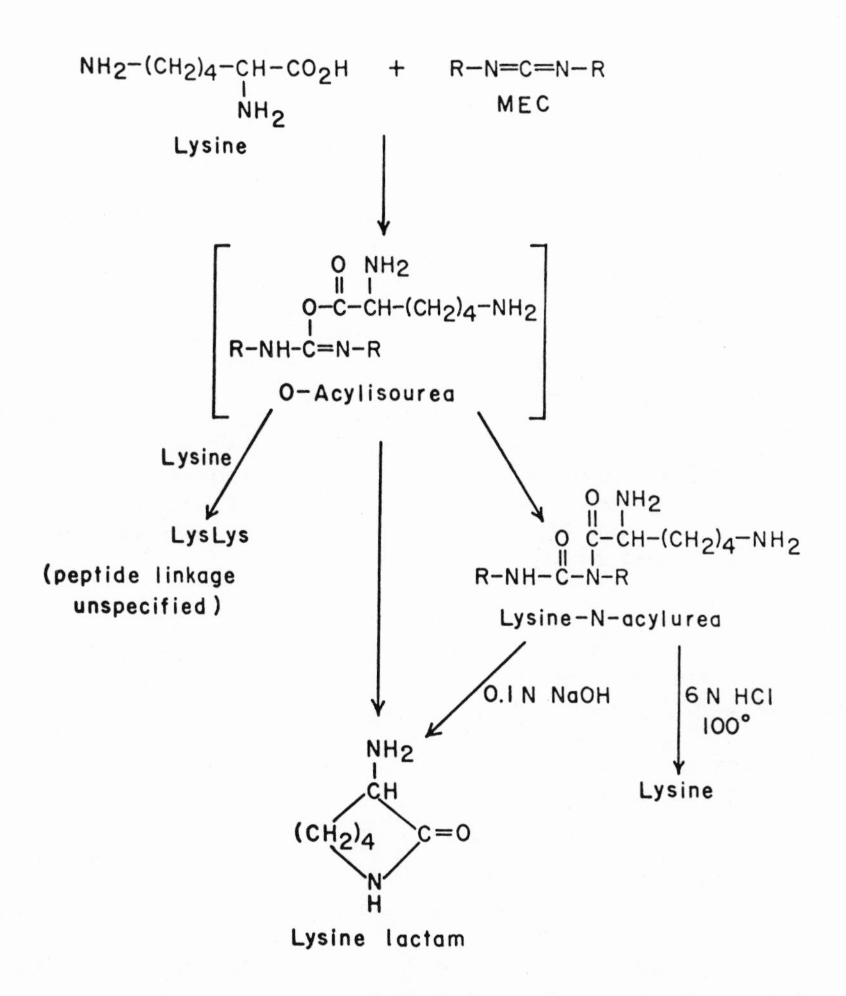

Figure 1. Reaction of lysine and 1,3-bis(2-methoxyethyl) carbodiimide. (R = $H_3COCH_2CH_2$-)

The MEC-induced coupling of lysine was carried out in dilute aqueous solution, at ambient temperature, for approximately 24 hours. In the reaction of ^{14}C-lysine (0.025 M) and MEC (0.05 M) at pH 2, the yield of the N-acylurea was about 30 percent, the lactam 8 percent, and dilysine 2 percent. The products of the reaction were best separated by thin-layer chromatography on silica gel. Typical autoradiograms of this separation are shown in Fig. 2, where lysine is ^{14}C-labeled in one reaction (A) and the carbodiimide labeled in the other (B). In Fig. 2A there are three products which contain ^{14}C-lysine. These are the N-acylurea, the lactam, and dilysine. The latter is hydrolyzed to lysine by heating in 6 N HCl. The peptide linkage in the dilysine is unspecified, although the product co-chromatographs with commercial (Miles Laboratories) L-lysyl-L-lysine. It is possible, however, that some of the ε-amino linked amide is also present.

In the reaction of lysine with ^{14}C-MEC (Fig. 2B) only two radioactive compounds are observed, one of which is the hydrolyzed carbodiimide, 1,3-bis-(2-methoxyethyl) urea (MEU), and the other the N-acylurea of lysine. Fig. 2C shows the ninhydrin positive compounds with commercial L-Lys-L-Lys co-chromatographed with the reaction mixture that contained ^{14}C-lysine. Evidence for the identity of lysine-N-acylurea was also obtained from an experiment where ^{3}H-lysine was reacted with ^{14}C-MEC. The molar ratio of ^{3}H-lys to ^{14}C-MEC was found to be 1.08 for the acylurea compound.

To test the hypothesis that polynucleotides could catalyze the polymerization of lysine, the reaction of lysine and MEC was carried out in the presence and absence of polyadenylic acid (Miles Laboratories). This polynucleotide was chosen because one of lysine's codons is AAA. In these experiments polyadenylic acid and lysine were precipitated in a 1 to 1 complex from a solution of 75 percent dimethylformamide-25 percent water. This precipitate was washed with acetone and dried *in vacuo*. It was then used in the experiments described above. A solvent of 80 percent acetone-20 percent water was used to reduce the dielectric constant of the medium so that there would be a maximum ionic binding between lysine and the polynucleotide. The results of a representative experiment are presented in Table I. The most pronounced effect of the polynucleotide is to triple the amount of dilysine that is formed. Thus, polyadenylic acid does exert a catalytic effect on the lysine condensation; the effect of poly G is not yet known. The other products are only slightly affected by the presence of the polynucleotide.

C. R. Woese has put forth the hypothesis (3) that evolution of the cell began with two very general kinds of polymers, polynucleotides that were purine-rich and polyamino acids composed largely, if not solely, of basic amino acids. The one type of polymer is viewed as catalyzing, in particular ways, the synthesis of the other, and *vice versa*. According to this model, translation began as a

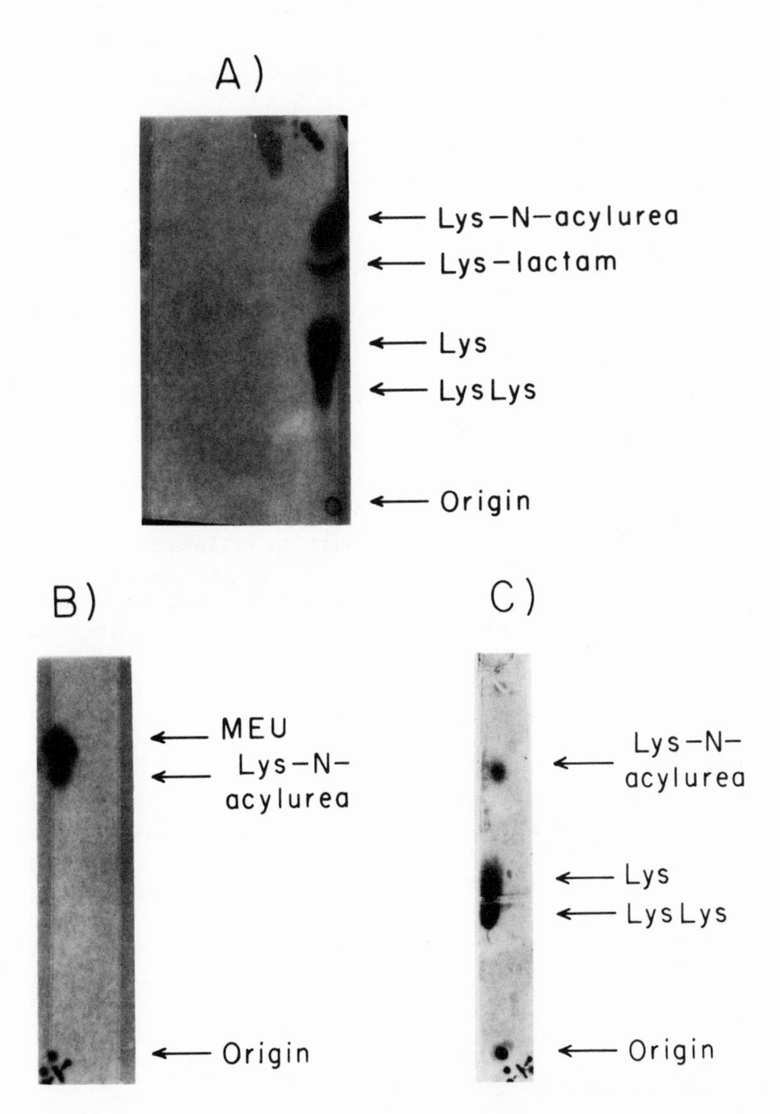

Figure 2. Autoradiograms from the TLC (silica gel, 95% EtOH:NH_4OH, 7:3) of the reaction of A) ^{14}C-lys + MEC, and B) lys + ^{14}C-MEC (MEU = methoxyethyl urea). C) shows ninhydrin positive products--carrier L-lysyl-L-lysine (Miles Laboratories) has been added.

Table I

Percent Yields of Products from the Reaction of ^{14}C-L-Lysine (0.002 M) and MEC (0.004 M) in the Presence and Absence of Polyadenylic Acid

(The solvent is 80 percent acetone-20 percent water, and the pH is about 7)

	LysLys	Lys	Lys Lactam	Lys-N-Acylurea
Polyadenylic Acid	2.5	62.2	5.3	30.0
----	0.8	58.0	4.2	37.0

"direct templating" and, in addition, "translation" was initially a reciprocal matter, not unidirectional as it is now. The evidence presented here, indicating that at least one polynucleotide catalyzes a particular peptide formation, tends to support such a model.

SUMMARY

The water-soluble, nonionic carbodiimide, 1,3-bis-(2-methoxyethyl) carbodiimide, a compound whose type was probably present on the prebiotic Earth, induces the coupling of lysine to dilysine in dilute aqueous solution. The yield of dilysine is increased by the presence of polyadenylic acid. Since ionic interactions are expected between lysine and poly A, and since AAA is a codon for lysine, these results support the concept that the polynucleotides played a role in the chemical evolution of the polypeptides.

*The work described in this manuscript was sponsored by the U.S. Atomic Energy Commission.

**Present address: New York University Medical Center, 550 First Avenue, New York, N. Y. 10016

REFERENCES

1. Calvin, M.,"Chemical Evolution," pp. 162-183, Oxford University Press, New York and Oxford, 1969.

2. Saxinger, C., and Ponnamperuma, C., paper presented at the Fourth International Symposium on the Origin of Life, Barcelona, Spain, June 24-28, 1973.

3. Woese, C. R., Proc. Nat. Acad. Sci. U.S. 59, 110 (1968).

4. Thomas, B. R., Biochem. Biophys. Res. Commun. 40, 1289 (1970).

5. Jukes, T. H., in "Prebiotic and Biochemical Evolution" (Kimball, A. P., and Oró, J., eds.), p. 122, North-Holland, Amsterdam, 1971.

6. Crick, F.H.C., J. Mol. Biol. 38, 367 (1968).

7. Orgel, L. E., J. Mol. Biol. 38, 381 (1968).

8. Busch, H., "Histones and Other Nuclear Proteins," Academic Press, New York, 1965.

9. Muller, W., and Widdowson, J., J. Mol. Biol. 24, 367 (1967).

10. Lacey, J. C., and Pruitt, K. M., Nature 223, 799 (1969).

11. Leng, M., and Felsenfeld, G., Proc. Nat. Acad. Sci. U.S. 56, 1325 (1966).

12. Nakashima, T., and Fox, S. W., Proc. Nat. Acad. Sci. U.S. 69, 106 (1972).

13. Schröder, E., and Lübke, K., "The Peptides" Vol. I, Academic Press, New York and London, 1965.

14. Barras, B. C., and Elmore, D. T., J. Chem. Soc., 4830 (1957).

15. Curraghand, E. F., and Elmore, D. T., J. Chem. Soc., 2948 (1962).

CONSIDERATIONS OF THE ORIGIN OF SPONTANEOUS MUTATIONS

S. E. Bresler

Leningrad Institute of Nuclear Physics
Academy of Sciences of USSR
Moscow, USSR

Spontaneous mutations are the main driving force of evolution. This is equally valid for the cellular or the prebiological period of life history. During the precellular prebiological period first considered by A. I. Oparin (1) and theoretically analyzed by M. Eigen (2), spontaneous mutations were the main cause of structural changes of information-containing molecules, that increased their "selective value." Gradually, the increasing sophistication of the self-instructional process led to supermolecular structures and finally to their most perfect combination--the cell. But the same mechanism preceded functioning and appeared as the basis of phylogeny of living matter. Therefore, it seems important to know the true origin of spontaneous mutations which, according to their name, arise without any visible reason. There were many attempts to explain spontaneous mutations by environmental factors like cosmic rays, or by the mutagenic action of some common metabolites, like hydrogen peroxide. All these ideas did not stand quantitative trial.

For instance, we can estimate the mutagenic action of the radiational background. It is well established that the mutation rate is proportional to the dose expressed in absorbed energy, i.e. in rads. The intensity of radiation at sea level is 10^{-5} rad per hr.

We can take the classical data of Demerec and Sams (3) or the more recent data of Kondo for reversions in bacteria at about 10^{-10} rad^{-1}. For an average gene, the rate of mutagenesis is by three orders of magnitude higher, because a reversion is due to a specific nucleotide insertion. Therefore, we find that the mutation rate due to the radiation background is $10^{-5} \cdot 10^{-7} = 10^{-12}$ $gene^{-1}$ hr^{-1}. But the average figure for spontaneous mutations in bacteria is 10^{-6}

$gene^{-1}$ generation $time^{-1}$ (4).

In this paper, we demonstrate that spontaneous mutations are absolutely legitimate and normal events, i.e. thermal fluctuations of "noises" of the process of DNA-replication. In other words, their source is the statistical nature of chemical reactions. We shall consider, for the sake of simplicity, our special case. During DNA-replication the enzyme DNA polymerase builds a Michaelis complex ES with one of the nucleoside triphosphates at every elongation step. If the complex does not involve a template, all four monomers participate equally well. But the complex is ternary; it contains a polynucleotide chain that selects each time one of the monomers, complementary to the polynucleotide template according to the principle of Watson-Crick. The formation of the Michaelis complex results in a decrease of free energy by $\Delta F°$ (Fig. 1), its dissociation constant is $K_m = e^{-\Delta°/RT}$, where T is the absolute temperature and R is the gas constant.

Of course to be precise, we must add to the free energy term a second one depending on the activity coefficients of the reagents Pi and the reaction products Pj,

$$RT\ln\left[\frac{\prod_i (Pi^{x_i})}{\prod_j (Pj^{y_j})}\right],$$

where x_i and y_j are

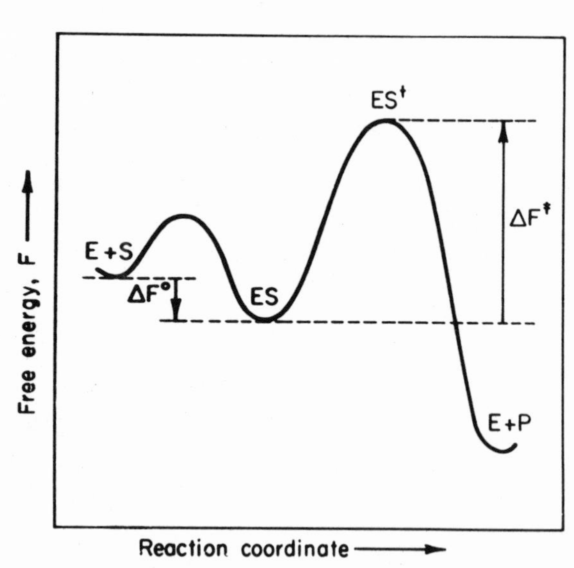

Figure 1. Free energy profile of an enzymatic reaction. Ordinate: free energy; abscissa: reaction coordinate.

the stoichiometric coefficients of the reaction. This term is unknown because we do not know the activities in the conditions of the cell. The same situation exists in all biochemical applications of thermodynamics when equilibria of enzymatic reactions are measured in solution, but applied to the cell interior. In our case, we are justified to do so because we shall consider only ratios of K_m, or differences of $\Delta F°$ values for similar substances, i.e. nucleoside triphosphates and DNA polymerase. For the differences of $\Delta F°$ values the additional terms cancel. Moreover, the only numeral estimate we shall make will use data measured in vivo, which include therefore the additional terms. The ternary complex is transformed via the activated state ES' into the reaction product. The free energy of the activated state $\Delta F'$ (according to the theory of absolute reaction rates) determined the maximal reaction rate

$$v_m = K'e^{-\Delta F'/RT} \quad .$$

It is clear that the value $\Delta F'$ and the corresponding v_m are similar for all four monomers because their addition rates are of the same order. Therefore, the reaction rate is approximately independent of the particular nucleotide. It is important only that the nucleotide added would conform to the Watson-Crick principle, i.e. to the complementarity rule. Now we shall consider the case of molecules having statistically distributed energies. Therefore, we must take into account the possibility of the formation of an erroneous noncomplementary Michaelis complex. For this erroneous complex ES" the change in the free energy will exceed the free energy of the regular complex by $\Delta F''$. Therefore, its dissociation constant will increase by the ratio

$$\frac{K_m''}{K_m} = e^{\Delta F''/RT}$$

The probability of the insertion of an erroneous monomer during DNA-replication will obviously depend on the ratio

$$\frac{K_m}{K_m''} \quad .$$

As a matter of fact, the rate of the elongation reaction is expressed by the Michaelis-Menten equation

$$v = \frac{v_m es}{K_m + s} \quad ,$$

where v is the reaction rate, e is the enzyme concentration, and s is the substrate concentration. At low substrate concentrations when $s \ll K_m$, we obtain simply

$$v = \frac{v_m es}{K_m}$$

The ratio of reaction rates of different monomers is approximately expressed by the reverse ratio of the dissociation constants

$$\frac{K_m}{K_m''}$$

We can make an estimate of the value $\Delta F''$--the difference of free energies of a regular and an erroneous Michaelis complex. For this purpose, we consider the experimental data for the level of spontaneous mutations in a population of Escherichia coli (4). The figures for different genetic loci diverge by several orders of magnitude. However, for a rough estimate, we can adopt an average value of 10^{-6} for one gene in one generation. If we take 10^3 nucleotide pairs as an average dimension of a cistron, it gives us the probability of a replication mistake on the order of 10^{-9} per nucleotide addition. In other words, once per billion elongation steps a thermal fluctuation results in a wrong nucleotide addition. For the ratio of complementary to noncomplementary rates of copying, we obtain

$$\frac{v}{v''} = \frac{K_m''}{K_m} = 10^9,$$

hence $\Delta F'' = RT\ln 10^9 =$ 13000 cal/mole. The value of $\Delta F''$ is the free energy barrier which protects the Watson-Crick helix from mistakes. We know that the enthalpy of hydrogen bonds between bases in DNA, measured in the calorimeter, is about 8000 cal/mole (5). It is not amazing that the calculated barrier is more than 1.5 times higher, because the melting of the helix in the calorimeter takes place in water solution. Therefore, H-bonds between bases are transformed into H-bonds with molecules of water. In case of an erroneous base insertion, the H-bonds in the Watson-Crick helix are broken as *in vacuo*. Therefore, the increase in energy must be much more pronounced than during the thermal helix-coil transition.

An important prediction can be deduced from the notion of spontaneous mutations as thermal replication errors. Both reactions of complementary and noncomplementary copying are independent. Each is characterized by its own rate. Therefore, if we decrease the reaction rate of regular DNA-copying by lowering the concentration of one of the indispensable precursors, i.e. one of the nucleoside triphosphates, we increase the replication time of the cell genome but the rate of erroneous copying remains constant because it does not need the deficient monomer. In other words, the probability of spontaneous mutations must be related to a time unit and not to the generation time. This conclusion was fully confirmed by Szilard and Novik in their chemostat experiments (6). It was shown to be independent of the growth rate of cells, which was changed manyfold by altered growth conditions. The rate of spontaneous mutant accumulation remained constant. Another important question is concerned with the properties of the replication enzyme. Does the probability of mutation depend on the quality of the enzyme? The answer is positive. The Michaelis complex contains the enzyme, substrate, and template. Therefore, the enzyme structure is crucial for the binding constant of the complex. It was found that bacteriophages code their own DNA polymerases and their level of spontaneous mistakes is two orders of magnitude higher than in bacterial polymerases. The opposite is found in eucaryotes. In case of their DNA polymerases, the fidelity of replication in general is more perfect and the level of errors is lower than in bacteria. Finally, there exist special mutations in the genes of DNA polymerases, the so-called mutators, which yield enzymes with increased noise level (7) and also antimutators (8) with low noise level.

An interesting model system, demonstrating the accumulation of spontaneous mutations as a result of replication errors is the well-known _in vitro_ replication of Qβ phage RNA (9).

In a solution of the phage RNA replicase and all four nucleoside triphosphates, unlimited multiplication of the phage RNA takes place. Gradually, some erroneous or mutated forms of phage RNA accumulate. Under the selective pressure of multiple transfers into fresh media those mutants are privileged, which reveal an increased replication rate. The "selective value" of templates is increased if they are copied faster. Therefore, shorter RNA variants accumulate. If analogues of nucleotides are introduced into the medium, they inhibit the replication reaction and gradually some mutated forms of RNA less sensitive to the antimetabolite action are selected. This experiment is a model of prebiotic evolution. One single enzymatic reaction of polynucleotide synthesis takes place and the selective value of the product is quite clear. In this case, it is obvious that mutations are the results of thermal fluctuations. No other sources of errors are present. Applying to bacteria the idea of spontaneous mutations as thermal noises, we designed a special experiment, which was performed in our laboratory (8) and was a direct support of the whole concept. We asked the question if it is possible to exaggerate the noise level in a bacterial population by slowing down its regular replication, say by a deficit of a monomer. We chose thymidine for this role because it is the only nucleoside which is not involved in RNA transcription and therefore not directly connected with protein synthesis. Taking a thymine dependent bacterial strain we know it must stop growing on a medium devoid of thymine, but will proceed at a very low rate in the presence of a small dose of thymine. We can bring this time of division to tens of hours instead of 20-30 min. Obviously, the rate of complementary copying will be low enough and the DNA polymerase will stop every time thymidine triphosphate is required by the complementarity rule. However, the reaction of noncomplementary copying is not affected, because instead of thymidylic acid another nucleotide may be inserted, for instance, cytidylic acid. The rate of erroneous DNA elongation remained unchanged under conditions of a thymine deficit. Under normal conditions, one elongation act takes place in about 10^{-3} sec. The probability of a wrong substitution is 10^{-9}. Therefore, one noncomplementary insertion will occur on the average once in $10^{9} \cdot 10^{-3} = 10^{6}$ sec = 10 days. If the time scale of DNA replication is lengthened to many days, we can expect enormous quantities of spontaneous mutants. Finally, an entire cell population will be composed of different mutants damaged in all possible genetic loci.

At first it seemed that there was an insurmountable barrier to the realization of this experiment. This was the so-called thymineless death of bacteria. In 1954, Cohen and Barner (11) observed the suspension of DNA synthesis in thymine-dependent cells in a thymine-deficient growth medium. RNA and protein synthesis proceeded and the cells lost viability in a short while (Fig. 2). The phenomenon of thymineless death was characteristic of any bacteria unable to synthesize thymidylic acid and deprived of exogeneous thymine. But we succeeded in finding conditions which preserved the viability of thymine-starved cells without inhibition of general metabolism. For this purpose thymine-starved cells must be plated on the surface of a supporting solid medium, i.e. agar. The filamentous bacterial cells, elongated because of unbalanced growth, are stabilized under these conditions against destructive mechanical perturbations, unavoidable in liquid medium. For the stabilization of cells against thymineless death, the agar medium must be supplemented with a minute amount of thymidine (10^{-3} µg/ml). This addition is insufficient for growth, but is used for repair of lesions, generated in DNA during thymine starvation (12). If the medium contains this stabilizing supplement of thymine, the viability of bacteria is preserved for many days (Fig. 2). If the concentration of thymidine (or thymine) is increased to 0.1-0.3 µg/ml, some growth of thymine-dependent cells occurs during the first 15-25 hr. The cell population is increased 50-100 times, depending on the density of the original innoculum. The cells divide some 5-10 times and then practically stop growing (Fig. 3a). During the subsequent 2-3 days, their viability is not affected and their number remains constant. Under these conditions, we observed an overall mutagenesis taking place in almost the whole bacterial population. This intense mutagenesis begins at the time of growth stoppage and develops gradually till 50-60% of the bacterial cells become mutated in different loci (Fig. 3b). In some experiments, the yield of mutants attained 70-80% of the cell count. The mutants are mostly auxotrophs with all sorts of nutritional requirements; some are resistant to antimetabolites; others are thermosensitive. The number of specific mutants

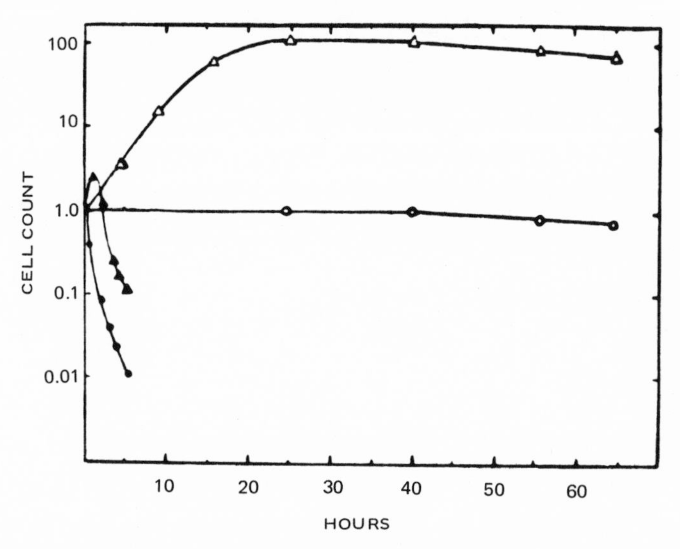

Figure 2. Thymineless death of cells and its prevention. Ordinate: $\frac{N}{N_o}$ relative number of cells (ratio to the initial).

Abscissa, t: time in hr

●-●-●-●-●-● Liquid medium devoid of thymine

▲-▲-▲-▲-▲-▲ Liquid medium, 0.3 μg/ml of thymine

o-o-o-o-o-o Solid medium, 0.001 μg/ml of thymine

△-△-△-△-△-△ Solid medium, 0.3 μg/ml of thymine

selected in one particular cistron increases with the same kinetics as the overall number (Fig. 3c). After 3 days the level of these particular mutations is 10^3-10^4 times higher than the background found in normal conditions. Many of the cells become double and triple mutants. Some of the mutants were identified. We found auxotrophs for histidine, arginine, biotin, nicotinamide, cysteine, uracil, and a mutant requiring both serine and threonine. Besides, we found plenty of streptomycin-resistant mutants (Fig. 3c) and also revertants to indole-independence and histidine-independence. This phenomenon has some very characteristic features: 1) it is observed only when some very slow vegetative growth is still possible. When thymine is totally absent or is present only as "supporting supplement" (10^{-3} - 10^{-2} μg/ml), the rate of mutagenesis decreases 1000 times and does not exceed the background more than 10 times. 2) There exists a concentration of thymine optimal for mutagenesis. In our conditions, which are self-adjusting, the optimal concentration is attained after some time of growth, when a considerable part of thymine is used by the cell population and further growth becomes extremely slow. These self-adjusting conditions are revealed by the

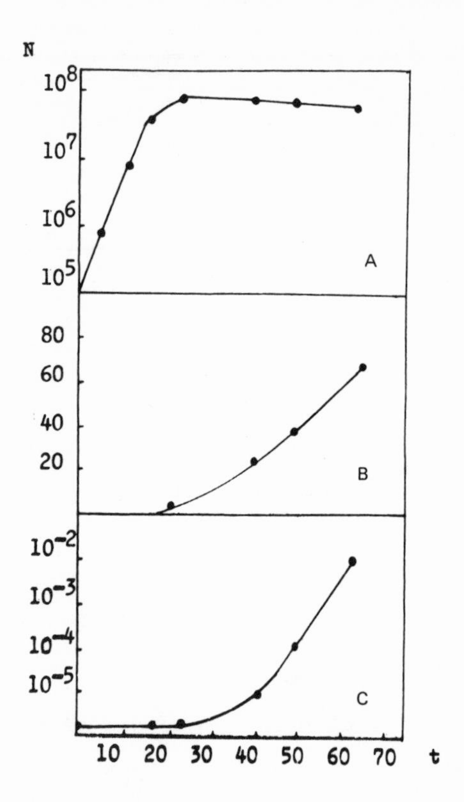

Figure 3. Overall mutagenesis in a Bacillus subtilis cell population, growing on agar supplied with a limiting amount of thymine (0.3 μg/ml).

Abscissa: time in hr
Ordinate: A, growth of cell population
B, total percentage of auxotrophic mutants in the cell population
C, percentage of streptomycin-resistant mutants

following facts. The attainment of optimal conditions depends on the density of the innoculum at the beginning of growth (Fig. 4). If we keep the time of incubation constant (46 hr), the denser the initial innoculum, the earlier mutagenic conditions are attained. Therefore, the yield of mutants increases with the density of seed. In case of a less dense innoculum, more hours of growth are needed and overall mutagenesis is observed later.

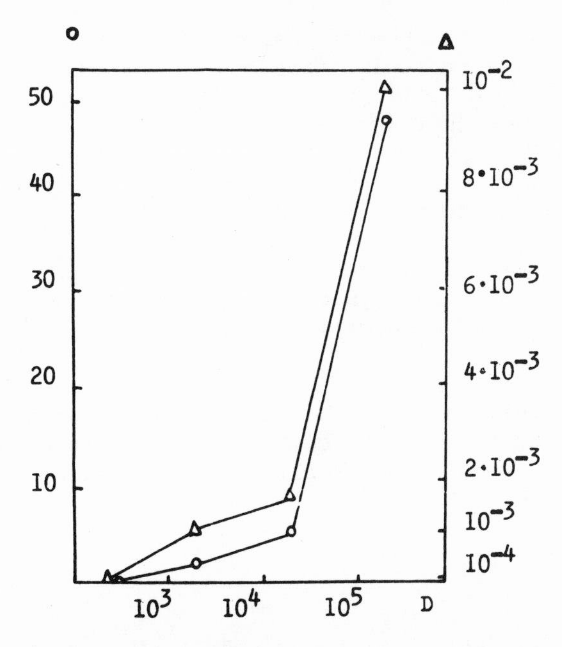

Figure 4. Efficiency of mutagenesis as a function of density of seed. Thymine-dependent cells (Escherichia coli) were grown on Petri dishes with agar supplemented with 0.3 µg/ml of thymine. The time of growth was 46 hr.

Abscissa: density of seed (per sq. cm.)

Ordinate: On the left, total percentage of auxotrophic mutants in the cell population. On the right, percentage of streptomycin-resistant mutants.

-Δ-Δ-Δ-Δ- Yield of streptomycin-resistant mutants

-o-o-o-o- Yield of auxotrophic mutants

Instead of adjusting optimal conditions in time, we can do it in space. For this purpose we use a Petri dish with a radial gradient of thymine (in the center of the plate a hole is filled with thymine solution). All mutants concentrate in a narrow ring zone on the plate and are easily found by replication on selective agar. 3) The question of the most probable chemical mechanism of mutations in case of growth on limiting thymine was solved by induction of reversions by means of mutagens with well-known properties. Most of the mutants were readily reverted by the action of 5-bromouracil. This is proof that they are transitions, i.e. simple replacements of C-G instead of T-A pairs. This fact is in perfect agreement with theoretical expectations.

It was important to show that spontaneous mutations are just the result of vegetative replication and not of DNA repair. This was proved by taking E. coli strains with different deficiencies in the repair enzymes (strains Uvr^-, $polA^-$, exz^-, rec A^-, rec B^-). All of them proved to yield mutants in case of limiting thymine growth with the same efficiency as wild type cells. All experimental details are to be found in ref. 10.

Finally, the overall mutagenesis was reproduced on a thymine-independent bacterial strain in conditions when thymidilate synthesis was inhibited by a specific poison, fluoruracildeoxyriboside (FUDR).

All these experiments showed unambiguously that the accumulation of replication errors in a cell population can be extremely exaggerated if the regular DNA-replication is slowed down by many orders of magnitude.

REFERENCES

1. Oparin, A. I., "Origin of Life," Macmillan, New York, 1938.
2. Eigen, M., Naturwissenschaften 58, 1 (1971).
3. Demerec, M., and Sams, J., in "Immediate and Low Level Effects of Ionizing Radiations" (Buzzati-Traverso, A., ed.), p. 283, L. Taylor Francis Ltd., 1960.
4. Freese, E., J. Theor. Biol. 3, 82 (1962).
5. Klump, H., and Ackermann, T., Biopolymers 10, 513 (1971).
6. Novik, A., and Szilard, L., Proc. Nat. Acad. Sci. U.S. 36, 708 (1950).
7. Hall, Z., and Lehman, J., J. Mol. Biol. 36, 321 (1968); Drake, J., and Greening, E., Proc. Nat. Acad. Sci. U.S. 66, 823 (1970).
8. Drake, J., and Allen, E., Cold Springs Harbor Sympos. Quant. Biol. 33, 339 (1968).
9. Levinsohn, R., and Spiegelman, S., Proc. Nat. Acad. Sci. U.S. 63, 805 (1969).
10. Bresler, S., Mosevitsky, M., and Vyacheslavov, L., Nature 225, 764 (1970); Bresler, S., Mosevitsky, M., and Vyacheslavov, L., Mutat. Res. 19, 281 (1973).

11. Cohen, S., and Barner, H., Proc. Nat. Acad. Sci. U.S. 40, 885 (1954).

12. Freifelder, D. J., J. Mol. Biol. 45, 1 (1969).

PRE-ENZYMIC EMERGENCE OF BIOCHEMICAL METABOLISM

R. Buvet
Laboratoire d'Energétique Electrochimique
et Biochimique
Centre pluridisciplinaire
Avenue du Général de Gaulle 94000, Creteil (France)

Since the concept of chemical evolution was expressed by Alexander Ivanovitch Oparin (1) for the first time fifty years ago, this concept has gradually changed its prime significance. Primarily, it was frequently understood as simply permitting the emergence of life on the Earth and any other proper planetary surfaces. It has become nowadays really fit for explaining the origin and nature of life.

The initial understanding is obvious from many papers presented in 1958 at the Moscow Symposium (2). At that time, the scarcity of experimental data on the formation of biochemical substances in conditions simulating the terrestrial periphery led to the conclusion that the initial occurrence of living cells could only result from a random succession of relatively rare macroscopic events, such as lightning, radioactivity, or volcanism, which bear hardly any relationship with the present functioning of the biosphere. But, starting from Miller's work (3), experimental data under widely varied conditions began to accumulate regarding the composition of the initial mixture and the choice of energy sources. This accumulation of data eventually established that the main categories of biochemical substance are formed when any sufficiently hard energy is fed into any reducing carbonaceous mixture in a very large range of pressures and temperatures enclosing ambient conditions. The extensive set of results obtained by G. Toupance et al. (4-7) for primary atmospheric products, together with data concerning the reactivity of these products in aqueous media (8,9), largely confirms the soundness of such statement.

It became possible, therefore, to examine to what extent the available theoretical or experimental data gave more insight about the emergence of metabolism from primitive chemical processes at the very beginning of the evolution of the Earth's surface.

In this respect, it can be assumed, as some biologists do (10), that the beginning of the biological era could only result from the random appearance of one special kind of molecule or polymolecular structure very precisely defined.

However, such an assumption, which is not accessible to scientific experimental control, loses its interest when the potentialities of the aqueous chemistry of carbon compounds are more closely compared with the achievements of enzymic biochemistry. A qualitative, though large, similarity is brought out by such comparison. It leads to the idea that metabolism emerged from a primitive set of processes which, in spite of their nonenzymic character, bore in germ the essential features of metabolic pathways as it concerns the kinds of processes which are involved, the criteria which rule their intervention, and the way they are assembled.

As we have presented elsewhere, the theoretical framework (11-14), mainly issued from the energetics of redox and condensation processes that underlie such a viewpoint; we shall shortly review here some of the experimental data supporting that viewpoint.

* * * * *

Since the energy requirements of the biosphere are nowadays met mainly from the photosynthetic activity of autotrophs, primitive models of this process have first to be defined (15). This photochemical behavior depends entirely on a primary act which, from the chemical standpoint, is an endergonic redox dismutation, coupled steochiometrically with the absorption of one or several photons. The following act is built up from elementary chemical processes which do not differ in principle from the other metabolic steps we shall be discussing later in this paper.

Under the most primeval conditions, i.e. in the atmospheric phase, the primary electronic act corresponded to the excitation, even possibly the ionization, of gaseous compounds, particularly CH_4, from the absorption of photons in the Schumann ultraviolet range.

The reactions of the high-energy species formed, such as CH_4^+, led, in the absence of solvent effects by water, to a large set of unsaturated products, mainly with C=C and C≡C bonds (7). In fact, this formation, occurring from fully hydrogenated compounds, is already an endergonic redox dismutation since it occurs with hydrogen release as in:

$$CH_4 + NH_4 \rightarrow HC{\equiv}N + 3\ H_2,\ \Delta G^{\circ\prime} > 0$$

However, photoelectronic processes occurring in aqueous media should be more representative of present-day biochemical photosynthetic activity. In this respect, let us recall (15,16) that the far ultraviolet absorptions of simple reducers (Red), such as halides or low-valency metal cations correspond to electron transfers to the solvent, in aqueous solutions, such as:

$$\text{Red} \rightarrow \text{Ox} + e_{aq}$$

where e_{aq} represents the hydrated electron, which should be only stable at potentials as low as -2.7 V/S,H,E, Then, the hydrated electron must rapidly disappear by forming other reducers, as in:

$$e_{aq} + H_2O \rightarrow H^{\bullet} + OH^-$$

$$2\ H^{\bullet} \rightarrow H_2\nearrow$$

If the standard redox potential of the oxidant obtained is sufficiently high, for example, when the nonbiochemical couple Ce^{4+}/Ce^{3+} is involved, the oxidant can produce oxygen, from water, as in the overall process:

$$4\ Ce^{4+} + 2\ H_2O \rightarrow 4\ Ce^{3+} + 4\ H^+ + O_2\nearrow$$

In such cases, however, the terminal absorption of Red must be located in the far ultraviolet range accessible in aqueous media, i.e. near 200 nm. Thus, for defining primordial models of the visible photochemical assimilation, we need to consider solvate-solvate charge transfers as Krasnovsky did, either with artificial porphyrins, which photochemically induce electron transfers between Fe^{2+} and organic dyes, or with transition oxides such as TiO_2, ZnO, and WO_3 (17).

Nevertheless, we must acknowledge that the attention that was paid until now to the possibility of re-exciting aqueous media obtained from evolution of solutions of primary atmospheric products is largely insufficient. Yet, we have recently shown (18) that the ultraviolet absorption of such solutions is strong into the visible range and is associated with reducing power in such a way that it implies the presence of many potential energy photoreceptors. In this respect, the observations which were made about the formation in aqueous solutions of biochemical substances, such as oses or osides, from the effect of light on solutions of atmospheric precursors such as formaldehyde (19), cannot be considered as relevant to the problem of the origins of photosynthetic activity in aqueous media. The observed processes do not seem to correspond to an overall accumulation of energy. New direct experimental studies must be made in this field. We reported at the Barcelona meeting (18) the principle of the CERES (Chemical Evolution or REactions into Solutions) program that we recently set up on these grounds.

* * * * *

Fortunately, it is not necessary to wait for such new data to examine further to what extent the other main categories of biochemical processes have occurred on the primordial Earth's surface before the occurrence of enzymic catalysts. Starting from the presence of redox energy in primeval aqueous media, we must first examine how condensed hydrolyzable energy-rich compounds have initially been produced during chemical evolution.

In this respect, an interesting result was mentioned by Calvin (20) concerning pyrophosphate condensation from orthophosphate by coupling with the oxidation of ferrous ions by H_2O_2 and other simple oxidants. In fact, this result was infirmed by Etaix (21). The pyrophosphate production observed by Barltrop was probably connected with the analytical procedure taken by this author. Several other nonenzymic processes leading to the storage of energy from redox reactions were proposed by Clark, Kirby, and Todd (22) and by Baüerlein and Wieland (23). Unfortunately, these results were obtained under conditions which probably have never occurred on the primeval Earth. Thus these experiments must be considered as not relevant to the origin of life on the Earth. However, in view of understanding the theoretical background underlying the biochemical energy-storage processes and to define possible archetypes of it, it is worth considering that these models respectively imply the oxidation of phosphoric esters of hydroquinone (22) and the bielectronic oxidation of thiols (23) in the presence of ADP and orthophosphate in nonaqueous media.

Confronted again with this relative irrevelancy of results collected until now, several authors have considered that the primitive source of energy-rich bonds should be located in some properties of unsaturated compounds which were formed during atmospheric processes. This hypothesis can be connected with the results obtained as early as 1960 by Jones and Lipmann (24), which showed that carbamyl phosphate is easily obtained in neutral aqueous solutions simply by reaction of orthophosphate on cyanate. Afterwards, the carbamyl phosphate can be nonenzymatically used as carbamyl donor.

More recently, several authors such as Beck and Orgel (25), Miller and Parris (26), Steinman, Lemmon and Calvin (27), and Ferris (28) obtained polyphosphoric compounds and peptides from aqueous solutions of the free monomers in the presence of cyanate, cyanamide, dicyandiamide, or cyanoacetylene. Likewise, polynucleotides were obtained by Ibanez, Kimball, and Oró at 40°C from mononucleotide monophosphates by action of cyanamide in the presence of polyornithine (29). All these reactions must in turn be considered as occurring through the formation of an energy-rich phosphorylated compound obtained from the addition of phosphate to a multiple bond of an atmospheric precursor. Since this precursor is eventually hydrated, it should be considered as the primary donor of energy.

Similarly, Degani and Halmann (30) studied the mechanism of phosphorylation of sugars by orthophosphate in the presence of $(CN)_2$. In this case, the primary step of the overall process is also an addition of phosphate on a multiple bond of the precursor, as:

$$CN - C \equiv N + HO - PO_3H_2 \rightarrow NH - C \begin{matrix} \nearrow\!\!\!\!\nearrow NH \\ \searrow OPO_3H_2 \end{matrix} \quad ,$$

the phosphorylated compound obtained acting nonenzymically as a phosphorylating agent of sugars.

This set of results clearly shows that the exergonic hydration of multiple bonds which are present in many atmospheric precursors should have played, at least to some extent, the role of coupling process in the primeval syntheses of condensed compounds. However, the theoretical explanations of energy storage processes based upon the energetics of redox processes that we recently presented (13) lead us to consider that experimental evidence of nonenzymic redox energy storage processes occurring in aqueous solutions should eventually be obtained, offering archetypic models more directly connected to present-day oxidative phosphorylations and other comparable biochemical processes.

* * * * *

The most impressive set of results related to archetypes of biochemical processes comes from the study of nonenzymic models of biochemical transacylations and to a lesser extent, transphosphorylations.

In this field, data have been collected partly by groups, such as Jencks _et al_. (31), and Bruice, Benkovic _et al_. (32), who are not directly centering their work on the problem of the origin of life. More recently, M. Paecht-Horowitz and A. Katzir-Katchalsky (36), and L. LePort, E. Etaix, and R. Buvet have more precisely considered the role of model transacylations and transphosphorylations in the course of chemical evolution. In our work, this quest was largely underlain by theoretical considerations related to an extensive discussion of the energetics of such processes (33,34).

Let us recall here that when the energetics of the over-all processes permit transfers of acyl, these are frequently observed by simple nucleophilic attack of an acylated compound by an acylable one with rapid kinetics and relatively high yields. Table I gives a brief view of such results for different kinds of acyl donors and acceptors.

Among other facts, these results show that the aminoacyl thiolesters which are slightly more energy-rich than aminoacyl-tRNA are, as assumed by Lipmann (35), quite capable of inducing the synthesis of polypeptidic chains without enzymic catalysts in slightly alkaline aqueous solutions.

M. Paecht-Horowitz and A. Katchalsky also presented evidence that relatively long polypeptidic chains are obtained from aminoacyl phosphorylated compounds, mainly in the presence of montmorillonite (36).

Many results are also available for nonenzymic transphosphorylation processes. Some of them have previously been discussed in this text when we considered the role of unsaturated atmospheric precursors. In addition, let us mention that the simple nonenzymic transfer of the terminal phosphate of ATP to labelled orthophosphate in the presence of simple divalent cations was studied as early as 1958 by Lowenstein (37). More recently, E. Etaix (38) showed that similar results are obtained from tripolyphosphate, which establishes that the adenosine group does not play a prominent part in this process. In both cases, marked influences of pH were noted but it should be mentioned that the rate of these reactions and their transfer yields are low.

Similarly, phosphorylations of nucleosides by polyphosphates, leading to nucleotides were observed by Ponnamperuma and Chang (39). Double transfers, ultimately ending with hydroxamates or peptides, were also observed (40,41).

We must note here that all available data on transphosphorylation converge to put into evidence that nonenzymic transphosphorylations generally occur with transfer yields and kinetics which are much less favorable than those observed in the case of transacylations. This comparison then leads us to think that transphosphorylations played a more limited role in the primeval metabolism, compared to that of transacylations in present-day living organisms.

* * * * *

Significant results were also obtained concerning the archetypes of present-day processes which led to the syntheses of carbonaceous skeletons.

Firstly, we must be aware of the fact that CO_2 fixation cannot be considered as akin to the primordial chemical metabolism of the

Table I

Acyl Transfer Yields Obtained in Nonenzymic Transacylations in Aqueous Solution at Ambient Temperature in Comparable Conditions

(Yields calculated as percentage of acyl groups transferred from the donor to the acceptor)

H labile on:	Acyl acceptors \ Acyl donors	$CH_3CO \sim OC_2H_5$	$CH_3CO \sim OCH_3$	$CH_3CO \sim OCH(CH_3)_2$	$H_2NCH_2CO \sim OC_2H_5$	$HSCH_2CO \sim OC_2H_5$	$CH_2(CO \sim OC_2H_5)_2$	$CH_3CO \sim SC_2H_5$	$CH_3CO \sim SCH_2CO_2H$	$CO_2H(CH_2)_2CO \sim SCH_2CO_2H$	$CH_3CO \sim OCOCH_3$	$CH_2CO \sim OCOCH_2$ (ring)	$CH_3CO \sim OPO_3H_2$
N	CH_3NH_2	57	59	51	34	45	35	89	89		69		88
N	$C_2H_5NH_2$	24				17	15	76			60		70
N	$HO(CH_2)_2NH_2$	19			15	11	14	60			59		55
N	$HS(CH_2)_2NH_2$				15	11					99 (b)		
N	$HO_2CCH_2NH_2$	9			3	7	5	65			65		60
N	$(CH_3)_2NH$	20			19	0	8	92			85		76
N	$HONH_2$	83			67	75	49	93			90		90
O	$(H_2O_3P)OH$										36 (b)		
O	$(H_3O_6P_2)OH$										42 (b)		
S	HO_2CCH_2SH										90 (b)	80	
S	$CH_3CONH(CH_2)_2SH$										57 (b)		
C	$(NC)_2CH_2$						5 (a')		100 (b')	100 (b')	50 (a)		

Experimental conditions:

Without subscript:	(Ac) = 1 M	(Don) = 1 M, (OH^-) = 2 M
(a):	(Ac) = 2.10^{-1} M	(Don) = 2.10^{-1} M
(a'):	(Ac) = 2.10^{-1} M	(Don) = 2.10^{-1} M, (OH^-) = 4.10^{-1} M
(b):	(Ac) = 4.10^{-1} M	(Don) = 4.10^{-1} M
(b'):	(Ac) = 4.10^{-1}M	(Don) = 4.10^{-1} M, slow

terrestrial surface. In fact, it is known that the primeval formation of skeletons by additions of monocarbonaceous fragments occurred mainly through exergonic additions on C≡N and aldehydic C=O bonds more than through endergonic additions to CO_2 coupled with hydrolyses of energy-rich bonds.

Results presenting valuable analogies with biochemical observations were obtained by Jencks et al. (42) and LePort (43,14) regarding the syntheses of linear carbon chains through the addition of C_3 fragments. Since the biochemical syntheses of lipid chains imply enzymic transfers of malonyl residues on C=O bonds of thiolester groups such as:

$$\text{(Enz)}\begin{cases}\text{S - CO - CH - CO}_2\text{H} \\ \qquad\qquad\ \ \text{O} \leftarrow \text{H} \\ \text{S - C - CH}_3\end{cases} \rightarrow \text{(Enz)}\begin{cases}\text{S - CO} \\ \qquad\quad \text{CH - CO}_2\text{H} \\ \text{S} \quad \text{HO - C - CH}_3\end{cases}$$

$$\rightarrow \text{(Enz)}\begin{cases}\text{S - CO - CH}(\text{CO}_2\text{H})(\text{CO - CH}_3) \\ \text{SH}\end{cases},$$

it was interesting to examine to what extent the use of some other malonic derivatives with more acidic methylenes could perform the same kind of process nonenzymically. In fact, this occurs perfectly well within pH 8 to 10, with rapid kinetics and practically quantitative yields, when dinitriles or other doubly condensed derivatives of malonic acid are used. For example, with malonodinitrile:

$$\text{CH}_3 - \underset{\text{SR}}{\text{C}} = \text{O} \;+\; (\text{CN})_2\text{HC - H} \rightarrow \text{CH}_3 - \underset{\text{SR}}{\overset{\text{CH(CN)}_2}{\text{C}}} - \text{OH} \rightarrow \text{CH}_3 - \text{CO} - \text{CH(CN)}_2 + \text{HSR}$$

Incidentally, the role that acyl thiolester plays in such a process led us to consider that insufficient attention has been paid to the destiny of sulfur in experiments modelling primitive chemical evolution.

* * * * *

Coming back to a more theoretical comparison between nonenzymic models of present-day biochemical processes and these biochemical processes, we must remark that, in all the aforementioned cases, the nonenzymic occurrence of the process was obtained at the expense of a slightly more favorable energy balance of either the overall reaction or one of its intermediate steps. From a general viewpoint nonenzymic processes should have belonged to the same categories as the later enzymic processes. The former depend on the thermodynamic balance of adequate substrates reacting with one another and of the presence in these substrates of activating groups in the close vicinity of their reacting sites, rather than on interactions with catalytic sites of an enzyme. Thus, the former processes required predominantly activation by intramolecular effects instead of activation by intermolecular reactions between substrate and enzyme.

The introduction of coenzyme molecules or of catalytic groups located in the side chains of amino acids should subsequently have led to an acceleration of the primeval processes. Completely new substrates turned sufficiently reactive to play their own game under these coenzyme catalytic effects. The sequential structure of enzymic catalyst should have played a role only when cooperation effects of several catalytic groups began to contribute significantly to the overall metabolism of the terrestrial surface.

REFERENCES

1. Oparin, A. I., "Proiskhozhdenie zhizni," Izd, Moskovskii Rabochi, Moscow, 1924.

2. Oparin, A. I., "The Origin of Life on the Earth" (Oparin, A. I. *et al*., eds.), Pergamon Press, London, 1959.

3. Miller, S. L., *Science* **117**, 528 (1953).

4. Toupance, G., Raulin, F., and Buvet, R., "Molecular Evolution," Vol. I (Buvet, R., and Ponnamperuma, C., eds.), p. 83, North-Holland, Amsterdam, 1971.

5. Toupance, G., Raulin, F., and Buvet, R., *Origins of Life* **5**, in press (1974).

6. Raulin, F., and Toupance, G., *Origins of Life* **5**, in press (1974).

7. Toupance, G., Thèse, Doctorat ès Sciences, Université de Paris VI, 1973.

8. Sanchez, R. A., Ferris, J. P., and Orgel, L. E., J. Mol. Biol. 30, 223 (1967).

9. Toupance, G., Sebban, G., and Buvet, R., J. Chim. Phys. 67, 1870 (1970).

10. Monod, J., "Le Hasard et la Necessité:" Essai sur la philosophie naturelle de la biologie moderne, p. 159, Seuil, Paris, 1970.

11. Buvet, R., Etaix, E., Godin, F., Leduc, P., and LePort, L., "Molecular Evolution," Vol. I (Buvet, R., and Ponnamperuma, C., eds.), p. 51, North-Holland, Amsterdam, 1971.

12. Buvet, R., Biological Aspects of Electrochemistry, Experientia supplementum 18 (Milazzo, G., Jones, P., and Rampazzo, L., eds.), p. 13, Birkhäuser Verlag, Basel, 1971.

13. Buvet, R., and LePort, L., "Bioelectrochemistry and Bioenergetics," Vol. I, Birkhäuser Verlag, Basel, 1974.

14. Buvet, R., "L'Origine des Etres Vivants et des Processus Biologiques," Masson, Paris, 1974.

15. Buvet, R., and LePort, L., Space Life Sci. 4, 434 (1973).

16. Orgel, L. E., Quart. Rev. 8, 422 (1954).

17. Krasnovsky, A. A., "Molecular Evolution," Vol. I (Buvet, R., and Ponnamperuma, C., eds.), p. 279, North-Holland, Amsterdam, 1971.

18. Buvet, R., Origins of Life 1, in press (1974).

19. Ponnamperuma, C., and Mariner, R., Radiation Res. 19, 183 (1963).

20. Calvin, M., in: "Horizons in Biochemistry" (Kasha, M., and Pullman, B., eds.), p. 51, Academic Press, New York, 1962.

21. Etaix, E., unpublished results.

22. Clark, V. M., Kirby, G. W., and Todd, A., Nature 181, 1650 (1958).

23. Bäuerlein, E., and Wieland, T., Abstr. Commun. 7th Meeting, Eur. Biochem. Soc. 57 (1971).

24. Jones, E., and Lipmann, F., Proc. Nat. Acad. Sci. U.S. 46, 1194 (1960).

25. Beck, A., and Orgel, L. E., Proc. Nat. Acad. Sci. U.S. 54, 664 (1965).

26. Miller, S. L., and Parris, M., Nature 204, 1248 (1964).

27. Steinman, G., Lemmon, R. M., and Calvin, M., Science 147, 1574 (1965).

28. Ferris, J. P., Science 161, 53 (1968).

29. Ibanez, J., Kimball, A. P., and Oró, J., in: "Molecular Evolution," Vol. I (Buvet, R., and Ponnamperuma, C., eds.), p. 171, North-Holland, Amsterdam, 1971.

30. Degani, C., and Halmann, M., in: "Molecular Evolution," Vol. I (Buvet, R., and Ponnamperuma, C., eds.), p. 224, North-Holland, Amsterdam, 1971.

31. Jencks, W. P., "Catalysis in Chemistry and Enzymology," McGraw-Hill, New York, 1969.

32. Bruice, T. C., and Benkovic, S., "Bioorganic Mechanisms," W. A. Benjamin, New York, 1966.

33. LePort, L., Etaix, E., Godin, F., Leduc, P., and Buvet, R., in: "Molecular Evolution," Vol. I (Buvet, R., and Ponnamperuma, C., eds.), p. 197, North-Holland, Amsterdam, 1971.

34. LePort, L., Thèse, Doctorat ès Sciences, Université Paris VI, 1974.

35. Lipmann, F., in: "Molecular Evolution," Vol. I (Buvet, R., and Ponnamperuma, C., eds.), p. 381, North-Holland, Amsterdam, 1971.

36. Paecht-Horowitz, M., in: "Molecular Evolution," Vol. I (Buvet, R., and Ponnamperuma, C., eds.), p. 245, North-Holland, Amsterdam, 1971.

37. Lowenstein, J. M., Biochem. J. 70, 222 (1958).

38. Etaix, E., and Buvet, R., Origins of Life 1, in press (1974).

39. Ponnamperuma, C., and Chang, S., in: "Molecular Evolution," Vol. I (Buvet, R., and Ponnamperuma, C., eds.), p. 216, North-Holland, Amsterdam, 1971.

40. Lowenstein, J. M., Biochim. Biophys. Acta 28, 206 (1958).

41. Etaix, E., and Stoetzel, F., unpublished results.

42. Lienhard, G. E., and Jencks, W. P., J. Amer. Chem. Soc. 87, 3863 (1965).

43. LePort, L., and Buvet, R., Fourth International Conf. Origin of Life, Abstract no. 83, Barcelona, 1973.

THE METHODS OF SCIENCE AND THE ORIGINS OF LIFE

A. G. Cairns-Smith

Chemistry Department, University of Glasgow

Glasgow, G12 8QQ, Scotland

Oparin brought the study of the origins of life into the domain of reasonable scientific enquiry. He showed protobiology to be thinkable in historical terms, that one can lay out a sensible set of alternatives and connect them with astronomical, chemical and biological ideas. Thus Oparin created a subject, not by solving its central problem, but by showing us that this problem could be solved and suggesting more or less where to look for a solution.

Probably many of the readers of this volume were, like me, convinced of the essential solubility of the problem of "the origin of life on the Earth" through reading Oparin's great book of that title (1). I was not, it is true, wholly convinced by the particular systems which Oparin considered, but this was a stimulus to think of alternatives. The real message driven home, first through general considerations and then through the consideration of particular hypothetical systems, was that there are some fairly particular solutions to be found that are bound to come nearer with increasing knowledge of chemistry and increasing understanding of the place of biological systems in the physicochemical world. Already with his 1924 paper (2), Oparin had opened out a new set of maps and clearly blocked out a strategy for tackling what was then generally taken to be an insoluble problem.

I would like to consider here briefly some questions of strategy and tactics in science and of the balance between these that may now be appropriate for protobiology.

"The methods of science" are not a fixed set of canons reflecting a constant attitude appropriate to all stages of scientific enquiry. There is not a scientific attitude, but many attitudes effective at different stages in the development of a science. This may seem obvious enough, but there is a widely held view that some subjects are intrinsically more "scientific" than others. For example, physics and chemistry are taken to be very scientific, biology rather less so, psychology hardly at all. This is to fail to see science in the time dimension. At first, even physics and chemistry were concerned with vague concepts like "motion," "fire" and "love and strife among the elements" (3). This was not because the very early natural philosophers were incompetent in that they failed to see some true method of science. On the contrary, their approach was a correct and necessary exploration of possibilities, a proper consideration of strategies before tactics. Indeed, the more practical-minded of the early scientists--for example, the Babylonian astronomers who could predict eclipses (4), or the alchemists with their considerable skills at matter manipulation--gave less in the end to the building of our now exact sciences than the early Greek philosophers with their vague and general speculations. When later the Greeks became more specific their contributions were, on the whole, less valuable. For example, Plato's attempt to account for the different "elements" by relating them to the five regular solids was an ultimately fruitful idea at the most general level; it is true that geometrical considerations at an invisible level are of great importance in accounting for the evident variety of material substances, but his specific assignments of substances to forms now seem absurd. Even much later when chemistry was just beginning to emerge from alchemy the specific ideas of, say, Paracelsus, Van Helmont or Descartes were of much less importance than the shifts of thinking at the most general levels which they encouraged.

When theory becomes too specific too soon, it can even have a crippling effect on further advance. Aristotle's detailed world view with its explanation for everything is a notorious example. More recently the concept of phlogiston had a similar if less prolonged effect, as did in the following century the too simple dualistic system of valency.

I am not suggesting that, to begin with, specific theories should not be attempted. They may serve, like Oparin's specific suggestions, to illustrate concepts, to help to make them thinkable. Even if, in the end, such a theory turns out to be inaccurate or even false it may have stimulated the experiments that produced the necessary realignment, or revolution. It can certainly be argued that much of eighteenth century pneumatic chemistry, which was to lead to the resolution of the central problem of fire, was stimulated by the evidently mind-catching idea of phlogiston. Again, the clear formulation of Berzelius' dualistic theory doubtless encouraged the certainty that some kind of precise explanation for chemical affinity was there to be found. So by all means let there be precise well worked out theories at all stages of enquiry; what must be avoided is a premature consensus, a premature universal commitment to a particular view.

The best way to avoid a premature consensus is to have many ideas around. Yet for an advanced subject, a plethora of opinions on fundamental matters could be a serious waste of time. At some stage it is sensible to agree; at some stage a consensus should emerge. The difficult question is when. As the history of science shows repeatedly, the fact that "everybody" agrees is not enough.

Now, how advanced is protobiology? Up to what level would it now be sensible to agree? After Watson and Crick (5), after Miller (6), after the subsequent work on abiotic syntheses (7,8), after such discoveries as the very early fossils (9) and of biochemical substances in meteorites (10), surely, you may say, something can be agreed about the nature of the first organisms beyond Oparin's original strategic ideas? Not yet, I think, anything very specific. We really have little idea about the actual chemistry of the very first life systems. In 1953, the year of Watson and Crick and of Miller, things looked very promising, but since then there have been two trends that make the problem of the nature of the first organisms now look more difficult. First, the universal modern life system seems now even more and necessarily sophisticated; it is becoming increasingly difficult to see how to formulate a sufficiently simplified version of a system whose components must be at least fairly complex and very competent to maintain the critical interdependence of functions on which the whole system seems to depend. The modern system is clearly the outcome of very considerable Darwinian processes. The gap between a primitive soup and anything like our system has widened. Second, the very variety of ways in which it is now known that modern biochemical substances can be kicked together under only loosely controlled conditions, although further tending to confirm the idea that modern biochemical substances could have been prominent among the substances around on the primitive Earth, diminishes the argument for any direct causal connection between this and the chemical constitutions of the first systems able to evolve indefinitely through natural selection. The general conclusion now from the abiotic syntheses is that organisms present on Earth now prefer molecules that are easy to make (10). But really this is what you would expect for an advanced life system. It could very well be the consequence of the considerable evolutionary processes that we know in any case must have taken place in creating the modern system. I have discussed this ambiguity recently in greater detail (11) and also, with G. L. Walker, the general question of the evolution of a multiply interdependent life system (12). The abiotic experiments clearly provide important evidence of some kind but it is not yet clear exactly where in the evolutionary story they fit.

I have suggested a particular example of a possible alien primitive life system both as a specific suggestion and to illustrate that such schemes can be formulated to the point at which experimental test becomes feasible (13,14). Indeed, Weiss (15) has recently produced evidence in favour of the central idea that layer silicates can be genetic materials. There may well be many other possibilities accessible to experimental test. Perhaps in the end it will become clear that only a nucleic acid-protein system will really work, but at the very least it is far too early to make any such assertion. Only a much wider view of the possibilities could begin to justify it. As I see it, the proper domain of protobiology is the totality of possible systems that are (a) capable of evolving indefinitely under natural selection and (b) belong to a reasonably large subset of the set of all possible physicochemical systems.*

To clarify the condition (b) let us start from nearer home. The relationship between the here-and-now biological world and the physicochemical world is that the former is a subset of the latter. That is, of all conceivable systems conforming to the laws of physics and chemistry some are organisms using our molecular biology. The problem of the origin of life arises at all because this subset is, relatively speaking, exceedingly tiny. To solve the problem, we must find some much larger kind of biological subset--the protobiological subset--large enough to overlap with reasonable probability another subset of the physicochemical world, namely that constituting all physicochemical systems that actually arose abiotically within a finite period, say a billion years, and a finite collection of atoms, say those constituting the crust of the primitive Earth. Oparin has convinced us that the protobiological subset exists. Our problem is to find it and show how members of it could have evolved in such a way as now to occupy the tiny region of here-and-now biology. I suspect that evolution switched systems, perhaps several times, from an initially "probable" one to the highly "improbable" one that we now see. Thus not only has the effective (occupied) biological set contracted, but it has shifted from its initial position in the field of physicochemical systems.

How, in the most general way, does one go about searching for these hypothetical protobiological systems if there were major biochemical discontinuities in evolution? Is there an appropriate "scientific method?"

One can use analogies. [See Hesse (16,17) for excellent discussions on the use of analogies in science.] To start with, one should use as many analogies as possible that are as diverse as possible both in their corresponding and non-corresponding aspects.

*Modifying Orgel's terminology, this is the study of SITROENS, where S stands for "Simple" (see ref. 7, p. 191).

In his 1924 paper Oparin used many general analogies to feel his way towards the ideas that he wanted to clarify. An organism was like a platinum catalyst (in one way), like a waterfall (in another way), like a growing sulphur crystal (in yet another way) and so on. The founders of modern science used similar analogies to give them a sense of the right direction. Descartes used many such (16), including his famous clock analogy, for organisms to illustrate the crucial idea that an organism is a machine. Newton's idea of forces operating across space could only be understood by analogy with direct experiences of pushing and pulling. Again the idea of chemical affinity started, with the Greeks, in rough analogies with human behavior to be refined later through use. Thinking broadly, like this, one should begin to grasp not a particular solution, not, as it were, a definite object, but a space of an increasingly particular shape--a region of overlap--that is to be filled. More or less slowly a set of requirements is realized--"What we really want is . . . " A specification for the unknown entity is built up. In principle, the superposition of a large enough number of broad considerations can specify an exact solution by specifying a narrow region of overlap. The Greeks were not to know that such a procedure would turn out to be ineffective for the final narrowing of possibilities where experiment and detailed observations are essential. But in the early stages of some quite new kind of enquiry, as now in our search for the first organisms on Earth, "the method of overlap" still has a place, and Oparin's 1924 paper remains a model for the kind of strategic thinking that is still needed.

REFERENCES

1. Oparin, A. I., "The Origin of Life on the Earth," 3rd ed., Oliver and Boyd, Edinburgh and London, 1957.

2. Oparin, A. I., Proiskhozdenie zhizny, Izd, Moskoski Rabochii, Moscow, 1924. Reprinted in English in Bernal, J. D., "The Origin of Life," pp. 199-234, Weidenfeld and Nicolson, London, 1967.

3. Russell, B., "A History of Western Philosophy," Chaps. 2, 4 and 6 on "The Milesian School," "Heraclitus," and "Empedocles," Allen and Unwin, London, 1946.

4. Toulmin, S. and Goodfield, J., "The Fabric of the Heavens," Chap. I, Hutchinson, London, 1961.

5. Watson, J. D., and Crick, F.H.C., Nature 171, 964 (1953).

6. Miller, S. L., Science 117, 528 (1953).

7. Orgel, L. E., "The Origins of Life: Molecules and Natural Selection," Chapman and Hall, London, 1973.

8. Lemmon, R. M., Chem. Revs. 70, 95 (1970).

9. Schopf, J. W., Biol. Revs. 45, 319 (1970).

10. Studier, M., in: "Extraterrestrial Matter" (Randall, C. A., ed.), p. 25, Northern Illinois University Press, Illinois, 1969.

11. Cairns-Smith, A. G., Origins of Life, in press.

12. Cairns-Smith, A. G., and Walker, G. L., BioSystems, in press.

13. Cairns-Smith, A. G., J. Theoret. Biol. 10, 53 (1966).

14. Cairns-Smith, A. G., "The Life Puzzle," Oliver and Boyd, Edinburgh, 1971.

15. Weiss, A., paper presented at the Fourth International Conference on the Origin of Life, Barcelona, June 1973.

16. Hesse, M. B., "Forces and Fields," Nelson, London, 1961.

17. Hesse, M. B., "Models and Analogies in Science," University of Notre Dame Press, Notre Dame, Indiana, 1966.

PHOSPHOLIPID MONOLAYERS--AS A PROTOTYPE OF BIOLOGICAL MEMBRANES

G. A. Deborin and A. D. Sorokina

A. N. Bakh Institute of Biochemistry
USSR Academy of Sciences
Moscow, USSR

One of the fundamental features of the theory of the origin of life on the Earth put forward by Academician A. I. Oparin is the assertion that phase-separated multi-molecular systems of organic substances should necessarily appear in the evolutionary process. This provided possibilities for further chemical and biological evolution of individual open systems, their natural selection, growth, and division (1).

Biological membranes, as the simplest form of spatial organization of matter, apparently, played a key role in the organization and evolution of the cell.

Surface mono- and poly-molecular films of lipid-like substances, and first of all phospholipids, should be considered as the primary, self-forming spatial structure, which is the molecular prototype of the biological membranes. Such films are capable of interacting with primary polypeptides, protein-like substances, oligo- and polysaccharides, nucleoproteins, and other biologically important compounds (2).

In view of this, great attention has been paid in recent years to the investigation of phospholipid surface films at the interfaces, interaction of phospholipid monolayers with biologically active substances, and the influence exerted on them by various chemical and physical factors. The results obtained produced valuable information about the formation conditions, packing, and basic parameters of such films, and allowed a number of inferences about molecular mechanisms of biological membranes.

We have studied the behavior at the liquid-air interface of the nucleoproteins of various forms and packing: rod-like tobacco mosaic virus (TMV) and spherical particles--ribosomes of germinated peas, as well as their interaction with a model phospholipid membrane.

In order to form a monolayer, 0.045 mg of egg lecithin dissolved in benzene was placed at the water surface of a Langmuir trough with the help of a microsyringe. The subsolution contained a 0.01 M solution of $MgCl_2$.

The solution of lecithin in benzene was presonicated for 45 min by cooling in a gaseous helium atmosphere at the piezocrystal oscillator (the ultrasonic wave frequency constituted 850 KHz, intensity 6 to 8 W/sm). The experiment was carried out by cooling to 0°C.

Suspended cytoplasmic ribosomes of germinated peas were carefully introduced under the phospholipid film into the subsolution volume.

The ribosomes were allowed to diffuse freely for 20 minutes, after which the film compression and decompression isotherms were taken by the Wilhelmy balance. The interaction of ribosomes and the phospholipid monolayer was evaluated by the changes in the compression isotherm and by electron microscopic observations.

The compression and decompression isotherms of lecithin monolayers are shown in Fig. 1. As seen from the figure (curves 1 and 2), the compression and decompression isotherms of the unsonicated lecithin films,and lecithin films with ribosomes introduced under them, coincide in shape and position, have the same inclination towards the x-coordinate, and similar values of maximum pressures. Based on this, it may be assumed that ribosomes do not enter the surface of the unsonicated lecithin film.

Electron microscopic photographs of such films (3) show single ribosomes under the monolayer on the liquid phase side. Obviously, ribosomes do not penetrate into the unsonicated lecithin film. Similar observations were carried out by other authors, who studied the interactions of phospholipid monolayers with certain proteins and toxins of protein nature. In the works by Tiffany and Blough (4), for example, the mixo- and paramixo-viruses (influenza A_2) adsorbed on the monomolecular films of the phosphatidylcholine only in case of its modification by protein. The penetration of the staphylococcus α-toxin into the egg lecithin monolayer (as shown by Bukelev and Colaccico, 5), occurred only in the presence of protein concentrated in the hypophase. As suggested by Quinn (6), the cytochrome "C" did not penetrate into the phospholipid monolayer, and such interaction occurred only under special conditions.

The authors assume that this is accompanied by the appearance of electrostatic and Vander Waals bonds, but do not disclaim the

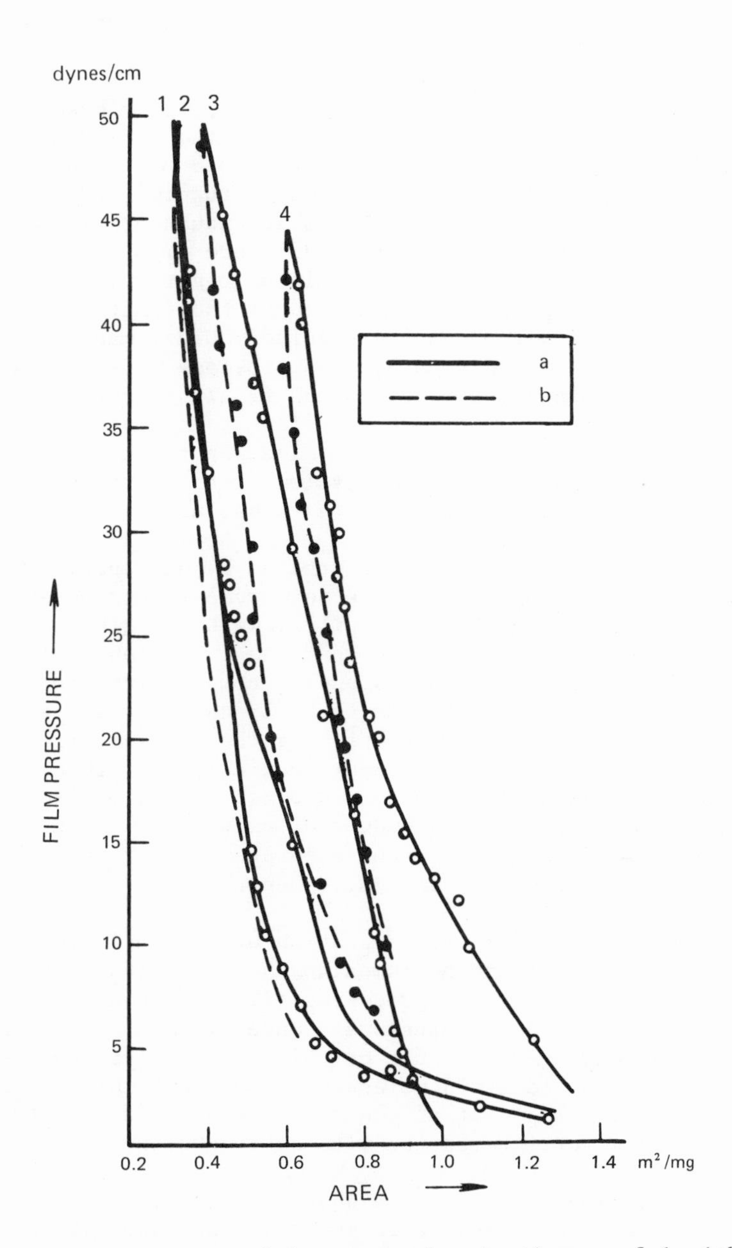

Figure 1. Compression and decompression isotherms of lecithin monolayers: (a) compression, (b) decompression. (1) Unsonicated lecithin monolayer, (2) ditto with ribosomes, (3) sonicated lecithin monolayer, (4) mixed monolayer of sonicated lecithin and ribosomes.

possibilities of hydrophobic interactions between protein and phospholipid molecules.

In the absence of the lecithin monolayer, the ribosomes themselves did not form a film on the surface of the 0.01 M $MgCl_2$ solution. The ribosome film at the interface could not be detected by surface pressure measurements. Electron microscopic (3) measurements showed that there was a considerable amount of ribosomes at the interface. Obviously, under the influence of surface forces, some of them were combined to form small aggregates of 2-4 ribosomes. In the check preparation, a drop of the fraction from the sucrose gradient was placed on the specimen holder made of formvar-plated copper grid. In this case, the ribosomes were single and had clear boundaries. Apparently, diffusing freely from the subsolution volume, the ribosomes do not form any film at the interface, as they do not undergo any structural changes, leading to the development of surface activity. During the compression of the surface with a barrier, they probably shift back into the subsolution.

The monolayer of lecithin, pre-treated by ultrasonic waves, possessed different properties. The measurements of the surface pressure of the sonicated lecithin films, with the ribosomes introduced into the subsolution, showed that the formation of a mixed monolayer is possible. The corresponding compression and decompression isotherms shown in the charts (Fig. 1, curve 4) was shifted to the side of the greater areas occupied by the lecithin-ribosome mixed film, as in the case of the pure sonicated lecithin. Besides, as shown in the figure, the isotherm did not have a liquid-expanded part, which testifies that the amount of particles in the film increased. Electron photomicrographs also proved that the ribosomes entered the monolayer of the sonicated lecithin.

As the conditions for the sonication of the lecithin-benzene solution, by cooling in the helium atmosphere, excluded the possibility of ultrasonic chemical processes (7), the changes obtained are, obviously, related to structural conversions of lecithin in the field of ultrasound. As a result of this, new active regions may get "exposed" in the lecithin molecules. Most probably, this happens in the hydrophilic part of the molecule which faces the liquid phase and is open for interaction with nucleoproteins.

In the experiments with TMV a phospholipid mixture, isolated from cattle brain and dissolved in benzene, was used. The surface concentration of phospholipids in all the experiments constituted 0.22 mg/m^2. In order to produce mixed films, 1 ml of virus suspension, containing 9×10^3 of particles, was introduced into the subsolution volume, after which a monolayer of phospholipid was spread onto its surface. The compression and decompression isotherms of the pure phospholipid and the phospholipid with TMV coincided well. The similarity of the curve shape, the same inclination towards the x-coordinate, and similar values of maximum pressures, allow us to assume that TMV does not enter the phospholipid surface film. What happens, apparently, is the adsorption of the TMV particles diffusing from the volume towards the surface of the phospholipid film from the liquid phase side. Different features were characteristic of the monolayer of phospholipid which was pre-treated by the ultrasound under the conditions described above.

The compression and decompression isotherms of the phospholipid, sonicated over a different period of time in the helium atmosphere, are shown in Fig. 2. It can be seen that the monolayer of the sonicated phospholipid is more condensed and more elastic than the monolayer of the unsonicated phospholipid. The compression and decompression curves of phospholipid films treated by ultrasound are shifted towards the side of the greater area values. With the increased duration of the sonication, the limiting area occupied by the phospholipid molecule increased as well. Also increased is the gap between the initial and ultimate values of the pressures applied. The tangent of the compression isotherm inclination angle increases as well. Peculiar is the fact that the sonicated phospholipid film can be compressed more than that of the sonicated phospholipid. This shows the growth of the aggregative forces between the monolayer-forming molecules. The fact that the sonicated phospholipid monolayers acquired the capability to interact with virus particles, forming a phospholipid-virus mixed film, indicated a change in the conformation of phospholipid molecules. In case the surface film is formed by the molecules of brain phospholipids, which did not undergo a structural rearrangement, a phospholipid-TMV mixed film is not formed.

The compression and decompression isotherms of the sonicated phospholipid films and the sonicated phospholipid with TMV are shown in Fig. 3. The B curve is shifted considerably towards the side of the greater areas compared to the A curve. This may testify to the increase in the amount of particles in the surface layer as a result of the TMV penetration into it. The collapse pressure of the phospholipid-TMV film (45-50 dynes/sm) is higher than that of the pure phospholipid film (37-40 dynes/sm). The increase in the collapse pressure is indicative of a greater stability of a mixed film. The phospholipid-TMV B curve has a flatter course than the A curve. This shows that the monolayer changes from the condensed to the liquid-expanded state. Similar phenomena were observed by studying the interaction of the TMV with the lipovitellin monolayers (8).

The films were also investigated by electron microscope (9). Seen clearly on the photos were the differences in the distribution of the hydrophilic and hydrophobic sites of the phospholipid monolayer. This indicates that under the influence of ultrasound the molecules are considerably rearranged. As the lyophobic part of the monolayer changed less than the lyophilic one, it may be assumed that the changes are related to a disturbing of the electrostatic inter-

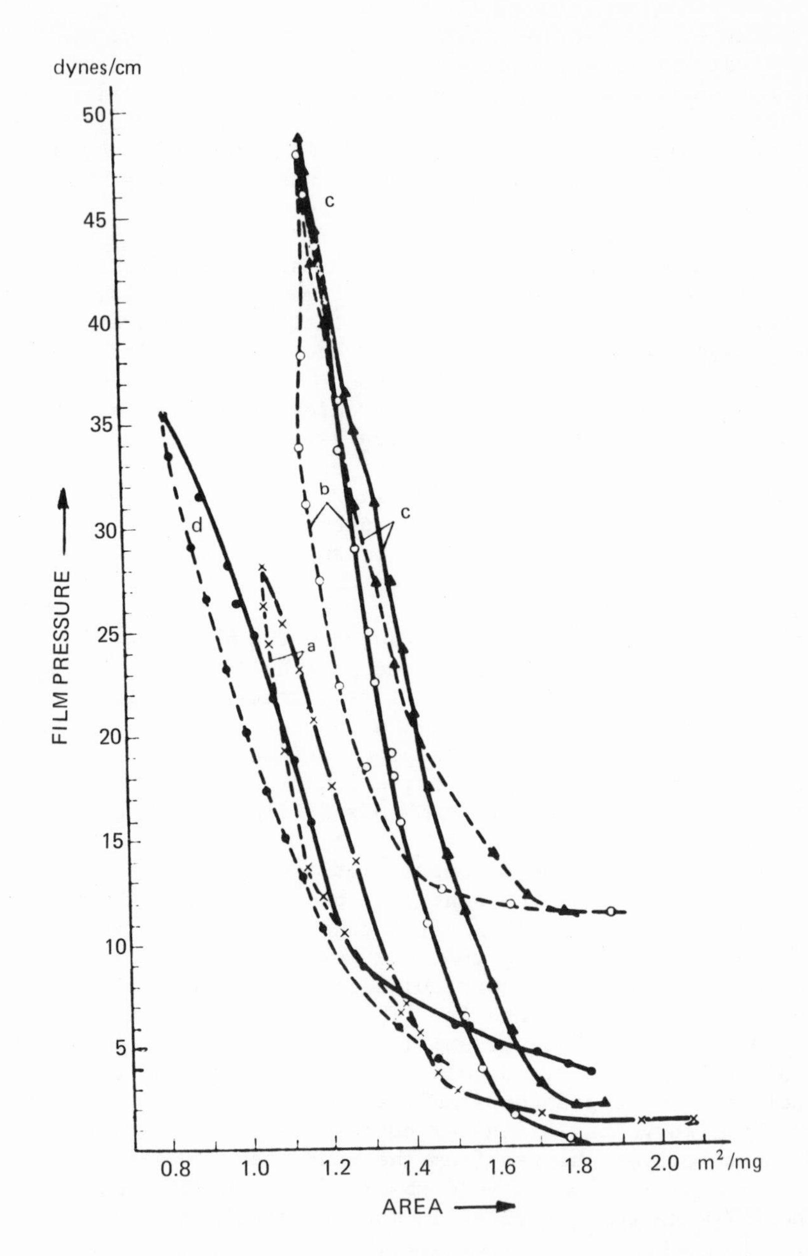

Figure 2. Phospholipid monolayer compression isotherms after treatment with ultrasound for 30 min (a), 45 min (b), and 60 min (c); unsonicated solution (d). Solid curves: compression; dotted curves: decompression.

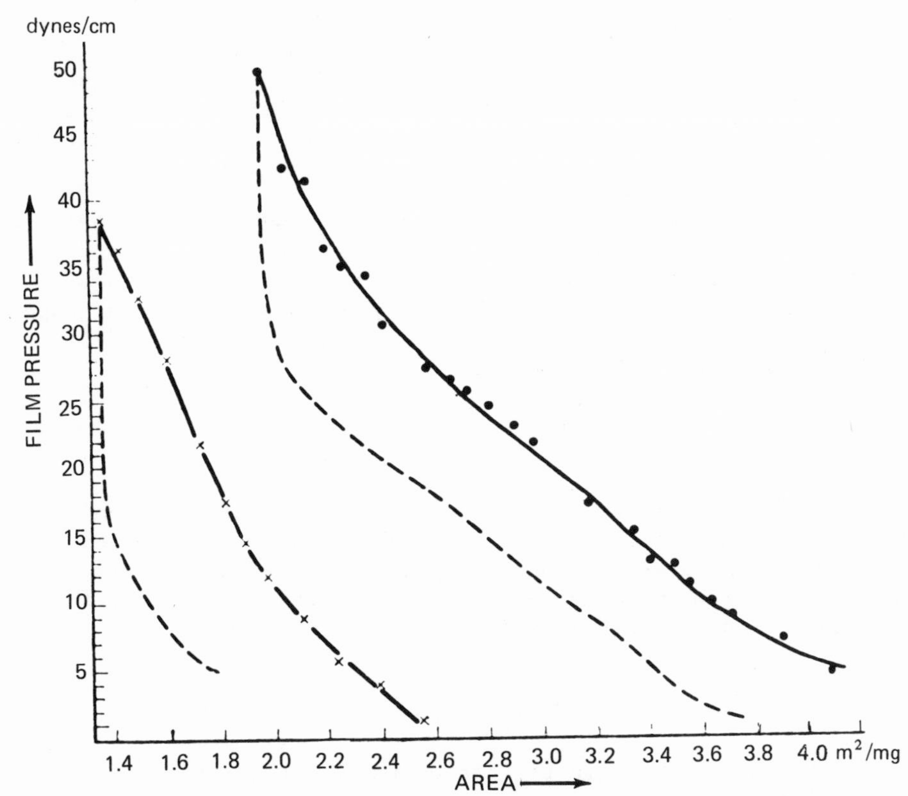

Figure 3. Compression isotherms of sonicated phospholipid (a) and sonicated phospholipid with TMV (b). Solid curves: compression; dotted curves: decompression.

actions in the lyophilic part of the phospholipid molecule placed into the interphase.

Structural rearrangements of the phospholipid molecules during the sonication were accompanied by change in behavior of the monolayer. The monolayer of the sonicated phospholipid acquired the capability to interact with nucleoproteins. It is natural to assume that ultrasound is only one of the factors modelling the complicated processes of regulating the structure of the phospholipid surface films. However, on the basis of these data, a conclusion may be drawn that the alterations in the state and the structure of the biological membrane or its components affect considerably its interaction with nucleoproteins and other substances vital for cells.

Under the conditions of the period of chemical evolution of our planet, various radiations, thermal factors, electric discharges, etc. could serve as energy sources altering the structure of lipid and protein surface films side-by-side with ultrasonic natural fluctuations. The possible role of ultrasonic energy on the primitive Earth is discussed by Elpiner (7). This view becomes even more likely, as it has been convincingly proved (10,11,12), that the above-mentioned energy factors ensured the abiogenesis of complex organic compounds and multimolecular systems under the conditions of the primary Earth.

In our experiments we suppose that the physicochemical principles of monolayer arrangement and interactions are the same as in the earlier stages of evolution. Therefore, it is permissible to use compounds, isolated from contemporary organisms for model experiments, that have the aim of demonstrating principles of such interaction both in primitive and modern systems.

ACKNOWLEDGMENTS

The authors gratefully acknowledge the gifts of TMV from T. Podjapolskaya and germinated pea ribosomes from I. Filippovich, and A. Tongur for assistance in electron microscopy of monolayers, and the Institute of Chemical Physics, USSR Academy of Sciences, for the sonication of our preparations.

REFERENCES

1. Oparin, A. I., "Origin of Life on Earth," U.S.S.R. Acad. of Sciences Publ., 1957.

2. Deborin, G. A., "Protein Lipid Surface Films--as a Prototype of Biological Membranes," U.S.S.R. Acad. of Sciences Publ., 1967.

3. Deborin, G. A., Sorokina, A. D., Tongur, A. M., and Fillippovich, I. I., Dokl. Akad. Nauk SSSR 206, 482 (1972).

4. Blugh, H. A., and Tiffany, J. M., FEBS Abstr. of Commun. 204, (1971).

5. Buckelev, A. R., and Collaccico, J., Biochim. Biophys. Acta 233, 7 (1971).

6. Quinn, P. J., and Dawson, R. M., Biochem. J. 113, 791 (1969).

7. Elpiner, I. E., "Biophysics of Ultrasound," Nauka Publ. House, Moscow, 1973.

8. Elpiner, I. E., Deborin, G. A., Baranova, V. Z., Sorokina, A.D., and Tongur, A. M., Biophysics 16, 44 (1971).

9. Deborin, G. A., Elpiner, I. E., Baranova, V. Z., Sorokina, A.D., and Tongur, A. M., Dokl. Akad. Nauk SSSR 198, 1445 (1971).

10. Calvin, M., "Chemical Evolution," Oxford University Press, London, 1969.

11. Fox, S. W., and Dose, K., "Molecular Evolution and the Origin of Life," Freeman and Co., San Francisco, 1972.

12. "The Origins of Prebiological Systems" (Fox, S. W., ed.), Academic Press, New York, 1965.

PEPTIDES AND AMINO ACIDS IN

THE PRIMORDIAL HYDROSPHERE

Klaus Dose

Institut für Biochemie
Universität Mainz
D-6500 Mainz BRD

INTRODUCTION

In this paper Oparin's original concept of the primordial ocean being a "primeval broth" in which "protobionts" have evolved is reviewed on the basis of recent and previous data on the formation and decomposition of peptides and amino acids. It will be shown that peptides and amino acids could never reach significant concentrations in the ocean. Locally, in pools of transient existence, however, amino acids, peptides, and other molecules of biological significance could have been concentrated by various geological processes to yield a "primeval broth" in which prebiotic systems such as proteinoid microspheres and coacervate droplets, could have evolved.

RESULTS AND DISCUSSION

Equation (1) describes the essential reaction of peptide-bond formation:

$$^{\oplus}H_3NCHRCOO^{\ominus} + {}^{\oplus}H_3NCHR'COO^{\ominus} \rightleftharpoons {}^{\oplus}H_3NCHRCONHCHR'COO^{\ominus} + H_2O \qquad \Delta G° = 2\text{-}4 \text{ kcal} \qquad (1)$$

Borsook and Huffman (1-3) published a first analysis of the thermodynamics of this reaction. Detailed thermodynamic data are available for the formation of simple peptides; e.g., in the case of the condensation of the two most common α-amino acids, glycine and alanine, according to equation (1a) and (1b):

$$\text{gly} + \text{gly} \rightleftharpoons \text{gly}\cdot\text{gly} + H_2O \qquad (1a)$$

$$\text{ala} + \text{ala} \rightleftharpoons \text{ala}\cdot\text{gly} + H_2O \qquad (1b)$$

the corresponding free enthalpies (at 37°C and pH 7.0) are 3.59 kcal and 4.13 kcal respectively (3). According to $\Delta G° = -RT \times \ln K$, the equilibrium constant of these reactions is only about 10^{-3}. The equilibrium concentration of these dipeptides for 1 M solutions of the free amino acids is thus only slightly above 10^{-5} M. The thermodynamic barrier is even larger for the formation of a polypeptide or a protein molecule in aqueous solution. Dixon and Webb (4) have pointed out that solutions 1 M in each of the twenty proteinous amino acids would at equilibrium yield a 10^{-9}gM concentration of a protein (M.W. = 12,000). The volume of this solution would have to be 10^{50} times the volume of the Earth to yield at equilibrium one molecule of this protein. To a small extent the requirement for energy is diminished by location of peptide bonds in the interior of the protein (5), but even under favorable circumstances the probability of generation of polypeptides in an aqueous system is extremely small.

The unlikelihood of spontaneous peptide-bond generation in the primitive ocean is in addition emphasized by the low equilibrium concentration of amino acids, as will be shown in the following paragraphs. Most of the amino acids available on the surface of the prebiotic Earth were formed from the constituents of the primordial atmosphere and hydrosphere. To promote this complex process of amino acid production, the starting material had to be activated by various forms of energy. After their formation in the atmosphere, the amino acids stayed there only for a relatively short time. Because of their low partial pressure and their high solubility in water, they were washed out and dissolved in the hydrosphere where they accumulated to a certain degree. The amino acids could reach only relatively low average concentrations, however, because of their decomposition by photochemical, radiationchemical, and thermal reactions. Data permitting estimates of the rate of production and decomposition of amino acids are available. These data also allow us to estimate the average equilibrium concentration of the amino acids in the hydrosphere.

The rate of amino acid production depends of course on the concentration of the raw materials and the energies available per unit time. The energies available from various sources to produce amino acids from the simple constituents of the primitive atmosphere and hydrosphere are summarized in Table I.

The amount of solar radiation below 2000 Å has been calculated from data presented by Jager (7) according to Fox and Dose (6). The actual values may have been lower or higher, but it is unlikely that they exceeded at any time the present value of 75 cal x cm^{-2} x y^{-1}. Only ultraviolet light below 2000 Å is directly absorbed by simple

Table I

Energies Available from Various Sources to Produce Amino Acids on the Earth 4×10^9 to 4.5×10^9 Years Ago

(Data after Fox and Dose, 6)

Type of Energy	Projected per cm^2 Surface and Year ($cal \times cm^{-2} \times y^{-1}$)
Solar radiation below 2000 Å	30
High-energy radiation (from the crust 35 km deep)	47
Heat from volcanic emissions (rocks and lava only)	$>0.15^*$
Electric discharges	4^*

*Found for the contemporary Earth. The values for the primordial Earth may have been significantly higher.

molecules such as H_2, H_2O, CH_4, CO, NH_3, and N_2, which may have served as the starting material for chemical evolution.

Only a small fraction of the high energy radiation within the Earth's crust may have actually contributed to the production of amino acids. The heat from volcanic emissions represents a lower limit. On the primitive Earth, the volcanic activity may have been significantly higher. The same applies to the electric discharges in the atmosphere, although in this case the whole amount of this kind of energy was released above the surface of the Earth and thus could directly contribute to amino acid production.

In Table II are summarized approximate values on the expenditure of energy to produce 10^{-9} mole of glycine from the constituents of simulated primordial atmospheres. The data have been deduced by Fox and Dose (6) from the results of a number of experiments carried out in various laboratories. Also presented in Table II are rates of glycine production calculated from the energy expenditures per 10^{-9} mole of glycine and the data on energies available from Table I. The actual rates of glycine production by the various kinds of energy were likely lower because (except for electric discharges) only a fraction of these energies was available for amino acid production on the surface of the Earth.

We shall assume that the total rate of amino acid production was about 7 times the rate of glycine production. This is an

Table II

Approximate Expenditure of Energy for the Production of Glycine from the Constituents of Simulated Primordial Atmospheres [after Fox and Dose (6) and calculated rates of glycine production on the primordial Earth]

Form of energy	Cal per 10^{-9} mole of glycine produced	Calculated rates of glycine production (moles of glycine x cm^{-2} x y^{-1})	
Ultraviolet light below 2000 Å	220	0.14×10^{-9}	
High-energy radiation	7.5*	6.26×10^{-9}	<u>Total</u>:
Heat	2	0.08×10^{-9}	7.28×10^{-9}
Electric discharges	5**	0.80×10^{-9}	

*Average from values for X-rays and β-rays

**Experimental values between 2 and 20 depending on kind of experiment.

optimistic estimate in view of the data from the laboratory because the yields of glycine are usually of the same order of magnitude as the total yield for the production of all other amino acids together (6). Thus the total rate of amino acid formation could have been 5×10^{-8} moles x cm^{-2} x y^{-1}.

To estimate the steady state concentration of amino acids, we must also consider the rates of amino acid destruction on the surface of the Earth. Hull (8) has estimated that only 3% of the amino acids formed in the upper atmosphere would have survived photochemical decomposition by solar radiation and reached the surface of the Earth, where they finally accumulated in the ocean. In our calculation, we shall neglect the destruction of amino acids occurring in the atmosphere, in order to approximate an upper limit of the steady state concentration of amino acids in the ocean.

The rate of amino acid destruction depends on the fluxes of the same kinds of energy which produce them. We shall in our consideration, however, neglect the effects of heat, radioactivity, and electric discharges. These forms of energy could reach only a tiny fraction of the amino acids once they were dissolved in the ocean. The ambient temperature of the hydrosphere (about 20°C) would have permitted thermal half-lives well above 1000 years (9-11). Thus we assume that only solar ultraviolet light below 3000 Å was significant in decomposing the amino acids of the primitive ocean. We estimate for this spectral range a flux, I, of 1700 cal x cm^{-2} x y^{-1} or 10^{22} quanta x cm^{-2} x y^{-1}, which is 1% of the total flux of optical solar radiations about 10^9 years ago (6,7). The quantum yield, ϕ, of amino acid destruction by this radiation is about 10^{-2} (12). With an absorption coefficient, ε, of about 0.15 lit. x $moles^{-1}$ x cm^{-1} and an absorption cross-section of $\sigma \simeq 6 \times 10^{-22}$ cm^2 x $molecule^{-1}$ for non-aromatic amino acids, the t_{37}-time of the amino acids can be calculated according to

$$\frac{A}{A_0} = e^{-\phi I \sigma t},$$

where A/A_0 is the fraction of molecules "surviving" the radiation time t. $A/A_0 = \frac{1}{e}$ yields $t_{37} \simeq 16.6$ years. The average equilibrium concentration of amino acids is thus 8.3×10^{-7} moles x cm^{-2} or, if the primitive ocean was on the average 100 m deep, 8.3×10^{-8} moles x $lit.^{-1}$. This concentration could have been somewhat higher because solar ultraviolet light probably penetrated only the upper 19 meters of the ocean (13,14), but such a correction does not significantly change the picture because the circulation of the ocean exposed the amino acids from greater depths to ultraviolet light also. Moreover, the rate of amino acid destruction in the atmosphere, which has been neglected here, would more than compensate for the partial protection of amino acids in greater depths of the ocean. Accordingly, we arrive at an estimated order of magnitude of 10^{-6} moles x cm^{-2} or 10^{-7} moles x $lit.^{-1}$ for amino acids in the primordial ocean. Such a concentration is by far too low for direct formation of polymers and more complicated structures. It may be compared with the concentration of free amino acids in the contemporary North Atlantic Ocean, which varies from 6 to 47 μg x $lit.^{-1}$ or from 5×10^{-8} to 4×10^{-7} moles x $lit.^{-1}$ according to Pocklington (15). The low values apply to samples from a depth of 3000 to 5000 meters; the high values have been found in samples from the surface. Zones of higher concentrations do not exist. It is therefore likely that the mean concentration of free amino acids in the primordial ocean was never significantly different from the present level.

The chances for polypeptides or polynucleotides to survive in any aqueous system are much smaller than those for the free amino acids. In this case, we may neglect the possible decomposition by the various forms of radiation (including ultraviolet light) and electric discharges because the thermal hydrolysis occurs relatively rapidly. As already indicated, peptide bond hydrolysis is an exothermic reaction. The activation energy in an aqueous solution can be 25000 calories per mole and higher (1,16). The activation energy of peptide hydrolysis corresponds with a half-life of the order of

ITHACA COLLEGE LIBRARY

days and months. In the presence of catalytically active minerals, these half-lives can be significantly lower. They will be larger, however, for native proteins and supramolecular assemblies of protein(oid)s, because in these cases the peptide links are better shielded against hydrolysis for steric reasons. No source of polypeptides can be visualized which could keep the peptides in the ocean at a level high enough, perhaps at least a 1% solution, to allow a spontaneous organization of prebiotic systems.

Hence, we arrive at the conclusion that the primitive ocean may have constituted a huge reservoir for all amino acids and other kinds of molecules of biological significance, but their concentration was too low to permit the direct selforganization of prebiotic systems. On the basis of the data supplied by Rubey (17,18) and Urey (19) less than one-tenth of the volume of water in the contemporary oceans was originally present on the surface of the Earth. The volume of the contemporary oceans is about 1.37×10^9 km^3; the primitive waters therefore had a volume of about 10^8 km^3 or 10^{18} lit. Hence, the average amount of amino acids present in this volume was about 10^{13} moles or as much as 10^9 tons. We arrive at similar values if we multiply the calculated equilibrium concentration of about 10^{-6} moles x cm^{-2} with the area of the surface of the Earth.

A variety of mechanisms could have operated to remove amino acids and other products from the ocean or from the zone of their formation, simultaneously concentrating them to levels where reactions between them would be possible and lead to more highly organized systems (6,20,21). These models of concentration include recessions of the seas, and drought and concentration by evaporation at elevated or ambient temperatures in lagoons (20,21). In a subsequent process the amino acids could have been condensed to yield protein(oid)s at temperatures above the boiling point of water* by volcanic heat (22). At such temperatures the amino acids are automatically dry. Once deposited with minerals at ambient temperature and in the absence of water amino acids, polypeptides, and proteins survive millions of years without considerable decomposition (23, 24). But in contact with water, the half-life of polypeptides and protein(oid)s is of no geological significance. As, in particular, Fox (25) has demonstrated, the selforganization of some proteinoids to microstructures (pre-organelles, pre- or proto-cells, microspheres) is a matter of minutes. Similarly, coacervate droplets are quickly formed from related starting materials (26-29).

In typical experiments to produce proteinoid microspheres, the concentration of proteinoids is initially well above 10%. Such high concentrations of organic materials could only exist in very limited

*Or at lower temperatures in the presence of phosphates.

parts of the primitive hydrosphere, e.g. in small pools, e.g. in a perivolcanic environment. It must be emphasized, however, that the formation of this kind of pool at any time in the history of the Earth is so far not backed by geological evidence, although pools which are rich in inorganic materials (e.g., tidal pools, pools at the Soufriere of St. Lucia, or in Yellowstone Park) are well-known geological phenomena.

The uneven distribution of polyatomic species such as formaldehyde and ammonia in interstellar space (30) and their possible role as precursors to prebiotic amino acids (31) supports the suggestion that on planets different from the Earth pools containing all compounds necessary to start the evolution of protobionts could have been formed with higher probability than on Earth. For such pools rich in organic materials, wherever they were formed, the Oparin-Haldane concept of the "primeval broth" or "thin soup" of organic molecules of biologically significant molecules ready to undergo self-assembly to more complex structures keeps its validity beyond doubt. The prebiotic polymers of this soup decompose within days and weeks. But as indicated above, under favorable conditions "protobionts" could have evolved within much shorter periods of time.

REFERENCES

1. Borsook, H., and Huffman, H. M., in: "Chemistry of the Amino Acids and Proteins" (Schmidt, C.L.A., ed.), p. 822, Charles C Thomas, Springfield, Ill., 1944.

2. Borsook, H., Advances Protein Chem. 8, 127 (1953).

3. Borsook, H., Chemical Pathways of Metabolism 2, 173 (1954).

4. Dixon, M. A., and Webb, E. C., "Enzymes," p. 666, Academic Press, New York, 1958.

5. Fruton, J. S., and Simmonds, S., "General Biochemistry," p. 712, 2nd ed., Wiley, New York, 1958.

6. Fox, S. W., and Dose, K., "Molecular Evolution and the Origin of Life," Freeman, San Francisco, 1972.

7. Jager, C. de, in: "Radiation Research" (Silini, G., ed.), North-Holland, Amsterdam, 1967.

8. Hull, D. E., Nature 186, 693 (1960).

9. Abelson, P. H., Mem. Geol. Soc. Amer. II 67, 87 (1957).

10. Conway, D., and Libby, W. F., J. Am. Chem. Soc. 80, 1077 (1958).

11. Vallentyne, J. R., Carnegie Inst. Wash. Yearbook 56, 185 (1956).

12. McLaren, A. D., and Shugar, D., "Photochemistry of Proteins and Nucleic Acids," p. 97, Pergamon Press, Oxford, 1964.

13. Berkner, L. V., and Marshall, L. C., J. Atmos. Sci. 22, 225 (1965).

14. Berkner, L. V., and Marshall, L. C., J. Atmos. Sci. 23, 133 (1966).

15. Pocklington, R., Nature 230, 374 (1971).

16. Abelson, P. H., in: "Organic Geochemistry" (Breger, I. A., ed.), Pergamon, Oxford, 1963.

17. Rubey, W. W., Geol. Soc. Amer. Bull. 62, 111 (1951).

18. Rubey, W. W., Geol. Soc. Amer. Spec. Paper 62, 631 (1955).

19. Urey, H. C., "The Planets," Yale University Press, New Haven, 1952.

20. Bernal, J. D., "The Physical Basis of Life," Routledge and Kegan Paul, London, 1951.

21. Bernal, J. D., in: "Proceedings of the First Intern. Symp. on Origin of Life on the Earth," Moscow, 1957; pp. 23-35, Pergamon, London, 1959.

22. Fox, S. W., Vegotsky, A., Harada, K., and Hoagland, P. D., Ann. N.Y. Acad. Sci. 69, 328 (1957).

23. Florkin, M., in: "Chemical Evolution and the Origin of Life," Molecular Evolution, Vol. 1 (Buvet, R., and Ponnamperuma, C., eds.), p. 10, North-Holland, Amsterdam, 1971.

24. Hare, P. E., in: "Organic Geochemistry," Results and Methods, p. 438, Springer Verlag, New York, 1969.

25. Fox, S. W., Science 132, 200 (1960).

26. Jong, H.G.B. de, Dekker, W.A.L., and Gwan, O. S., Biochem. Z. 221, 392 (1930).

27. Oparin, A. I., in: "The Origins of Prebiological Systems" (Fox, S. W., ed.), p. 331, Academic Press, New York, 1965.

28. Oparin, A. I., "The Origin and Initial Development of Life," Meditsina Publishing House, Moscow, 1966.

29. Evreinova, T., and Bailey, A., Dokl. Akad. Nauk SSR 179, 723 (1968).

30. Donn, B., Science 170, 1116 (1970).

31. Fox, S. W., and Windsor, C. R., Science 170, 984 (1970).

AMINO ACIDS AND CARBOHYDRATES IN PRECAMBRIAN ROCKS

I. A. Egorov and I. Z. Sergiyenko

Bakh Biochemical Institute
USSR Academy of Sciences
Moscow, USSR

The theory of origin of life on the Earth formulated by A. I. Oparin more than 50 years ago for the first time raised a problem of a possible abiogenesis of organic substances in the early period of the Earth's existence.

In this connection, of considerable interest are the investigation of the Earth's oldest rocks in the search for some remains of the earliest living forms or their molecular imprints, detection of biologically important organic substances in them, and determination of the mechanisms of their origin, i.e., whether they were formed abiogenetically or they were only the remains of the primitive organisms existing long ago. Searches for the traces of life during the earliest epoch of the Earth's history are being carried out on both the morphological and molecular levels.

As a result of investigating the Earth's oldest rocks, a great amount of experimental material has already been accumulated which makes it possible to assert that algae, bacteria, and simple and complex organic molecules can be preserved in the Earth's oldest sedimentary rocks from the geological epochs; these can be isolated and identified by the modern methods. Thus the algae, bacteria, and other microorganisms were isolated and studied in the sedimentary deposits of the Early (2), Middle (3), and Late (4) Precambrians, many of these organisms being morphologically similar or identical to the contemporary algal thallophytes (5).

The earliest work reporting study of free and bound amino acids was the paper of W. A. Schmidt (6) who isolated free and peptide-bound amino acids from Egyptian mummy tissue in 1908. In 1922, W. Gothan (7) described some cellulose available in the Early-Tertiary lignin, and in 1933, E. Abderhalden and K. Heyns found chitin remains in the bug wings from Eocenic brown coals (8).

In recent years the investigation of such organic matter has especially expanded to include different archeological, paleontological, and geological objects as well as meteorites. For instance, it is possible to point to the papers of P. H. Abelson (9,10) on the detection of amino acids in the Ordovician trilobites and mineralized fossils, a paper of E. S. Barghoorn (11) on the investigation of the oldest cherts, a paper of J. W. Schopf (12) on the investigation of the chemical composition of fossils and amino acids in the Precambrian sediments, papers of T. V. Drozdova (13), S. M. Manskaya and T. V. Drozdova (14) as well as many others.

A number of investigations are growing on the hydrocarbons (15), porphyrins (16), amino acids (17), carbohydrates (18), fatty acids (19), and a number of other biologically important compounds in the recent and oldest sediments of the Earth's rocks and meteorites (20).

However, as A. I. Oparin (21) points out: "Our knowledge in the field of the investigation of organic matter of the Earth's oldest rocks and its evolution accumulated at present is still contradictory and requires a considerable expansion especially in the investigation of carbon compounds in the deepest parts of the crust and upper mantle."

We studied the Precambrian rocks for their amino acid and carbohydrate contents to expand our knowledge of the organic matter in the Earth's oldest rocks with a view to understanding its history and evolution.

The Early-Cambrian skeletal remains of the Tuva tenialic Archaeocyatha (age exceeding 500 million years; 17), black crystalline schists of the Contiosari formation in the Layzhskaya series of the Early-Proterozoic era (1.8-2.0 billion years ago; 18), and cyanite schists of the Kiev formation on the Kola Peninsular (2.6 billion years ago; 22) were studied.

The method of isolating "free" and "bound" amino acids for all the materials under consideration was generally the same: samples were ground, a weighed portion (10-50 g) was extracted with acidified 80% ethanol (15-200 ml) at 100°C in a water bath with continuous stirring (1-3). "Free" amino acids and sugars were determined in this extract after its centrifugation from the deposit, cleaning, and separation in ion-exchange columns. The deposit produced as a result of centrifugation was hydrolyzed in 6 N HCl (150-300 ml) for 10-12 hrs at 100°C with intensive stirring. "Free" amino acids and sugars were determined in this hydrolysate after its cleaning, neutralization, and separation in ion-exchange columns.

The morphology of the tenialis archaeocyathean skeletons was studied (23). The chemical composition of the organic remains in the skeletons has not yet been investigated. Below are given the data on the tenialic archaeocyathean skeletons, investigated for the first time, to determine the content of amino acids and carbohydrates by the paper chromatographic method; at the same time free and bound amino acids were also determined.

As a result of the paper chromatographic quantitative analysis it was possible to establish that the skeletons of the tenialic archaeocyathean cubes had preserved free and bound amino acids and sugars in their composition. At the same time, the free amino acids on the chromatograms were represented by the following six amino acids: arginine, valine, threonine, phenylalanine, lysine, and a nonidentified amino acid with the lowest R_f. The bound amino acids determined in the hydrolyzate manifested themselves as five amino acids on the chromatogram; only arginine, threonine, leucine, and phenylalanine were identified. Free and bound sugars represented glucose, ribose, and one nonidentified sugar.

Remains of unicellular and colonial algae and initially unclear remains of invertebrate animals were revealed as a result of the morphological investigations in the black crystalline schists of the Contiosari formation of the Karelian Lower-Proterozoic era (absolute age 1.8 -2.0 billion years). Remains of the algae of the family Ptiloptonceae, Vologdin, and spongelike organisms of the family Ladogaellaceae, Vologdin, were isolated and described from such rocks. Soluble bitumens were found by the method of luminescent analysis. The paper chromatographic investigation was carried out for the first time to determine the content of amino acids and carbohydrates; it showed that the alcoholic extracts and acidic hydrolysates isolated from the samples of these schists had free and bound amino acids and sugars in their composition. Thus the free amino acids (alcoholic extract) are represented by the following amino acids: glycine, threonine, valine, phenylalanine, and leucine.

The bound amino acids were found as the following eight amino acids; besides the five already indicated, there were also revealed glutamic acid, methionine, and one nonidentified amino acid. The free sugars are found on the chromatogram as maltose, glucose, mannose, and ribose; the bound sugars as glucose and mannose only, plus one nonidentified carbohydrate.

The content of amino acids and carbohydrates was first investigated in the kyanite schists of the Keyvy formation dated from the Precambrian period and aged 2.6 billion years. The formation stratigraphy and history of this suite are given in detail in the papers by P. V. Sokolov (25) and I. V. Bel'kov (26) where a high level of saturation with the carbon-bearing matter is indicated in the kyanite schists of this formation. The particular formation belongs to sedimentary formations and its sedimentation took place under the conditions of a shallow basin with a comparatively stable structural environment during the rock deep-weathering in the areas of erosion by water and a high degree of sedimentary differentiation in the formation of sediments.

We succeeded in establishing the presence of free and bound amino acids and carbohydrates in the cyanite schist samples of the Kiev formation by using the method of paper chromatography and amino acid analyzer.

Fig. 1 shows a chromatogram of the free and bound carbohydrates isolated from the alcoholic extract (free sugars) and hydrolysate (bound sugars). The free sugars yielded four stains: 1. nonidentified sugar; 2. glucose; 3. mannose; 4. nonidentified sugar. The same four sugars were found in the hydrolysate (bound sugars) with the only difference that the fourth stain of the hydrolysate was much larger in size and more intensive in color.

Figs. 2 and 3 give the results of paper chromatography (Fig. 2) and amino acid analyzer (Fig. 3) in the determination of free and bound amino acids.

The free amino acids appeared on the chromatogram as six stains (Fig. 2): 1. nonidentified stain; 2. serine-aspartic acid; 3. leucine-threonine; 4. alanine; 5. valine-methionine; 6. phenylalanine. The bound amino acids produced seven stains on the chromatogram (Fig. 2): 1. nonidentified amino acid; 2. serine-aspartic acid; 3. lysine; 4. threonine-leucine; 5. alanine; 6. valine-methionine; 7. phenylalanine.

The graph of the amino acid analyzer shows the free amino acids (Fig. 3) as 13 amino acids: 1. aspartic acid; 2. threonine; 3, serine; 4. glutamic acid; 5. proline; 6. glycine; 7. alanine; 8. valine; 9. methionine; 10. isoleucine; 11. leucine; 12. tyrosine; 13. phenylalanine; and the bound amino acids are represented by eleven amino acids except for aspartic acid and methionine.

The results of investigating the composition of free and bound amino acids with the help of paper chromatography and the amino acid analyzer, when compared with each other, are well coordinated and supplement each other. It should be added that depending on the sample in determining bound amino acids some noticeable quantities of aspartic acid are recorded on some graphs of the amino acid analyzer not presented here. Taking into consideration this additional material, it is possible to say that the free and bound amino acids have the same composition, considering that methionine is usually destroyed in hydrolysis.

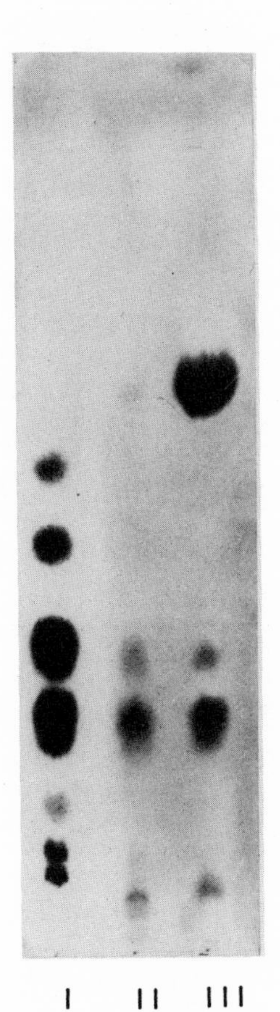

Figure 1. Chromatogram of "free" and "bound" carbohydrates.

I. Evidences: 1. lactose; 2. maltose; 3. saccharose; 4. glucose; 5. mannose; 6. ribose; 7. rhamnose.

II. "Free" carbohydrates: 1. nonidentified sugar; 2. glucose; 3. mannose; 4. nonidentified sugar.

III. "Bound" carbohydrates: 1. nonidentified sugar; 2. glucose; 3. mannose; 4. nonidentified sugar.

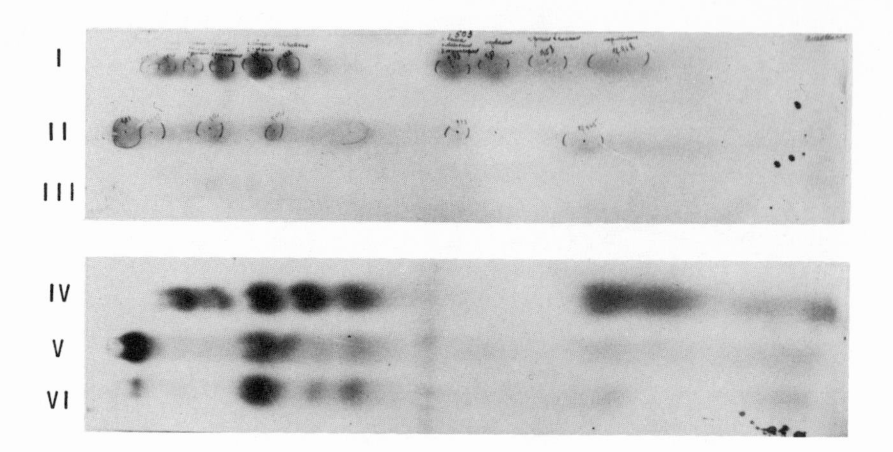

Figure 2. Chromatogram of free and bound amino acids.

I. Evidences: 1. cysteine; 2. ornithine-lysine; 3. histidine-asparagine-arginine; 4. serine-aminoacetic acid; 5. leucine-threonine; 6. alanine; 7. valine-methionine; 8. norvaline; 9. phenylalanine; 10. norleucine.

II. Bound amino acids: 1. nonidentified amino acid; 2. lysine; 3. serine-aspartic acid; 4. threonine-leucine; 5. alanine; 6. valine; 7. phenylalanine.

III. Control of reagents and purity of operation.

IV. Evidences: 1. cysteine; 2. ornithine-lysine; 3. histidine-asparagine; 4. serine-glycine; 5. leucine-threonine; 6. alanine; 10. norleucine.

V. and VI. Free amino acids: 1. nonidentified amino acid; 2. serine-aspartic acid; 3. leucine-threonine; 4. alanine; 5. valine-methionine; 6. phenylalanine.

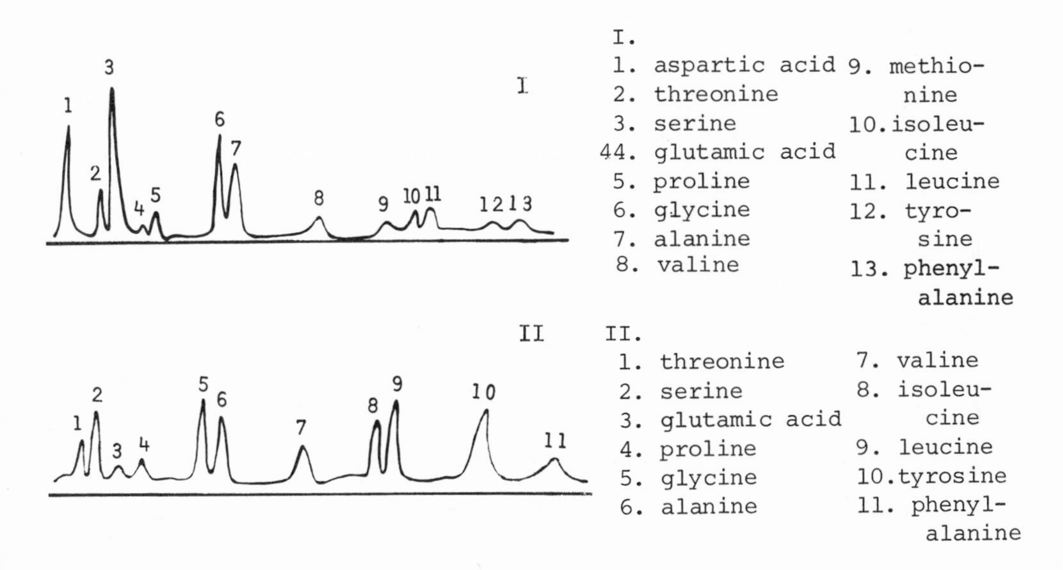

Figure 3. Chromatogram of free and bound amino acids obtained with the "Hitachi" amino acid analyzer.

I. Free amino acids.

II. Bound amino acids.

Thus, the investigations of the Lower-Cambrian skeletal remains of the tenial archaeocyatheans (500 million years ago), the black crystalline schists of the Contiosari formation in the Ladozhskaya series of the Early-Proterozoic era (1.8 - 2.0 billion years ago), and the cyanite schists of the Kola Peninsular Kiev formation (2.6 billion years ago), for the content of amino acids and carbohydrates with the method of paper chromatography and amino acid analyzer have shown that amino acids and carbohydrates can be preserved in the Earth's sedimentary rocks in the fossil state during the geological periods and can be isolated and identified by modern methods. Based on these data it is impossible to judge for certain the origin of the compounds found, as recently the investigations in abiogenic synthesis have shown a possible formation of organic compounds abiogenically, even the most complex such as porphyrins or isoprenoids (phytane and prystane) which until recently have been considered as signatures of biological origin. Nevertheless, based on the literature data which indicate that the organisms have already existed over 3 billion years ago it is possible to suppose that these amino acids and carbohydrates are of biological origin.

REFERENCES

1. Oparin, A.I., "Life, its Nature, Origin, and Evolution", Izd. AN SSSR, Moskva, 1960.

2. Schopf, J.W., and Barghoorn, E.S., Science 156, 508 (1967).

3. Cloud, P.E., and Hagen H., Nat. Acad. Sci. U.S. 54, 1 (1965).

4. Barghoorn, E.S., and Schopf, J.W., Science 150, 337 (1965).

5. Schopf, J.W., in McGraw-Hill Yearbook of Science and Technology, p. 46 (1967).

6. Schmidt, W.A., Z. Allgem. Physiol. 7, 369 (1908).

7. Gothan, W., "Non Arten der Braunkohlen", Ezsuchung-IV Braunkohle. Vol. 21, 8, 400 (1922).

8. Aberhalden, E., and Heyns, K., Biochem. Z. 259, 320 (1933).

9. Abelson, P.H., Science 119, 576 (1954).

10. Abelson, P.H., Sci. Amer. 195, (I), 83 (1956).

11. Barghoorn, E.S., and Tyler, S.A., Science 147, 563 (1965)

12. Schopf, J.W., Kvenvolden, K.A., Barghoorn, E.S., Proc. Nat. Acad. Sci. U.S. 59, 639 (1968)

13. Drozdova, T.V., Trudy Biochem. Labor GEOkhN AN SSSR im. V.I. Vernadskogo 12, 333 (1968).

14. Oro, J., Nooner, D.W., Zlatkis, S.A., Wikstrom, S.A., and Barghoorn, E.S., Science 148, 77 (1965).

15. Manskaya, S.M., Drozdova, T.V., "Organic Matter in Recent and Fossil Sediments, Moskva, "Nauka", 1971.

16. Barghoorn, E.S., Meinschein, W.G., and Schopf, J.W., Science 148, 461 (1965).

17. Sergiyenko, I.Z., Bobyleva, M.I., Egorov, I.A., and Fokin, V.D. DAN SSSR 190, 725 (1970)

18. Vologdin, A.G., Sergiyenko, I.Z., Egorov, I.A., and Bobyleva, M.I., DAN SSSR 191, 11 (1970).

19. Hoering, T.C., and Abelson, P.H., Carnegie Institution of Washington Year Book, 64, 218 (1965).

20. Oparin, A.I., Report at International Geological Congress in Canada, 24 session, 106 symposium, 1972.

21. Sergiyenko, I.Z., Bobyleva, M.I., Sidorenko, C.A., and Egorov, I.A., DAN SSSR, in press.

22. Vdovykin, G.P., "Carbon Substance in Meteorites", Izd. "Nauka", Moskva, 1967.

23. Vologdin, A.G., "Archaecyatheans. Principles of Paleontology" Vol. 2, Moskva, Izd. AN SSSR, 1962.

24. Vologdin, A.G., DAN SSSR 175,1143 (1967).

25. Sokolov, P.V., USSR Geology, V.XXVII (1958).

26. Bel'kov, I.V., "Cyanite Schists of Kiev Formation", 1963.

THE SURFACES OF COACERVATE DROPS AND THE FORMATION OF COLONIES

T. N. Evreinova, B. L. Allakhverdov, and V. I. Peshenko

Department of Plant Biochemistry, Biological Faculty

Moscow University, Biophysical Institute, USSR Academy of Sciences (Pushchino)

In his theory of the origin of life, Academician A.I. Oparin attached particular importance to the cooperation of molecules from the water of the primordial ocean in the form of coacervate drops, which were later transformed into the probionta (1). According to J. D. Bernal, the sediments of the organic compounds that occurred between the aluminosilicate clay layers in the ocean might have also served as material for concervate drops. The hydrophilic coacervate systems are made up of the coacervate drops 0.5-640 microns in diameter. Molecules are concentrated in the drops. The coacervate systems, mostly consisting of gelatin and gum arabic, were investigated by the Dutch Professor Bungenberg de Jong in the first half of this century (2).

A. I. Oparin published his first book "On the Origin of Life" in 1924; 25 years later appeared a cinema film that showed the process of formation of the coacervate drops from sediments. The first literature in the field appeared in 1954 (2).

A coacervate system can be obtained from synthetic polymer molecules, e.g., from polyadenylic acid and polylysine (according to Oparin and Serebrovskaya), and also from biopolymer molecules: proteins, nucleic acids, carbohydrates, and other compounds (2). To date, over 300 chemically different coacervate systems are known. Enzymic reactions have been carried out in many of these. The coacervate drop is one of the probable models of the protocell. Figure 1 shows the place of these structures in the evolution of matter towards life (3).

The aim of this paper is to discuss the part played by the

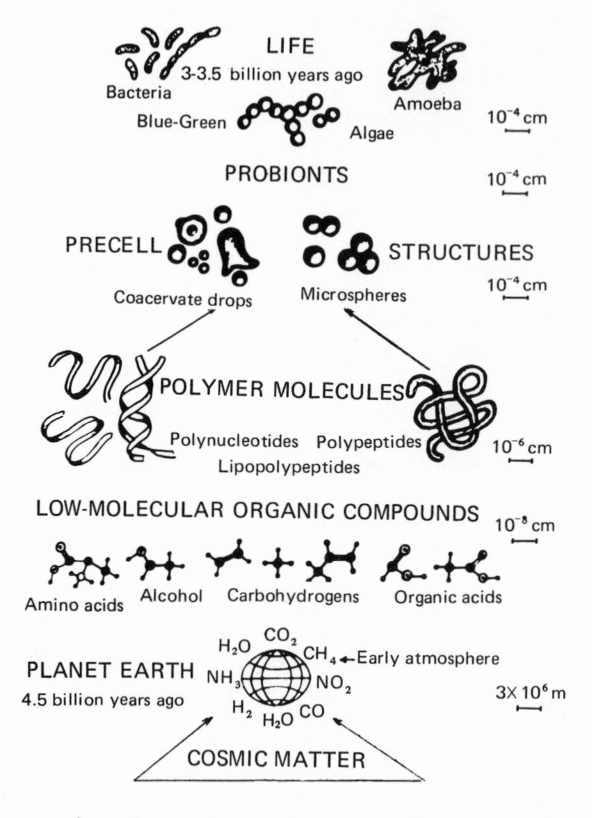

Figure 1. Evolution of matter (schematic).

surface of the coacervate drops of stable systems and the formation of colonies of such drops.

METHODS OF INVESTIGATION OF INDIVIDUAL COACERVATE DROPS

Interference microscopy - for determining the dry mass of molecules (excluding water) in a living cell and in a coacervate drop. The limit of the method is 10^{-14}g. The error of the method is 2 to 5 per cent of the quantity measured.

Cytospectrophotometry - for measuring the content of quinones and oxidized o-dianisidine. The device MUF-4, MUF-5, SIM (scanning integrating microphotometer); measurement limit of 10^{-13} - 10^{-15}g. Error of 2 to 4 percent of the quantity measured (5).

Electron microscopy - Standard Soviet UEM-100B microscope; the drops were examined at magnifications from 4,000X to 50,000X. Japanese YEOL YEE YB scanning electron microscope, magnifications from 5,000X to 20,000X.

Coacervate systems vary greatly in stability. In the unstable systems, the drops settle on the bottom of the vessel under the action of gravity, are destroyed, and form a coacervate layer. For example, in the system "serum albumin-gum arabic" at pH-6,0, the coacervate drops survive for 30 minutes. In the stable systems, the drops also settle on the bottom but do not disappear. They keep their individuality for years. The first system was obtained in 1968, and the drops are still intact (4). Stable protein-carbohydrate and protein-nucleic acid coacervate systems were obtained by including in them oxidizing enzymes and their substrates at pH-6,0 and 16-30°. The enzymes used were polyphenoloxidase (1.10.3.1) and peroxidase (1.2.1.7).

Polyphenoloxidase oxidized pyrocatechol to quinones, peroxidase oxidized, with participation of H_2O_2, pyrogallol to purpurogallin, and o-dianisidine (amino group) to o-dianosidine (imino-group). The oxidized products were accumulated in the drops and the systems became stable. The quinone content in the individual drop ranged from 0.001 to 1.4 percent, that of o-dianisidine, from 0.003 to 6 percent (5). This shows that even a small amount of oxidized products had a stabilizing effect. Therefore, the dry mass of the stable drops was practically the same as in all coacervate systems studied earlier. The data on the dry mass are given in Table I (2).

Table I

Weight and Concentration of Dry Mass in Various Objects

Objects	Diameter (cm)	Volume (cm^3)	Weight (g)	Concentration (%)
Bacteria	2.5×10^{-4}	8.5×10^{-12}	2.5×10^{-12}	30-25
Mammalian cells	2.5×10^{-3}	8.5×10^{-9}	2.1×10^{-12}	25-10
Amoeba	1.0×10^{-2}	5.2×10^{-7}	7.8×10^{-8}	15-10
Coacervate drops	2.42×10^{-4} to 1.02×10^{-2}	7.4×10^{-12} to 5.3×10^{-7}	2.5×10^{-12} to 3.5×10^{-8}	34-7

Obviously, the quinones and the oxidized o-dianisidine produce bridges linking the polymer molecules (histone, DNA). The surface of the drops gains in stability. The drops are not only preserved but become able to join by their surfaces, giving rise to colonies. Mechanical shaking may lead to destruction of the colonies, but after a time they arise again. The structure of the surface layers varies (6-8). Figures 2-3 present the photographs of the drop in the scanning and in the transmission electron microscopes (3). The data suggest that the oxidized low-molecular compounds are of considerable importance for the formation of new structures from the polymer molecules in the coacervate drops. We suppose that the emergence of the colonies of protocell structures (in particular, those of the coacervate drops) had been an indispensable stage in their further development and becoming more complicated on the way towards the appearance of life.

The authors deem it a pleasant duty to congratulate Academician A. I. Oparin on the occasion of the festive anniversary and to thank him cordially for his kind attention to their endeavours.

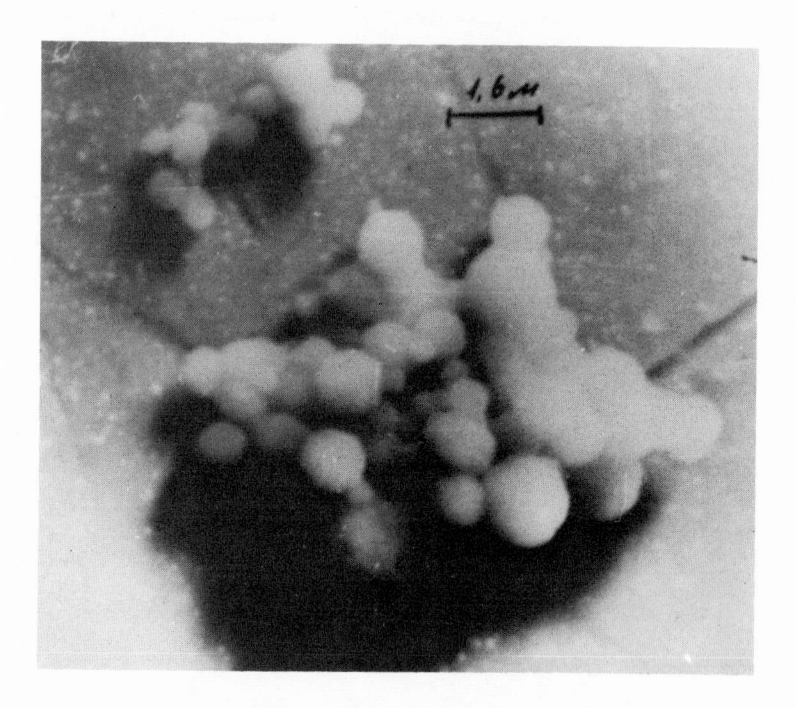

Fig. 2a. Coacervate drops in the scanning electron microscope: polyphenoloxidase-histone-DNA-quinones.

Fig. 2b. Coacervate drops in the scanning electron microscope: polyphenoloxidase-histone-gum arabic-quinones.

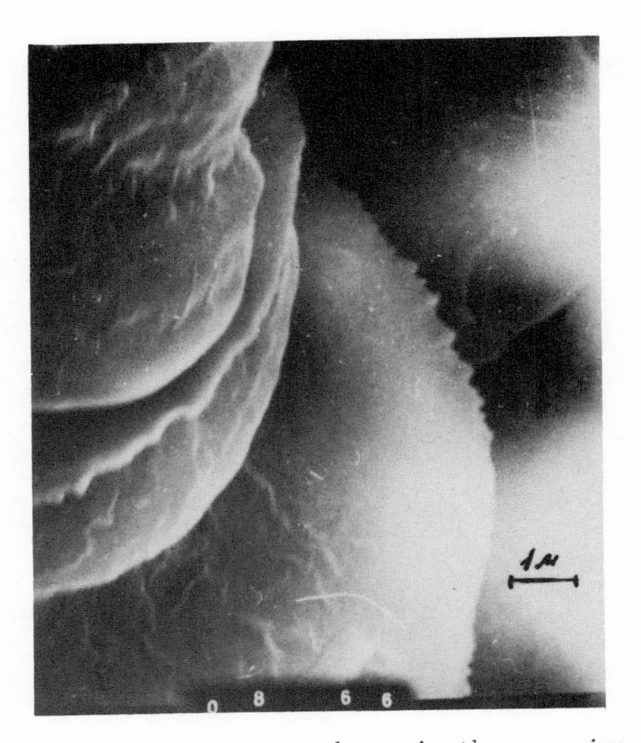

Fig. 2c. Coacervate drops in the scanning electron microscope. Surfaces of drop parts. Coacervate system: peroxidase-histone-gum arabic-purpurogallin.

Fig. 3. Part of drop surface in transmission electron microscope. Coacervate system: polyphenoloxidase-histone-quinones.

REFERENCES

1. Oparin, A. I., "Life, Its Nature, Origin and Development," "Nauka," Moscow (in Russian), 1968.

2. Evreinova, T. N., "Concentration of Substances and Action of Enzymes in the Coacervates," "Nauka," Moscow (in Russian), 1966.

3. Evreinova, T. N., "Coacervate Drops and the Origin of Life," "Novoe v zhizni, nauke i tekhnike," Znanie Publsh Evolusionnaya Biokhimiya, 12 (in Russian), 1973.

4. Evreinova, T. N., and Bailey, A., Dokl. Akad. Nauk SSSR 179, 723 (1968).

5. Evreinova, T. N., Mamontova, T. V., and Khrust, Y. R., Dokl. Akad. Nauk SSSR 204, 991 (1972).

6. Evreinova, T. N., Stephanov, S. B., and Mamontova, T. V., Dokl. Akad. Nauk SSSR 208, 242 (1973).

7. Evreinova, T. N., Mamontova, T. V., Karnauhkov, W. N., Stephanov, S. B., and Khrust, U. R., Coacervate systems and origin of life. Abstracts Fourth Internat. Conference on the Origin of Life, Barcelona, Spain, June 25-28, 1973.

8. Evreinova, T. N., Gladilin, K. L., Stephanov, S. B., and Karnaukhov, W. N., "Action of oxidoreductases in protein-nucleic acids and other coacervates." Abstracts of Nineth International Congress of Biochemistry, p. 435, Stockholm, 1-7 July, 1973.

ON SOME PROBLEMS OF EVOLUTION OF THE PHOTOSYNTHETIC PIGMENT APPARATUS

V. B. Evstigneev

Institute of Photosynthesis, USSR Academy of Sciences

Moscow, Putschino, USSR

According to the theory of origin and evolution of life on the Earth, first proposed and developed by A. I. Oparin (1-4), at a certain period of the evolution some organisms in their struggle for existence were forced to evolve such metabolic pathways, which not only permitted them to assimilate external (exogenous) organic substances in a more effective way but also gave them the possibility of utilizing other means of extracting energy from the environment and of assimilating the simplest forms of carbon-containing compounds.

Of the various possibilities available, utilization of solar light energy turned out to be the most effective. Solar radiation is the most powerful and practically inexhaustible source of energy on the surface of the Earth. Furthermore, the absorption of light quanta resulting in formation of excited states of molecules gave the simplest possibility for generation of a system possessing a high energetic potential that could be subsequently utilized to create energy-rich chemical compounds necessary for the metabolism of organisms. At the aforementioned time, the photochemically most active radiation, the ultraviolet rays, ceased reaching the Earth's surface as a result of a number of processes taking place in the upper layers of the atmosphere (formation of the ozone layer); the visible and near infrared spectral regions therefore became the main source of photochemically active light energy. It is precisely to the light of these spectral regions that organisms had to adjust themselves during their transition to a photoautotrophic mode of life.

First of all, an apparatus had to be evolved which could capture and utilize light energy. Colored compounds which strongly absorb light quanta could meet the requirement. The transition period from heterotrophic to autotrophic life evidently should have equalled the period required for formation of the pigment apparatus and selection of the pigments most suitable for photosynthesis. It would of course be incorrect to assume that colored substances absorbing in the visible and infrared regions of the spectrum appeared only during this period. It seems more probable that such substances existed at an earlier period; moreover, even during the period of chemical evolution, when the first "protobionts" had not yet appeared and only the formation of carbon and other chemical compounds occurred, the primeval ocean also contained various types of colored compounds (1-4). The colors of such compounds were simply a consequence of their chemical structures and did not serve any special purpose. There exist today many strongly colored natural substances whose light absorbing ability is of no physiological significance (for example hemoglobin).

Nevertheless, the color of some of these compounds could be an essential factor in why they at first found themselves in the composition of the primary structures of the coacervate type (1) and then in primary organisms. Their ability to absorb light could have had the consequence that they were more active in the excited state chemically than uncolored compounds and could themselves more easily enter into interaction with other substances or act as more effective catalysts for other reactions.

It is possible that further improvement of the catalytic apparatus and appearance of more perfect catalysts made unnecessary their light activation, but in the initial period the new apparatus could have played an essential role. It seems very likely that the chromophore groups of cytochromes or some other compounds playing an important role in contemporary plant metabolism can be regarded as examples of such colored substances.

In any case there can be no doubt that bodies of primary heterotrophic organisms, at least of some among them, contained various colored substances side by side with colorless ones. It cannot be excluded that the larger quantity of colored compounds and their greater variety could have played an essential role in why just certain organisms became an initial stage in the long development of a photoautotrophic mode of life.

During the transition to a photoautotrophic mode of life, organisms had to select among the colored compounds which they contained those which were suitable in their properties as effective photocatalysts in the developing photosynthetic apparatus and which were able to catalyze reactions involving the storage of light energy.

Not all intermediate stages of development of the photosynthetic pigment system, beginning from inclusion of the earliest pigments of this type and ending with modern photosynthetic pigment complexes, can be traced. We have no concrete data on the existence of contemporary organisms of an intermediate type, containing simple systems corresponding to an intermediate stage of evolution. In all modern pigmented organisms, including the blue-green algae and photosynthesizing purple and green bacteria, the pigment systems are highly developed and contain many different components. The intermediate forms were probably nonviable and therefore disappeared during evolution.

However, it is hard to imagine that from the very beginning of development of pigment systems, the only pigment that organisms made use of was chlorophyll or similar compounds. It seems more natural to assume that on transition to photoautotrophy organisms initially "tested" various types of colored substances present in their tissues. Those compounds which were unsuitable were rapidly (certainly on an evolutionary scale) rejected by natural selection, whereas others could have played an active role over a certain period of time only to be forced out by pigments which turned out to be more suitable.

On the grounds of the chemical structure of the major contemporary photosynthetic pigments we can think that those pigments which could not compete with chlorophyll, but probably functioned as photosensitizers of photosynthesis over a certain period of time, possessed a structure akin to that of tetrapyrrole compounds.

As is well known, tetrapyrroles with a closed porphyrin ring (derivatives of porphin) and in particular those which contain metal atoms in the center (iron, cobalt, manganese, etc.) are widely distributed and are frequently vital components of many dark biocatalysts or enzymes. Such compounds apparently originated at the very earliest period of emergence of life. Their synthesis from simpler compounds under conditions imitating those of the primitive Earth has been demonstrated in laboratory experiments.

However, it should be emphasized that not all metal derivatives of porphin are photoactive and can be employed as photocatalysts. Thus, the very abundant iron or copper porphyrins are photochemically inactive as are a number of porphyrins containing other metals with varying valency. On the other hand, porphyrins which do not contain metals, or contain magnesium or zinc, are quite active as photosensitizers. Nature had to make an initial choice and then continue seeking pigments most suitable for photocatalysis of reactions proceeding with storage of light energy and capable of being built into the photosynthetic electron transfer chain (5).

In spite of the fact that at present all photosynthetic pigments (chlorophyll, bacteriochlorophyll, and bacterioviridin) possess closed tetrapyrrole rings, it is quite possible that during the initial transition period open-system tetrapyrrole pigments similar to the modern phycobilins played some role.

In particular, primitive photosynthetic algae, in which biliprotein-like compounds played a major role, could have appeared in the primeval ocean. Data obtained in our and other laboratories show that the contemporary phycobiliproteins, phycoerythrin and phycocyanin, are photochemically active in vitro. In particular, they are capable of participating in photochemical reactions involving electron transfer and hence can sensitize oxidation-reduction reactions (6-8). On this, one may suggest that, not only at an early evolutionary period but even at present, phycobiliproteins may play a definite role as direct participants in the photosynthetic reactions and not only as carriers of energy to chlorophyll. However, the biliproteins were probably less suitable than pigments of the chlorophyll type and could not compete with the latter, especially when living organisms began to populate the land after emerging from the ocean. The biliproteins are completely absent in modern terrestrial plants but exist in significant amounts in a number of algal species.

One of the interesting advantages of tetrapyrroles with closed rings is their greater ability to form complexes with metals. It is well known that inclusion of such a metal as magnesium in the center of porphyrin molecules appreciably raises the photochemical activity of these molecules. It has been demonstrated in model experiments, for example, that the magnesium complexes of some porphyrin compounds donate an electron with greater ease on illumination, that is, are more readily photooxidized, than the respective compounds without metal atoms (9).

Moreover, inclusion of the metal usually enhances the coloring of porphyrin, that is, its ability to absorb light. The presence of such metals as magnesium also increases the ability of the pigment to form the coordinational bonds required for linkage of the pigment to other components of the photosynthetic complex such as water or proteins.

Tetrapyrrole pigments with closed rings of conjugated double bonds thus became, probably as a result of the advantages mentioned above, and perhaps of others as well, the choice of evolution. Consequently, chlorophyll and related pigments are now the most widespread and most important pigments in the photosynthetic pigment complex. The existence of several modifications of chlorophyll in different modern photosynthetic organisms evidently reflects the differences in chemical evolution pathways of the pigments.

According to modern concepts (1-5), an intermediate step in the transition from heterotrophic metabolism to complete photo-autotrophy with water and carbon dioxide serving as initial reagents was the appearance of organisms which stored light energy but were not capable of removing electrons from water; various organic or inorganic compounds were used by these organisms as reductants (sources of electrons or hydrogen atoms). Some photosynthesizing bacteria (purple sulphur or nonsulphur and green sulphur bacteria) are representatives of this type of autotrophy which have persisted to the present time. Much less energy is required for removal of an electron from, say, hydrogen sulphide, than from water. Therefore the oxidizing potential of the excited state of the pigment photosensitizing this process may be less than required in the case of true photosynthesis. Investigations indeed show that the bacterial pigments, bacteriochlorophyll and bacterioviridin, possess both weaker oxidative and reductive possibilities. In experiments carried out by us, it was found that chlorophylls a and b photosensitized photooxidation and photoreduction reactions of bacterial pigments; this means that the oxidative and reductive potentials of intermediate labile forms of chlorophyll responsible for photosensitization are greater than the potentials of the corresponding forms of bacterial pigments (10,11). It follows from this that differences in properties of bacterial pigments and chlorophyll exist at the molecular level and this explains the substitution of one pigment for another during the evolutionary transition to the utilization of water as electron donor and the impossibility, also, of further perfection of the photosynthetic apparatus without substitution of the pigments.

It is interesting to note that bacterioviridin (chlorobium chlorophyll) is almost identical with chlorophyll a, as far as the absorption spectra are concerned, but photochemically is closer to bacteriochlorophyll (10,11). This again shows that pigments participate in photosensitization of photosynthetic processes more as photochemical catalysts than as simple physical sensitizers.

Regarding the formation of the photosynthetic pigment system, it should be noted that a biochemical apparatus was required which could ensure the synthesis of sufficient amounts of pigments at least much more than would be required for dark catalysts or for photocatalysts accelerating energy-evolving, instead of energy-storing, reactions. Very small amounts of enzymes are sufficient for dark biocatalysis, providing the enzymes are capable of undergoing rapid reversible transformations. The energy required for driving the reactions is derived from the reacting compounds themselves and not from any external source.

In the case of photosynthesis, there must be a continual supply of large amounts of energy to the sites at which the reaction with the pigment actually takes place. The amount of such an active pigment required for the reaction may be small, but much larger amounts of pigment are required for its continual operation in order to absorb and transfer the energy to the reaction center.

A result of this requirement was apparently the formation of photosynthetic organelles containing large quantities of pigments which absorb practically all incident light. On the other hand, the high concentrations of the pigments led to strong interactions between them, irrespective of whether they were free or complexed with proteins or lipids, and it was inevitable that association or aggregation of the molecules should result. This is manifest in a change in optical properties (9,10).

Careful measurements of the absorption spectra of chlorophyll *in vivo* have revealed existence of a rather large number of various forms of chlorophyll in the chloroplast. These forms differ in the position of the major absorption band and hence in degree of packing of the molecules. Evidently it would be wrong to assume that each of the forms has a special physiological function and is required for photosynthesis as such (12,13).

In our opinion, the multiplicity of pigment forms *in vivo* with differing absorption spectra is mainly due to physical causes and is not conditioned by physiological requirements. It seems to us very probable that the number of pigment forms with different functions in photosynthesis is appreciably smaller than the number of forms which differ physically, and in particular with respect to their optical properties. This problem requires further study.

On the other hand, there is no doubt that in modern photosynthesizing organisms there exist two types of chlorophyll which functionally are very different; however, it is still not clear to what extent these differences correspond to those in the spectral properties of the "physical" pigment forms mentioned above. The bulk of the chlorophyll in chloroplasts, very likely not connected directly with the photosynthetic electron transfer chain, serves simply to absorb light and to transfer the absorbed energy to the reaction centers where it is used in photochemical reactions by another, "active" fraction of chlorophyll (11-14).

During the evolution and perfection of the photosynthetic apparatus and enhancement of its efficiency, a "division of labor" gradually occurred between the various chlorophyll molecules which initially, when there were few of them, probably all fulfilled an identical function consisting in direct absorption of light and of photochemical catalysis. As the plant organisms became more complex, this state of affairs became progressively less efficient and therefore inpracticable. As the capacity of the apparatus for biosynthesis of chlorophyll and the concentration of chlorophyll increased, more and more of the pigment molecules began to fulfill an accessory role, namely that of energy transfer to the reaction centers.

Apparently, it was most effective that the major role in energy absorption and transfer to the reaction centers was played by chlorophyll itself, that is, by the same pigment which was located at the reaction centers. Indeed, due to almost complete overlapping of the absorption spectra, the transferred energy could excite the "active" chlorophyll with maximal efficiency. Transfer of energy from other pigments is undoubtedly less efficient and in contemporary organisms is required only under special circumstances, for instance under weak illumination or when it is of value to utilize light weakly absorbed by chlorophyll molecules.

In this connection it seems to be rational to say some more words about the so-called accessory pigments which include chlorophylls *b*, *c*, and *d*, the phycobiliproteins, phycoerythrin and phycocyanin, and carotenoids.

An auxiliary role in photosynthesis is ascribed to all these pigments. They are assumed to transfer the light energy absorbed by them to the "major" pigment which is chlorophyll *a* in higher plants and algae or bacteriochlorophyll and bacterioviridin in bacteria. The latter pigments are true sensitizers of photosynthesis in these organisms (12,13).

Now, if an accessory pigment is defined as that which transfers energy and does not participate directly in reactions at the reaction center, then the bulk of major pigments should also be regarded as accessory pigments. Moreover, the composition of the reaction centers of plants and bacteria is not known in sufficient detail. There is no ground to assert that the reaction centers contain only the "major" pigment. A small fraction of accessory pigments may be a part of the reaction center and may play a role differing from that of the bulk of pigment. Evidently it will be difficult to assess the role of an accessory pigment simply by studying the relation between the bulk of it and bulk of chlorophyll *a* which essentially is also an accessory pigment.

More information on the composition of the reaction center is required. On the other hand, one should not exclude the possibility that such pigments as the phycobiliproteins have their own reaction centers which, as in the case of the chlorophyll centers, contain only a small amount of the total pigment.

It seems to us that considerations based on an evolutionary approach may be of help when trying to assess the role of the accessory pigments in the photosynthetic apparatus. First of all, the accessory pigments can apparently be divided into two groups.

Carotenoids comprise one group and all other accessory pigments the other. All living organisms contain carotenoids and these pigments were evidently abundant in organisms at the very beginning of development of the photosynthetic apparatus. It is still not clear what their role is in the photosynthesis of present-day organisms and it may be that they play a truly accessory role such as that of protecting chlorophyll from destruction by light. For many carotenoids it is not clear what role they play in transfer of energy to chlorophyll. From the foregoing it may be concluded that even if carotenoids were of any significance in the initial period of transition to autotrophy as direct photosensitizers of photosynthesis, they rapidly disappeared from the scene because of the low efficiency of the systems containing them and also because they absorbed only in the shortwave range of the spectrum.

Accessory pigments of the second group are contained only in photosynthesizing organisms and apparently they began to be formed in significant quantities only after the beginning of development of photoautotrophs. The possible role of phycobiliproteins in the evolution of photosynthesis was briefly considered above. At any rate, there are reasons to believe that during the transition period phycobiliproteins, being strongly colored substances and possessing the ability to photosensitize oxidation-reduction reactions, could have played an independent role over a long period of time, as sensitizers of photosynthesis in certain algae. In this role they were gradually replaced by chlorophyll which turned out to be a more suitable pigment.

This viewpoint is supported by electron microscopic investigations which show that in contemporary red and blue-green algae phycobiliproteins are contained in special granules (called phycobilisomes) which are attached to the surface of chloroplast lamellae but separately from chlorophyll (15-17). The fact that phycobiliproteins are not a structural part of the lamellae testifies once more that there cannot exist a close complex between them and chlorophyll, as was confirmed by results of our investigations (18).

We shall not consider in greater detail this question as well as that concerning the evolution of accessory pigments of the chlorophyll-type. Moreover, there are practically no data pertaining to the origin and evolution of the functions of chlorophylls *b*, *c*, and *d*. It seems to us, however, to be more correct to consider these pigments as "rudiments" of evolutionary development of the photosynthetic pigment apparatus, which played in the transition period a role equal to that of the subsequent major pigments and then gradually lost their significance, rather than as accessory apparatus which appeared only after creation of modern photosynthesis with participation of chlorophylls.

The points considered above are not the only ones relevant to the problem of the origin and evolution of the photosynthetic pigment apparatus. Of great importance, for example, is the development of the spatial organization of the pigments, the coupling of the latter to the protein-lipid carriers and many other problems. Nevertheless, on the basis of what has been said it may be asserted that the pigment composition of the photosynthetic apparatus of present-day organisms is the result of a long period of evolutionary selection of the most suitable major pigment and of gradual creation of the best relationship between this pigment and others, under the given conditions, some of which may have played a major role but gradually lost this leading position.

It is worth noting that a rise of interest in the hypothesis of symbiotic origin of chloroplasts of higher plants, suggested already in the beginning of our century, can be observed in recent time (19,20). According to this hypothesis, the chloroplasts originated from blue-green algae which had penetrated in ancient time into some heterotrophic organisms and had begun a symbiosis with them.

This hypothesis does not certainly remove the question about the primary origin and evolution of photosynthetic apparatus inasmuch as the question will concern the blue-green algae themselves, in case it is correct. At the same time, the further development of the photosynthetic apparatus can be regarded in this case somewhat differently; as parallel with its direct improvement, this apparatus under the action of newly facilitated conditions of existence had to undergo simplification as well. It can explain, for instance, the rudimentary character of functions or even disappearance of some pigments.

The study and discussion of evolution of the pigment apparatus of higher plants from this point of view seems to us to offer promise.

REFERENCES

1. Oparin, A. I., "The Origin of Life," Oliver and Boyd, London, 1957.

2. Oparin, A. I., "Genesis and Evolutionary Development of Life," Academic Press, New York, 1968.

3. Ponnamperuma, C., "The Origins of Life," E. P. Dutton, Inc., New York, 1972.

4. "Chemical Evolution and the Origin of Life" (Buvet, R., and Ponnamperuma, C., eds.), Vol. 1, North-Holland, Amsterdam, 1971.

5. Krasnovski, A. A., in: "Vosniknovenie zhisni na Zemle" (Origin of Life on the Earth).

6. Evstigneev, V. B., and Gavrilova, V. A., Biofizika 9, 739 (1964).

7. Evstigneev, V. B., and Bekasova, O. D., (a) Molekularnaya Biologia 2, 380 (1968); (b) Biofizika 15, 807 (1970); 17, 997 (1972).

8. Evstigneev, V. B., and Bekasova, O. D., Izvestia Acad. Nauk SSSR, ser. biol., No. 3, 344 (1973).

9. Evstigneev, V. B., and Gavrilova, V. A., Doklady Acad. Nauk SSSR 74, 781 (1950).

10. Evstigneev, V. B., and Gavrilova, V. A., Doklady Acad. Nauk SSSR 141, 477 (1961); 154, 714 (1964).

11. Evstigneev, V. B., and Gavrilova, V. A., in: "Problemy evolucionnoi i teknicheskoi biohimii," p. 232, Nauka SSSR, 1964.

12. Fogg, G. E., "Photosynthesis," The English Universities Press Ltd., London, 1969.

13. Rabinowitch, E., and Govindjee, R., "Photosynthesis," John Wiley and Sons, Inc., New York, 1969.

14. Vernon, L. P., and Ke, B., in: "The Chlorophylls" (Vernon, L. P., and Seely, G. R., eds.), p. 569, Academic Press, New York, 1966.

15. Gannt, E., and Conti, S. F., in: "Energy Conversion by the Photosynthetic Apparatus," Brookhaven Symposium in Biology No. 19, 393 (1966).

16. Gannt, E., and Conti, S. F., J. Cell Biol. 29, 423 (1966); J. Bact. 97, 1486 (1969).

17. Gannt, E., and Lipschultz, G. A., J. Cell Biol. 54, 313 (1972).

18. Evstigneev, V. B., and Bekasova, O. D., Biofizika 17, 997 (1972).

19. Breslavetz, L. P., Trudy Instituta istorii estestvoznania i tekhniki, Ed. Acad. Nauk SSSR M. 23, No. 4, 257 (1959).

20. Ostroumov, S. A., Priroda, No. 3, 21 (1973).

AMMONIA: DID IT HAVE A ROLE IN CHEMICAL EVOLUTION? (1)

J. P. Ferris and D. E. Nicodem

Department of Chemistry

Rensselaer Polytechnic Institute, Troy, New York

The possibility that biomolecules necessary for life were formed in a reducing atmosphere was first suggested by Oparin in his classic treatise on the origins of life (2). Urey calculated from equilibrium thermodynamic data that CH_4 and NH_3 were the predominant compounds of carbon and nitrogen in the atmosphere of the primitive Earth if a minimum of 10^{-3} atmosphere of H_2 were present (3). Rasool and McGovern postulated, on the basis of an equilibrium model for the primitive Earth, that the exospheric temperature was between 500-1000°K, so that the rate of H_2 loss was relatively slow (4,5). They calculated that, with this slow rate of loss, there would have been sufficient H_2 to maintain NH_3 in the primitive atmosphere for 10^8-10^9 years.

The postulated CH_4-NH_3 atmosphere was consistent with Miller's observation in 1953 of the formation of amino acids when an electric discharge was arced through a mixture of CH_4, NH_3, and H_2O vapor (6). Since 1953, there has been a multitude of reports of the abiotic synthesis of amino acids and other prebiotic molecules, using NH_3 as one of the starting materials (7). Other abiotic syntheses that use NH_3 include porphyrins (8), the pyrophosphate linkage (9), purines (10,11), nicotinamide (12,13), and the nucleotide condensing agent cyanamide (14,15).

Bada and Miller investigated the possible role of NH_3 in chemical evolution in detail and concluded that the NH_3 on the primitive Earth was either present mainly as NH_4^+, which was free in solution or bound to clay minerals, or was covalently bound in organic compounds. They estimated that the concentration of dissolved NH_4^+ would have been in the 10^{-2} to 10^{-3} M range (16).

The proposed CH_4-NH_3 atmosphere for the primitive Earth has been challenged. Rubey emphasized the volcanic origin of the primitive atmosphere and claimed that H_2 would not have been outgassed from the Earth's crust sufficiently rapidly to maintain a steady state pressure of 10^{-3} atmosphere (17). Since H_2O, CO_2, and N_2 are the major gaseous compounds emitted from contemporary volcanoes, he reasoned that the primitive atmosphere contained mainly those compounds.

The concept of the volcanic origins of the primitive atmosphere was extended by Holland (18). He postulated that the primitive atmosphere was in equilibrium with molten iron for 1/2 billion years before the Earth's core formed. He estimated that initially the Earth's atmosphere contained mainly CH_4, with lesser amounts of H_2, NH_3, and H_2S. Once the core formed, N_2 was the predominant compound of nitrogen because there was no pathway for the reduction of N_2 to NH_3. While Abelson agrees with the models of Rubey and Holland, he proposed that any NH_3 present in the primitive atmosphere would have been photolyzed to N_2 and H_2 in 3×10^4 years or less (19). Photolysis of NH_3 is likely because short wavelength (<300 nm) light penetrated the atmosphere of the Earth in the absence of the contemporary ozone shield. Cloud (20) concluded from geological arguments that there was no NH_3 in the primitive atmosphere.

This problem merits further investigation because of the central role of NH_3 in many prebiotic syntheses and because we feel that a number of the previous workers omitted important consideration in their estimates of the levels of ammonia that were present in the atmosphere of the primitive Earth. The rates of formation and photodestruction of NH_3 were not considered by Urey (3) nor by Rasool and McGovern (4,5). Although equilibrium considerations favor NH_3 formation, the strong uv flux may have dissociated the NH_3. Bada and Miller did not consider the photolysis of NH_3 in their calculations of the concentration of NH_4^+. As a consequence, their estimates of NH_4^+ concentration may be too high. On the other hand, Abelson (19) may have overestimated the rate of NH_3 photolysis. He assumed that all of the NH_3 would have been in the atmosphere and he did not consider the effect of its dissolution as NH_3 and NH_4^+ in the primitive ocean. Dissolved NH_3 would have been protected from photolysis by the absorption of light by other dissolved compounds. Furthermore, dissolved NH_3 and NH_4^+ absorb light at shorter wavelengths than does gaseous ammonia. Since most of the NH_3 would have been dissolved, only the small amount of gaseous NH_3 would have been subjected to photolysis. Therefore, the rate of photolysis might have been much less than Abelson's estimate. Furthermore, NH_3 may have been protected from photodestruction by water vapor. Gaseous H_2O absorbs light strongly over much of the same wavelength region where NH_3 absorbs (20).

Two approaches were used in an attempt to make a detailed evaluation of the role NH_3 played in chemical evolution. First, a computer program was devised to calculate the time needed to photodestroy NH_3, if one assumes that all the nitrogen in the contemporary atmosphere had been present on the primitive Earth as NH_3 and NH_4^+. The pressures of NH_3 used in this calculation are those which would have been present in the atmosphere if the ocean were pH 8.1 and if the concentrations of NH_4^+ were those given in Table I. Although the total concentration of NH_4^+ was varied from 1 $\underline{M}$ to 10^{-4} $\underline{M}$ in the calculation, the total amount of NH_3 photolyzed remained constant. Input into the program included the light emission from the Sun divided into 1 nm bands from 100-400 nm. The average light intensity at a distance equal to the average radius of the orbit of the Earth was used in units of photons cm^{-2} sec^{-1} nm^{-1} (21). The absorption coefficients of NH_3 and the other gases used in this study were also divided into the same 1 nm bands (20). The computer program calculated the length of time it would take for the NH_3 to absorb an equivalent number of photons in the presence of the other gases. This calculated lifetime was then increased by 16; the ratio of the area of the surface of a sphere to its cross-sectional area is 4 and the quantum yield for NH_3 photolysis is 0.25 (22).

This calculation assumed that the amount of light absorbed by the atmospheric NH_3 is constant, even though photodestruction is taking place. This is a valid assumption for two reasons: (a) atmospheric NH_3 destroyed by photolysis would be replenished by a shift in this equilibrium, and (b) NH_3 has a very intense and continuous uv absorption which starts abruptly at about 220 nm and extends below 100 nm (21). Even very low partial pressures of NH_3 will absorb the bulk of the uv radiation in this region.

The calculated lifetimes of NH_3 (Table I) are some 10 times greater than Abelson's value of 3 x 10^4 years. These calculations do show that, when the partial pressure of NH_3 is decreased, its lifetime is increased somewhat (compare 1 and 4 in Table I); however, the effect is not large. These data demonstrate that even if the bulk of the NH_3 were dissolved in water or bound to clay minerals its rate of photolysis would still be appreciable. This finding is a consequence of the intense uv absorption of NH_3. Partial pressures of NH_3 as low as 10^{-4} torr would still result in the absorption of all the solar radiation below 220 nm.

We expected to find that H_2O vapor protected NH_3 from photolysis; however, it exerted a negligible protective effect (compare 2 and 8, Table I; 23). Further consideration revealed that this protective effect is small because of the rapid decrease in the Sun's radiation around 200 nm (21). Even if H_2O absorbed all the sunlight from 190-100 nm, this radiation would only amount to 2.5%

Table I

Calculated Lifetimes for Ammonia Photolysis

	Aqueous NH_4^+ (molarity)[a]	Potential Atmospheric Gases (torr)[b] NH_3	H_2O	CH_4	H_2S	CO_2	NH_3 lifetime (years x 10^6)
1.	1	0.51	0.76	1.3	0	-	0.4
2.	10^{-2}	0.0051	0.76	1.3	0	-	0.6
3.	10^{-2}	0.0051	0.76	13	0	-	0.6
4.	10^{-4}	0.000051	0.76	1.3	0	-	1.8
5.	10^{-2}	0.0051	0.76	0	0	-	0.6
6.	10^{-2}	0.0051	2.4	130	-	-	0.6
7.	10^{-2}	0.0051	7.6	130	-	-	0.6
8.	10^{-2}	0.0051	24	130	-	-	0.6
9.	10^{-2}	0.0051	0.76	130	-	1	0.6
10.	10^{-2}	0.0051	0.76	130	-	100	0.6
11.	10^{-2}	0.0051	0.76	130	0.00063	-	1.4
12.	1	0.51	0.76	13	0.0063	-	0.6
13.	10^{-2}	0.0051	0.76	13	0.0063	-	4
14.	10^{-2}	0.0051	0.76	130	0.0063	-	26
15.	10^{-2}	0.0051	0.76	130	0.63	-	220
16.	10^{-2}	0.0051	0.76	130	6.3	-	2000

a) Bada and Miller (16) calculated that the NH_4^+ concentration would be 10^{-2} - 10^{-3} M.

b) It was assumed that the remainder of the atmosphere consisted of inert gases that did not have significant ultraviolet absorption.

of the total solar output in the 220-100 nm region where NH_3 absorbs light.

It was anticipated that the rate of NH_3 photolysis would decrease in the presence of H_2S because of its intense absorption in the 240-100 nm region where NH_3 absorbs light (24). We found, however, that a constant pressure of 6.3 torr H_2S (1000 times the maximum NH_3 pressure) is required to protect NH_3 for 2×10^9 years while an H_2S pressure equal to that of NH_3 results in only a 6-7 fold increase in the lifetime of NH_3.

The above calculations consider only the amount of light absorbed by potential atmospheric constituents with no consideration of the chemical reactions between photochemically excited molecules and other atmospheric constituents. In our second approach, the quantum yield for NH_3 decomposition was measured in the presence of other potential constituents of the atmosphere of the primitive Earth. Mixtures of NH_3 with other gases were irradiated with a 206.2 nm light source in a 10 cm cylindrical quartz cell (25). The rate of photolysis was followed by the decrease in the uv absorption of NH_3 at 209, 213, or 217 nm. The lamp intensity was measured concurrently in each experiment. This was done by measuring the rate of photolysis of a standard mixture of NH_3 (10 torr) and argon (740 torr). The quantum yield for NH_3 photolysis in the simulated primitive atmosphere was calculated from the ratio of the first order rate constants for the simulated atmosphere and the NH_3-argon standard, and 0.25,the quantum yield for the photolysis of NH_3 in the standard mixture (27).

Hydrogen was the only gas which markedly decreased the decomposition of NH_3 (Table II). However, this effect is only apparent at pressures greater than 50 torr. Approximately the same pressure of hydrogen is needed to inhibit the photolyses of either 10 torr or 1 torr of NH_3, a result which shows the inhibition is only proportional to the pressure of H_2. This inhibition is undoubtedly due to reaction of the photochemically formed NH_2 with H_2 to regenerate NH_3.

$$NH_3 \xrightarrow{h\nu} NH_2\cdot + H\cdot$$

$$NH_2\cdot + H_2 \longrightarrow NH_3 + H\cdot$$

The gases H_2S, CO, or O_2 each accelerate NH_3 photolysis. The quantum yields for the former two gases are tentative because solids were produced which coated the windows of the photolysis cells (Table II). Sulfur is undoubtedly the solid formed in the presence of H_2S; however, this had not been proved (29). Ammonium cyanate is the major solid product formed in the presence of CO. Lesser amounts of urea, biurea, biuret, formamide, HCN, and semicarbazide

Table II

Effect of Other Gases on the Quantum Yield of Ammonia Photolysis

	NH_3[a]	Gas A[a]		Gas B[a]		ϕNH_3
1.	12.5	CH_4	736.5			0.25
2.	9.0	CO_2	92.0	Ar	660	0.23
3.	11.0	H_2O	21,2	Ar	714.8	0.16
4.	20.0	He	737.4			0.25
5.	52	H_2S	0.45	Ar	708	∿0.5[b]
6.	14.4	CO	743.1			∿1[b]
7.	14.7	N_2	572	O_2	150	0.85
8.	10	H_2	740			0.03
9.	1	H_2	749			0.04
10.	1	H_2	99	Ar	650	0.07
11.	1	H_2	60	Ar	689	0.06
12.	1	H_2	45	Ar	704	0.10
13.	1	H_2	40	Ar	709	0.14
14.	1	H_2	30	Ar	719	0.10
15.	1	H_2	20	Ar	729	0.24

a) concentration is given in pressure (torr).

b) a precipitate formed in the irradiation cell.

were also isolated from preparative scale runs (1). The urea is probably formed from the thermal rearrangement of the ammonium cyanate in the hot uv lamp in the preparative scale runs.

Our calculations suggest that NH_3 could have been protected from photodestruction on the primitive Earth by H_2S. Light is absorbed by H_2S in the 240-100 nm region; it could have shielded NH_3 from solar radiation. However, the pressure of H_2S in the atmosphere would have exceeded 1 torr in order for it to serve as an effective shield. Appreciable amounts of H_2S probably would not have been present in the primitive atmosphere, because H_2S is photolyzed to H_2 and S with a quantum yield of one over a wide pressure range (29). We have calculated, by the same procedure used for the calculation of NH_3 lifetimes, that H_2S would have been photolyzed in about 10^4 years. In addition, the total pressure of H_2S would have also been attenuated by its dissolution and ionization in H_2O (the pK of H_2S is 7.24) and by its precipitation as FeS.

There are other possible mechanisms by which NH_3 may have been shielded from uv photolysis on the primitive Earth. Brinkmann suggested that sufficient O_2 (5 torr) was produced by H_2O photolysis to form a protective ozone shield (30). However, this conclusion has been challenged (31) and, furthermore, appreciable amounts of ammonia would not have been formed nor would appreciable amounts of NH_3 have coexisted with such a large amount of O_2 (3).

It has been suggested that the pH of the primitive ocean was 6-7 with high salt concentrations (low volumes), rather than the contemporary value of 8.1 (32). This would have shifted the NH_3-NH_4^+ equilibrium in favor of dissolved NH_4^+ and would have thereby decreased the rate of NH_3 photolysis. However, high salt concentration would have displaced the NH_4^+-clay mineral equilibrium in the direction of free NH_4^+ and NH_3, which would have partially compensated for the increased NH_4^+ concentration of lower pH (16). Furthermore, our calculations demonstrate that even if the NH_4^+ concentration were one-tenth of Bada and Miller's lower limit, the NH_3 would still have been photochemically destroyed in 1.8×10^6 years (Table I, run 4).

The observation that ammonium cyanate and urea are formed by the photolysis of CO-NH_3 mixtures (1) suggests that NH_3 may have been preserved as urea and that NH_3 may have been released slowly into the primitive atmosphere by the hydrolysis of urea. The hydrolysis of urea is a unimolecular process which proceeds at the same rate in acidic, basic, and neutral solutions (33). The first order rate constants for urea hydrolysis at 30° and 0° can be obtained by extrapolation of the Arrhenius plots (33). Half-lives of 10^2 years were calculated at 30°, and 10^4-10^5 years at 0°. This rapid hydrolysis suggests that NH_3 would have been protected only if there were sufficient CO present in the primitive atmosphere to assure the efficient photochemical conversion of NH_3 back to ammonium cyanate and urea. It has been noted that CO is a reactive substance which would have been depleted from the atmosphere by reaction with OH^- to give formate (19,34). We have not determined the limiting amount of CO necessary for the efficient photochemical conversion of NH_3 to urea.

Our observation that high pressures of H_2 decrease the rate of NH_3 photolysis suggests that H_2 may have preserved NH_3 from photochemical destruction in the atmosphere. The model of Rasool and McGovern (4,5) postulates that pressures of H_2 greater than 50 torr were present in the primitive atmosphere for about 5×10^7 years. If this is correct, NH_3 could have been preserved for a maximum of 10^8 years.

Finally, the possibility exists that NH_3 was continually formed during the first 10^8-10^9 years of the Earth's history. It could have been generated during the outgassing of the Earth's crust (18), or it could have resulted from the hydrolysis of the HCN formed by the action of electrical discharges on a CH_4-N_2 atmosphere (35). We feel that the rate of photolysis of NH_3 would have been much greater than its rate of synthesis by these pathways because it has been estimated that the energy form of short wavelength uv light was much greater than that in the form of electrical discharges and heat (36). Furthermore, the thermal synthesis of NH_3 from N_2 and H_2 requires special catalysts and high pressure to proceed at a significant rate. The syntheses of NH_3 from N_2 and H_2 has an activation energy of 55 Kcal mol^{-1} in the absence of catalysts (37).

If there were no high pressure (>50 torr) of H_2 on the primitive Earth, any NH_3 present in the primitive atmosphere would have been photolyzed to N_2 in 10^6 years. For NH_3 to have had a significant role in chemical evolution, life must have formed during that geological short time period. Otherwise one must conclude that NH_3 had no role in chemical evolution (38).

ACKNOWLEDGMENTS

We thank Professors R. Reeves and P. Harteck of Rensselaer Polytechnic Institute for invaluable assistance with the construction and use of the iodine lamp. Support by NASA research grant 33-018-148 and USPHS Career Development Award GM 6380 (to J.P.F.).

NOTES AND REFERENCES

1. Chemical Evolution XX. Previous paper in this series: Ferris, J. P., Williams, E. A., Nicodem, D. E., Hubbard, J. S., and

Voecks, G. E., Nature, in press (1974). A preliminary report of some of these data has been published by Ferris, J. P., and Nicodem, D. E., Nature 238, 268 (1972).

2. Oparin, A. I., "The Origin of Life on Earth," Macmillan, London, (Reprinted, Dover, New York, 1953).

3. Urey, H. C., Proc. Nat. Acad. Sci. U.S. 38, 351 (1952).

4. Rasool, S. I., and Mc Govern, W. E., Nature 212, 1225 (1966).

5. Mc Govern, W. E., J. Atmos. Sci. 26, 623 (1969).

6. Miller, S. I., Science, 117, 528 (1953); Miller, S. L., J. Am. Chem. Soc. 77, 2351 (1955).

7. Lemmon, R. M., Chem. Revs. 70, 95 (1970).

8. Hodgson, G. N., and Ponnamperuma, C., Proc. Nat . Acad. Sci. U.S. 59, 22 (1968).

9. Lohrmann, R., and Orgel, L. E., Science 171, 496 (1971); Handschuh, G. J., and Orgel, L. E., Science 179, 483 (1973).

10. Ponnamperuma, C., Lemmon, R. M., Mariner, R., and Calvin, M., Proc. Nat. Acad. Sci. U.S. 49, 737 (1963).

11. Oro, J., and Kimball, A. P., Arch. Biochem. Biophys. 94, 217 (1961).

12. Dowler, M. J., Fuller, W. D., Orgel, L. E., and Sanchez, R. A., Science 169, 1320 (1970).

13. Friedman, N., Miller, S. L., and Sanchez, R. A., Science 171, 1026 (1971).

14. Ibanez, J. D., Kimball, A. P., and Oro, J., Science 173, 444 (1971).

15. Schimpl, A., Lemmon, R. M., and Calvin, M., Science 147, 149 (1964).

16. Bada, J. L., and Miller, S. L., Science 159, 423 (1968).

17. Rubey, W. W., Geol. Soc. Amer. Spec. Paper 62, 631 (1955).

18. Holland, D. H., in "Petrologic Studies: A Volume to Honor A. F. Buddington " (Engel, A. E. J., James, H. L., and Leonard, B. F., eds.), p. 447, Geological Society of America, New York, 1962.

19. Abelson, P. H., Proc. Natl. Acad. Sci. U.S. 55, 1365 (1966).

20. Cloud, P. E., Science 160, 729 (1968).

21. Goody, R. M., "Atmospheric Radiation I, Theoretical Basis," p. 421, Oxford University Press, London, 1964.

22. Noyes, W. A., and Leighton, P. A., "The Photochemistry of Gases", p. 374, Dover, New York, 1966.

23. The value of 2.4 torr in Table I for the partial pressure of water vapor was based on the assumption that the primitive Earth had a relative humidity of 50% at 0°. The value of 24 torr is equivalent to a relative humidity of 100% at $25^{o}C$.

24. Watanabe, K., and Jursa, A. S., J. Chem. Phys. 41, 1650 (1964). Values for the absorption coefficient in the 200-300 nm region were measured in this laboratory.

25. Harteck, P., Reeves, R. R., Jr., and Thompson, B. A., Z. Naturforschg. 19a, 2 (1964); Thompson, B. A., Reeves, R. R., Jr., and Harteck, P., J. Phys. Chem. 69, 3964 (1965). It has been reported that 20% of the emission from the iodine lamp is at wavelengths less than 206.2 nm (26). The quartz window on our lamp filtered out the bulk of this shorter wavelength light. Furthermore several of the photolyses were performed with a 1 cm water filter which isolated the 206.2 nm line with no observed effect on the quantum yield or reaction.

26. Buschman, N. W., and Groth, W., Ber. Bunsenges, Physik. Chem. 73, 859 (1969).

27. The quantum yield for ammonia photolysis is reported to be 0.25 (22). We obtained a value of 0.27 $\pm$ 0.03 using an HBr actinometer (28), and have adopted the 0.25 value.

28. Calvert, J. G., and Pitts, Jr., J. N., "Photochemistry," p. 782, John Wiley, New York, 1966.

29. de Darevent, B. and Roberts, R., Proc. Roy. Soc. A216, 344 (1953).

30. Brinkmann, R.T., J. Geophys. Res. 74, 5355 (1969).

31. Van Valen, L., Science 171, 439 (1971).

32. Sillen, L. G., Arkiv. Kemi 24, 431 (1971).

33. Eloranta, J., Soumen Kemistilihti 34, 107 (1961); Shaw, W. H. R.,

and Bordeaux, J. J., J. Am. Chem. Soc. 77, 4729 (1955).

34. Van Trump, J. E., and Miller, S. L., Earth and Planet. Sci. Lett. 20, 145 (1973).

35. Sanchez, R. A., Ferris, J. P., and Orgel, L. E., Science 154, 784 (1966).

36. Horowitz, N. H., and Miller, S. W., Prog. Chem. Org. Natural Products 20, 423 (1962).

37. Vancini, C. A., "Synthesis of Ammonia" (Bogars, B. J., ed., transl. by Pirt, L.), p. 60, Macmillan, London, 1971.

COACERVATE DROPLETS, PROTEINOID MICROSPHERES, AND THE GENETIC APPARATUS

Sidney W. Fox

Institute for Molecular and Cellular Evolution
University of Miami
Coral Gables, Florida 33134, U.S.A.

INTRODUCTION

In this paper, coacervate droplets and proteinoid microspheres are compared, both as laboratory models and as sources of inference. Depending upon the definition used for coacervate droplets, proteinoid microspheres themselves constitute a type of coacervate droplet. Both in the Soviet Union* and the United States, this author has been asked to compare and contrast the two. The context of those questions indicated that coacervate droplets from gelatin and gum arabic were usually meant on the one hand, whereas microspheres assembled from proteinoid alone were meant on the other. Difficulties stem from the fact that either coacervate droplets or proteinoid microspheres are produced in many types. Coacervate droplets so designated are complexes of two or more polymers of markedly different composition. The simplest proteinoid microsphere consists mainly of a family of compositionally closely related copolyamino acids, rather than of two colloidal types of markedly different sources and polarities. It is also true that coacervate droplets of, for example, polyvinyl sulfonic acid (34) alone have been reported, but this type is not relevant to a question of associations of functions related to vital properties, associations such as are found in copolyamino acid preparations. A fundamental difference between the typical coacervate droplets of the Oparin school and the typical proteinoid microspheres

---

*This question arose frequently during an interacademy cultural lecture exchange of Academician Oparin and this author in 1969. The answers at that time were discussed by the author for a paper translated into Russian (14).

is that the former are produced from polymers obtained from contemporary organisms, whereas microspheres are aggregates of proteinoid arising in turn from monomeric amino acids under one set of geologically relevant conditions [occasionally in the presence of material such as seawater salts (48), basalt (13), or a variety of other minerals (20, p. 150].

Discussions of each type of structure carry some of the same implicit emphases. One emphasis is that of the need for holistic organization, which is destroyed when a cell is disassembled for analysis. The utility, or necessity, of the early appearance on the scene of a (primitive) cell has been stressed also by Ehrensvärd (8) and Onsager (38), among others. The experiments with proteinoid microspheres have shown that, once proteinoid was formed, only the presence of liquid water was necessary to produce such minimal cells; the latter would have emerged easily. Experiments with basic proteinoids and polynucleotides have yielded, also, models for protoribosomes. These have enabled, in turn, inferences about the origin of the nucleic acid-controlled genetic apparatus.

The fact that inferences from (a) the kind of coacervate droplet specified and from (b) the proteinoid microspheres have much in common when broadly viewed, is reflected in the frequency with which the two are discussed in close proximity in recent textbooks (e.g. 5, 7, 10, 28, 30, 33, 44, 45; compare 1, 31, 47). The similarity in emphasis is, for example, explained in Lehninger's book (33) in a section titled Life without Nucleic Acids: Coacervate Droplets and Microspheres. Lehninger refers to Oparin's coacervate droplets as models for protobionts and then states that "A corollary of his (Oparin's) hypothesis is that the genetic apparatus which yielded precise self-replication of cell catalysts may have been a later event in biological evolution." In this and other treatments, each kind of structure serves conceptually to bring a cell early onto the scene, since it provides a favorable environment for catalyzed reactions, reproduction, and the evolution of metabolism. Indeed, my view (18,21,23) is that evolution of molecules and their functions would have early arrived at a dead end until and unless a cell formed.

The science of protobiogenesis (spontaneous generation, abiogenesis, biopoesis) especially owes to Oparin a naturalistic explanation that reversed the negative view bequeathed to us by Louis Pasteur and his disciples (20,39, pp. 37-38). Oparin, himself, however made what seems to be a curious statement (translation by A. Synge) for 1957, "However, as the methods of scientific investigation of living nature become more and more precise, spontaneous generation was gradually relegated to simpler and simpler organisms. Finally the sudden appearance of even the most primitive organisms from inanimate material was shown to be impossible. Thus, to-day, the theory of spontaneous generation has no more than a historical interest and cannot serve as an approach to the problem with which we are concerned." Our interpretation of the experimental results from microspheres agrees fully with the first of these three sentences, but not with the last two. Nor is it apparent to me how one would demonstrate that spontaneous generation of even the most primitive organism is impossible. Even Pasteur, in his later years, is reported to have said, "Spontaneous generation? I have been looking for it for twenty years but I have not yet found it, although I do not think that it is an impossibility" (37).

Despite this seeming inconsistency in Oparin's evaluation of 1957 on this one issue, his writings contributed critically to reversing the general thinking against spontaneous generation (of the modern kind) which persisted for so many decades. Oparin's contributions were however much more than that. Among the other emphases that he made was that (39, p. 392) of the importance of organization of the living systems, e.g. "a very highly developed organisation of the metabolic network."

THE PRIMORDIAL SEQUENCE

Oparin (39, p. 392) said, for example, "First there was the formation of individual catalysts of great reactivity and specificity. Later the activity of these catalysts was co-ordinated and there arose the whole chains and cycles of enzymic reactions which form the basis for the separate departments of metabolism. Still later came the spatial organisation of the system and the localisation of processes and the rationalisation of the interacting energic and structural branches of metabolism. This guaranteed, within limits, the continual self-preservation and self-reproduction of living systems."

The above states that enzymes, compartmentalized cells, and reproduction arose in that order.* With most of this our experiments are in agreement; I would however separate reproduction into cellular reproduction and molecular reproduction, a classification required as the result of combined inferences from experiments (16,17).

This last-mentioned view follows from the need for ribosomes to provide a contemporary type of peptide bond synthesis governed by polyribonucleotides (17), and the fact that a minimal cell would arise quickly and often as soon as proteinoid appeared and came into contact with water.

---

*It also resembles remarkably the inferences of Eigen (9) arrived at by a quite different line of reasoning.

The enlarged flowsheet of concepts is as follows:

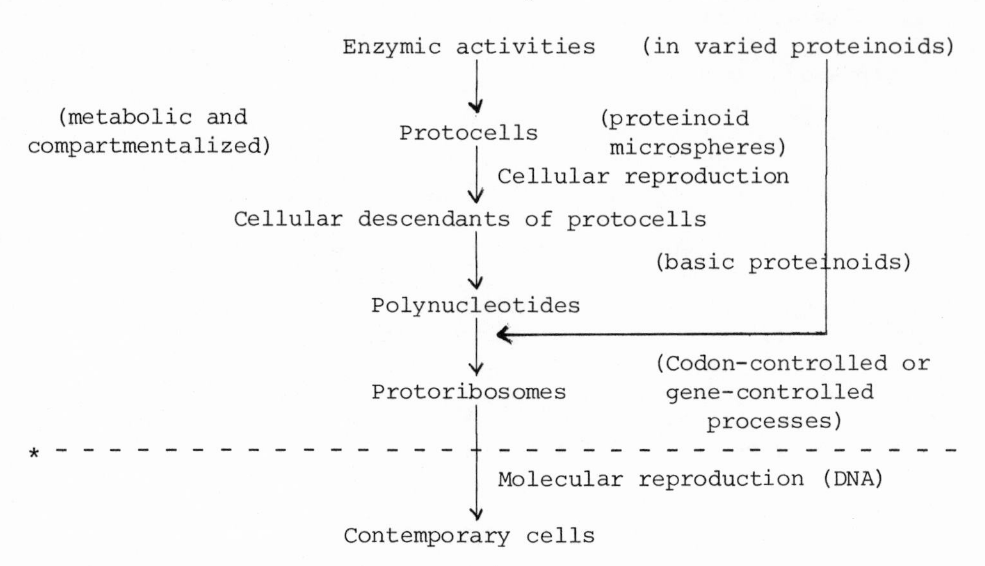

*Physical models to this point

Although the ribosome concept is included in this sequence, awareness of the ribosome as such was essentially absent in 1957. The awakening to the role of the ribosome, therefore, came about since Oparin wrote the quoted edition of his work and since we began our experiments.

COMPARISONS OF CONCEPTS

The crucial refinements of Oparin's ideas made possible by the proteinoid microspheres are, in my view, two. For the first of these (39, p. 290), the conceptualization of "organic polymers . . . having, as yet, no orderly arrangement of amino acid and nucleotide residues" was replaced by the finding that, in the thermal polycondensation, amino acids order themselves.

The other crucial refinement flows from the obtaining of models for the protocell from polymers produced in turn from monomers under geologically relevant conditions. The coacervate droplets of Bungenberg de Jong and of Oparin have been made from polymers extracted from evolved organisms, an approach that, as earlier indicated, is anachronistic for investigation of origins (even though important physical principles may be revealed).

Neither of the two crucial advances just referred to could be fully visualized from reductionistic approaches; what was necessary were experiments in the production of more complex systems from simple components, those components being primordial instead of contemporary in character.

Oparin's emphases upon organization of open systems are epitomized (39, p. 332) in his statement,

"Thus the simplest abiogenic system which could have served as the starting point for the evolutionary process which led up to the appearance of life must already have had the organisational features characteristic of open systems, in which the separate reactions form a network of chemical transformations which are coordinated in time."

While we have carried out all of our key processes, in the making or converting of proteinoid, in open systems, in agreement with this statement of Oparin, with the emphases of Prigogine (41), and with concepts of irreversible thermodynamics (18) we have also extended the principle back to the production of micromolecules (20). The chemist is conditioned to the use of closed flasks* in studying his reactions, but the planetary surface is an open realm. The open state has been brought into focus dramatically by analysis of the failure of the Apollo program to find on the surface of the Moon a "bonanza of organic molecules," an expectation based in part on the results of organic synthesis in closed flasks (15).

Our emphases on open systems have been further supported by the studies of Orgel (25), who showed that adenosine could be synthesized from adenine and ribose in the solid state in the obligatory absence of near-absence of water. One special virtue, thus highlighted for open systems, is that they permit the volatilization of water.

The abiotic synthesis of amino acids, proteinoids, polyphosphates, adenosine, and ATP, each under hypohydrous conditions in open systems, are thus indirect support of the early emphases on open systems (43).

Although Oparin recognized the need for a synthesis of proteins early in evolution, he conceived only of disordered polymers, as mentioned earlier. He did not suggest an explanation of how the amino acids would be polymerized. This was accomplished in our program by simple anhydrization of amino acids, first by the open-system heating of amino acids. This simple and geologically

*Energy is transferred, but matter is not, into and out of closed flasks.

relevant process could have been missed in the thinking at the time because of a general feeling of the futility of obtaining polymers of α-amino acids by heating the latter, as expressed, for example, by Emil Fischer (11) and by W. H. Carothers (3) who produced nylon from ω-amino acids.

The successful polymerization of α-amino acids is due to the special effects of copolymerization. Most α-amino acids decompose on heating, as Carothers knew, but inclusion in the mixture of a sufficient proportion of nonneutral amino acid (aspartic acid, glutamic acid, or lysine) bring into largely linear copolymers all of the amino acids that would otherwise suffer decomposition on extended heating (20).

An equally significant addition from these same experiments was the finding that amino acids order themselves during thermal copolymerization. The thermal program, in fact, evolved from an earlier interest in learning about the evolution of order, expressed as primary sequence, in protein molecules. After initial study of sequences in evolved proteins, beginning in 1944 (12,17, 42), it became clear that a fundamental understanding or the appearance of order in protein molecules, i.e., the first events, required assembling protein molecules rather than disassembling them.

When one thinks of the origin of enzymes, instead of the origin of proteins, the perspective enlarges. Of course, principal properties of proteins include both enzymic activity and structural function. The proteinoids have both qualities and, by inference, the protoproteins had both qualities. The question emphasizing enzymes has been eloquently expressed by Dixon and Webb (6) as:

"Given the pre-existing enzymes, the formation of other enzymes is understandable; but the difficulty arises: if enzymes are formed only by enzymes, how were the first enzymes formed?"

The studies of Oparin and his school have vividly explained the biological benefit of containing enzymes within a micropackage. High concentrations of both enzymes and substrates are maintained, a sequence of reactions is aided by being held in proximity, and the system is protected from thermodynamically unfavorable effects of the surrounding fluid.

The proteinoid microsphere also provides these advantages plus the fact that the material of the microsphere itself, like the membrane and protein contents of contemporary cells, can itself be enzymically active (20).

ORIGIN OF THE GENETIC APPARATUS AND CODE

Various aspects of the theoretical basis for the origin of the genetic code, mostly as formulated in advance of most of the constructionistic experiments, have been analyzed by Woese (51), Ycas (52), Crick (4), Calvin (2), and others (8,9,19,20,23,24,26,32,35, 36). What has especially been needed is an understanding of the origin of cellular protein, of cellular nucleic acids, and of selective interactions between nucleic acids and amino acids.

The approach toward these goals has recently been advanced experimentally and summarized to a late date (17). Serebrovskaya, in Oparin's institute had, however, in 1964 shown that ADP was converted to polyadenylic acid in (RNA-histone) coacervate droplets by polynucleotide phosphorylase (46), and proteinoid microspheres have been shown to have the ability to produce internucleotide bonds from ATP (27). A model for the original ribosome, a protoribosome from basic proteinoid and polynucleotide, was reported in 1968 (20); it has the approximate composition, pH-sensitivity, and salt-sensitivity of nucleoprotein organelles. The model protoribosome has been shown to have the fundamental ability to produce peptides from amino acid and ATP (21).

The need for polymer-polymer interactions to chart the stereochemistry of the genetic code has been inferred with proteinoids (53,20) in interaction with polynucleotides. These have yielded results not obtained in searches for polymer-monomer interactions.

Codonic and anticodonic preferences, varying with conditions such as lysine content, have been established in such interactions (24). A set of codonic preferences has been identified in a dynamic system using the model protoribosomes and preformed aminoacyl adenylates (35). It is these results, especially, that have first yielded the inference that the genetic code was determined by stereomolecular constants rather than by chance statistical occurrences (4).

The overview derived from all of the experimental results cited is consistent with a nearly simultaneous origin of protocellular protein and protocellular nucleic acid, especially in view of the fact that the aminoacyl adenylate consists of a monomer of each of the two classes of polymer, adenine for nucleic acid and amino acid for protein (36).

THE CHICKEN-EGG QUESTIONS

The proteinoid approach permits uniquely a discussion of the chicken-egg problem, a question that has titillated many viewers of the scene (e.g. 8,9,33,49,50).

In 1957, Oparin (39, p. 287) analyzed the question as follows:

"In this case, which came first, nucleic acids or proteins?

"This question reminds one somewhat of the scholastic problem about the hen and the egg. The problem is insoluble if we approach it metaphysically in isolation from the whole previous history of the development of living matter. Nowadays every hen comes from an egg and every hen's egg from a hen. Similarly, nowadays proteins can only arise on the basis of a system containing nucleic acids while nucleic acids are formed only on the basis of a protein-containing system. The hen and its egg developed from less highly organised living things in the course of their evolution. In the same way, both proteins and nucleic acids appeared as the result of the evolution of whole protoplasmic systems which developed from simpler and less well adapted systems, that is to say, *from whole systems and not from isolated molecules*. It would be quite wrong to imagine the isolated primary origin either of proteins or of nucleic acids (italics added).

"Many contemporary authors do, however, follow this line of thought. They take the view that in the first place nucleic acids arose in some way and that at once, simply by virtue of their intramolecular structure, they were able both to synthesize proteins and to multiply themselves spontaneously. It is, however, clear from all our previous discussion that a hypothesis of this sort is in direct opposition to the facts as they were at present known."

Out total experimental results are in accord with this judgement of Oparin's, for as far as the latter goes, and especially for its emphasis on the need for whole systems for synthesis of nucleic acids and proteins (21). In one respect, the analysis does not go (back) far enough.

The initial "whole system" had to arise from simpler components--macromolecules. (It is not sufficient to use systems to understand cellularly synthesized macromolecules without explaining the origin of the systems, much as it is not sufficient to use polymers from contemporary organisms to explain the origin of those systems.)

The experiments indicate that the initial "systems" were preceded by precursor proteinoids (17,21). The chicken-egg problem now needs to be rephrased and then ramified:

For the first stage, the question now is: which informational macromolecule came first? The macromolecule need not have been nucleic acid to have been informational. The thermal experiments specify the answer to that question as proteinoid-first (not protein-first), and that such polymer was highly ordered internally, and also had arrays of enzymic activity. The protocell resulting from such polymer by contact with H_2O would have initiated reproduction, at the the protocellular level.

In the second stage, protonucleic acid was produced (8,21,27); it combined with basic preprotein (proteinoid) to yield protoribosomes.

In the third stage, protein and nucleic acids were produced in approximate simultaneity. Synthesis of protein by protoribosomes would have had to take over for the protocell the heterotrophic obtainment of preformed proteinoid.

The flowsheet, which will undoubtedly have to be ramified as new experiments are completed, now appears as follows:

FIRST STAGE
amino acids $\xrightarrow{-H_2O}$ acidic preprotein, basic preprotein $\xrightarrow{H_2O}$ protocell

SECOND STAGE
preprotein $\xrightarrow{\text{nucleoside triphosphates}}$ protonucleic acid (+ basic preprotein) $\xrightarrow{H_2O}$ protoribosomes

THIRD STAGE
protoribosome $\xrightarrow{\text{ATP, amino acids}}$ protein

While the sequence given is not entirely modelled by laboratory demonstrations, it is to a considerable degree in that stage.* Formation of protoribosomes in a protocell has not yet been demonstrated. All steps are, however, operationally simple and geologically relevant (17). Since proteinoid, in appropriate systems, can synthesize internucleotide and peptide bonds, it provides the possibility of evolvability from a protocell (21).

The first preprotein, as modelled by proteinoid, contained information derived from diverse mixtures of amino acids, making the presence of nucleic acids superfluous at that point. The experiments of the first stage also show that reproduction occurred easily at the protocell stage, and was inevitably at first a manifestation of systems (16), as Oparin (39, p. 357) has also emphasized in another context.

---

*Especially needed in models are: a surer understanding of the origin of basic amino acids, a demonstration of the production of a family of purines and pyrimidines under a single set of open-system geochemical conditions, cellular production of large polynucleotides, and the protected functioning of protoribosomes within a protocell.

A protoribosome of the type described would have met the requirement for synthesis of peptide bonds in the presence of water (cf. 54). The model protoribosome however arises from prior proteinoid, itself synthesized in the presence of much less water. The recognition that nucleic acid was not needed at first for either ordering of amino acids in proteinoid nor for reproduction at the level of microsystems is crucial. But Oparin's "first systems" were, according to the microspheric experiments, preceded by partly ordered, catalytically active proteinoids, not proteins (the latter defined here as polymers synthesized by ribosomes).

According to my interpretation of the flowsheet, the first proteins did not get their information indirectly from earlier proteinoids. The instructions for the proteins were generated by the nucleic acids within the protoribosomes. However, stereochemical and other forces operative in the synthesis of proteinoid and in the ribosomally guided synthesis of protein must have, according to a number of indications, yielded similar products even though the forces were manifested at different stages.

OVERVIEW

In order to answer the complex question of the origin of life, Oparin has posed a number of questions and answers for the contributory materials, systems, and processes. Some of these have had to be rephrased. From our experimentally disciplined perspective, Oparin's answers were mostly not incorrect; the questions, rather, were in some instances incomplete, as exemplified by the contrast between the source of coacervate droplets and that of proteinoid microspheres. We can anticipate further testing of the newer answers, and further refinement of the questions and their selection, all based on Oparin's pioneering concepts. Basically, however, the questions have had to be asked experimentally rather than merely deductively and we can now see that the experimental approach has had to precede the appropriate questions as well as the answers (16). As one example, classification of the early chemistry into three stages appears not to have been feasible from the perspective provided by analysis of the contemporary.

The course of the research on proteinoid and microspheres (and the organization of the Wakulla Springs conference in 1963) reflect Oparin's conceptual modulation from micromolecule to macromolecule to prebiological system. We could not at that time foresee all of the properties later to be found in proteinoid microsystems. When the "remarkable" properties of proteinoid microsystems (29,33,44,45) were found to include also the ability to make internucleotide bonds and peptide bonds, the term prebiological system became less definitive than it appeared to be earlier.

ACKNOWLEDGMENTS

Research on proteinoid microspheres has been aided by Grant NGR 10-007-008 of the National Aeronautics and Space Administration. Contribution no. 272 of the Institute for Molecular and Cellular Evolution.

REFERENCES

1. Ambrose, E. J., and Easty, D. M., "Cell Biology," pp. 479-480, Addison-Wesley, Reading, Mass., 1970.
2. Calvin, M., "Chemical Evolution," pp. 162-183, Oxford University Press, 1969.
3. Carothers, W. H., Trans. Faraday Soc. 32, 39 (1936).
4. Crick, F.H.C., J. Mol. Biol. 38, 367 (1968).
5. CRM Books, "Biology--An Appreciation of Life," pp. 51-52, Del Mar, Calif., 1972.
6. Dixon, M., and Webb. E. C., "Enzymes," p. 666, Academic Press, New York, 1958.
7. Dowben, R. M., "Cell Biology," pp. 528-531, Harper and Row, New York, 1971.
8. Ehrensvärd, G., "Life: Origin and Development," University of Chicago Press, 1962.
9. Eigen, M., Naturwissenschaften 58, 465 (1971).
10. Etkin, W., Devlin, R. M., and Bouffard, T. G., "A Biology of Human Concern," pp. 502-503, J. B. Lippincott Co., Philadelphia, 1972.
11. Fischer, E., Chem. Ber. 39, 609 (1906).
12. Fox, S. W., Advances Protein Chem. 2, 155 (1945).
13. Fox, S. W., Nature 201, 336 (1964).
14. Fox, S. W., J. Evol'y Biochem. Physiol. 6, 131 (1970).
15. Fox, S. W., Bull. Atomic Scientists 29(10), 46 (1973).
16. Fox, S. W., Naturwissenschaften 60, 359 (1973).

17. Fox, S. W., Molec. Cell. Biochem. 3, 129 (1974).

18. Fox, S. W., in: "Unity in Natural Science" (Mintz, S., ed.), in press.

19. Fox, S. W., Amer. Biol. Teacher 36, 161 (1974).

20. Fox, S. W., and Dose, K., "Molecular Evolution and the Origin of Life," W. H. Freeman, San Francisco, 1972.

21. Fox, S. W., Nakashima, T., and Jungck, J. R., Fourth International Conference on Origin of Life, Barcelona, Abstract 24, 1973; ibid., Origins of Life, in press (1974).

22. Fox, S. W., Lacey, J. C., Jr., and Mejido, A., unpublished experiments, 1972.

23. Fox, S. W., McCauley, R. J., and Wood, A., Comp. Biochem. Physiol. 20, 773 (1967).

24. Fox, S. W., Yuki, A., Waehneldt, T. V., and Lacey, J. C., Jr., in: "Chemical Evolution and the Origin of Life" (Buvet, R., and Ponnamperuma, C., eds.), p. 252, North-Holland, Amsterdam, 1971.

25. Fuller, W. D., Sanchez, R. A., and Orgel, L. E., J. Mol. Biol. 67, 25 (1972).

26. Jukes, T. H., in: "The Origins of Prebiological Systems" (Fox, S. W., ed.), p. 407, Academic Press, New York, 1965.

27. Jungck, J. R., and Fox, S. W., Naturwissenschaften 60, 425 (1973).

28. Keeton, W. T., "Biological Science," pp. 694-695, W. W. Norton, New York, 1972.

29. Kenyon, D. H., Science 179, 789 (1973).

30. Koob, D. B., and Boggs, W. E., "The Nature of Life," pp. 348-350, Addison-Wesley, Reading, Mass., 1972.

31. Korn, R. W., and Korn, E. J., "Contemporary Perspectives of Biology," pp. 456-459, John Wiley, New York, 1971.

32. Lacey, J. C., Jr., and Pruitt, K., Nature 228, 799 (1969).

33. Lehninger, A. L., "Biochemistry," pp. 782-784, Worth, New York, 1970.

34. Millich, F., and Calvin, M., J. Phys. Chem. 66, 1070 (1962).

35. Nakashima, T., and Fox, S. W., Proc. Nat. Acad. Sci. U.S. 69, 106 (1972).

36. Nakashima, T., Lacey, J. C., Jr., Jungck, J. R., and Fox, S. W., Naturwissenschaften 57, 67 (1970).

37. Nicolle, J., "The Life of Pasteur," p. 75, Basic Books, New York, 1961.

38. Onsager, L., in ref. 17.

39. Oparin, A. I., "The Origin of Life on Earth," Academic Press, New York, 1957.

40. Orgel, L. E., J. Mol. Biol. 38, 381 (1968).

41. Prigogine, I., "Thermodynamics of Irreversible Processes," 2nd ed., Interscience, New York, 1961.

42. Rosmus, J., and Deyl, Z., J. Chromatography 70, 333-335 (1972).

43. Ryan, J. R., and Fox, S. W., BioSystems 5, 115 (1973).

44. Salthe, S. N., "Evolutionary Biology," pp. 31-33, Holt, Rinehart, and Winston, New York, 1972.

45. Samuel, E., "Order: In Life," pp. 261-268, Prentice-Hall, Englewood Cliffs, N.J., 1972.

46. Serebrovskaya, K., in "Problems of Evolutionary and Technical Biochemistry" (Kretovich, V. L., ed.), p. 127, Nauka, Moscow, 1964.

47. Smallwood, W. L., and Green, E. R., "Biology," p. 234, Silver Burdett, Norristown, N.J., 1968.

48. Snyder, W. D., and Fox, S. W., Federation Proc. 32, 640 abs. (1973).

49. Tatum, E. L., in: "Evolving Genes and Proteins" (Bryson, V., and Vogel, H. J., eds.), p. 10, Academic Press, New York, 1965.

50. Thimann, K. V., "The Life of Bacteria," 2nd ed., p. 834, Macmillan, New York, 1963.

51. Woese, C. R., "The Genetic Code," Harper and Row, New York, 1967.

52. Ycas, M., "The Biological Code," North-Holland, Amsterdam, 1969.

53. Yuki, A., and Fox, S. W., Biochem. Biophys. Res. Commun. 36, 657 (1969).

54. Wald, G., Origins Life 5, 7 (1974).

ASPECTS OF THE EVOLUTION OF SOME SYSTEMS AND PROCESSES OF SELF-REGULATION OF PLANT METABOLISM

G. Georgiev and N. Bakardjieva

Department of Plant Physiology, University of Sofia Bulgaria and Institute of Plant Physiology, Bulgarian Academy of Sciences, Sofia

The evolution of living matter is connected with the appearance of increasingly complex and differentiated systems, which display firm coordination and sequence of functions. One of the main characteristics of living systems is their ability, in the process of interaction with the environment, to preserve the metabolic parameters specific for the given type of organism and the kind of elements they are composed of--low molecular compounds, polymers and organelles. The realization of the genetic program proceeds in the ontogeny. The higher the level of evolution of the organism, the more complex are the systems of genetic and physiological control. This ensures the coordination of the vital functions in time and space.

A. I. Oparin (1) remarks that the enormous quantity of descriptive facts on living organisms accumulated prior to Darwin has been interpreted and summarized by the idea of evolution. Oparin emphasizes that, by analogy, the great amount of data gathered on the organization of metabolism and on the structure of molecules, membranes and organelles in organisms at different evolutionary stages will gain a new significance when interpreted from the point of view of evolution.

The question raised by Oparin inevitably leads to the problem of the evolution of the control mechanisms of the metabolic processes. The existence of systems with many structural and functional levels, and with a clearly expressed morphological differentiation, presupposes that they should display more complete coordination of the structural-functional relations, i.e. a system of self-regulation. All this poses a new task for scientific investigation--the comparative study of the systems, principles and mechanisms of the self-regulation of metabolism.

This conclusion is logically derived from a conception in Oparin's theory on the origin of life. According to this conception, a characteristic feature of living systems is that the network of metabolic reactions is not only coordinated, but is also directed toward self-preservation and self-reproduction of the system as a whole. This purposefulness has not come about by chance, but has been formed in the process of constant improvement (1,2).

In connection with this conception an attempt is made in this paper to discuss certain aspects of the evolution of some systems and processes of self-regulation of the metabolism in plant organisms. This article marks the fiftieth anniversary of Oparin's first publication on the origin of life and the eightieth anniversary of his birth.

It is known that the coordination of living processes is a result of the coordinated unity of specific structural and functional elements of the biological system. Organizational and functional subordinated systems may be distinguished in a complex living system. The great variety of the structural elements they are composed of--molecules, molecular aggregates, membranes, cell organelles, tissues and organs--is doubtless a prerequisite and an evolutionary adaptation for achieving greater metabolic regulation.

Functional systems unite the biological activity of different structures by means of a common function. This assumption becomes clearer if based on Anokhin's (3) definition. This states that the functional system is a complex of components included on a selected basis, the interaction and inter-relation of which acquires the character of a mutual concerted action of the components. This mutual concerted action is aimed at achieving a useful result. The effective participation of the functional system in the self-regulation of the metabolism depends on the degree of mutual concerted action achieved.

If the evolution of living systems is interpreted on the basis of cooperation and mutual concerted action among different organizational and functional systems, we may then search for the specific systems and processes which bring about the self-regulation at the different levels of evolution of the metabolism. There are two main questions to be dealt with--the evolution of the functional systems and their hierarchical subordination. The evolution of the functional system is expressed in the improvement and specialization of the function itself, usually connected with changes in the structural complexity of the system.

As yet, two stages in the evolution of the organism have been fairly well elucidated: the evolution of macromolecules and the evolution of the species. They are interdependent, but little is known about the systems and the processes which bring about the connection between them. This is probably the place to search for the self-regulation mechanisms.

There is quite a considerable amount of data on evolution at the molecular level. Two general tendencies may be observed: 1) unity and uniformity of the basic molecules (nucleic acids, proteins) in all organisms from the simplest to the superior ones, and 2) multiplicity and diversity in the forms of these compounds. It may be supposed that the velocity of evolution at the molecular level is greater than the speed of the evolution of the species. However, very little is known about the real relation between the two evolutionary processes. An attempt to find a link between them has been made by Soran (4).

From the point of view of evolution the changes in the molecule of the nucleic acids are connected with the coding and translation of genetic information and with the functional organization of the genetic material. Ohta and Kimura (5) emphasize that the transition from lower to higher groups of organisms is characterized by differences in the nucleotide sequence, as well as by an enlargement of the molecule (mammalian DNA being 1000 times greater than bacterial DNA). The larger DNA molecule allows greater possibilities for multiplicity in the organization of its structural elements.

More indices may be used in order to comprehend this variety in a comparative study of DNA in different systematic groups of organisms. These may include not only nucleotide composition and sequence, but also the presence and composition of satellite DNA, the percentage of homology and other indices (6). Belozersky and Mednikov (6) point out that the existence of a DNA specificity in the different species is completely understandable. One must bear in mind that in these substances, which are so important for the organism, the evolution of the species must be reflected. The evolution of the gene and the genetic code is a question widely studied and discussed nowadays (7-9).

The evolution of protein may also be thought of as an evolution of biological catalysis. The simplest enzyme is much more complex than the non-protein type of catalysts. Perhaps the most important advantage of the biocatalytic systems widely used in evolution lies precisely in the multiplicity of the protein molecule. For this reason special importance in the adaptation of the activity of the catalysts to the peculiarities of the regulation in a given organism, organ, or cell is attached to the protein component of the biocatalyzer (10). In this case the size of the molecule and its structure also changes. While preserving its characteristic function, the protein molecule may pass through different microevolution states, which will determine the ascending and divergent evolution.

Changes in the amino acid residue sequences of a number of enzymes in phylogenetically different plants and animals have been widely investigated in recent years. It has been established that enzymes with an equal catalytic function in different groups of organisms have specific and different primary structures. A typical example is provided by the ferredoxins (11-13), and by cytochrome C (14,15). According to Webb (16), in the protein component of the enzymes there may be a difference not only in the sequence of the amino acid residues, but also in the molecular weight, the sedimentation constant, the immunobiological character, and other properties.

It is logical to deal with the question of the evolutionary meaning of this multiplicity of the molecular forms in parallel with the rather conservative and slow change in the function. This multiplicity gives the possibility for formation of the specific type of metabolism.

Apart from the nucleic acids and proteins, other components of the plant system also undergo evolutionary changes. In plants, as autotrophic organisms, has appeared and developed a photoreactive system specialized in absorbing and utilizing light in the biosynthetic processes. We may now speak definitely of evolution of the pigments, of photochemical reactions, of photosynthetic apparatus, of the evolution of the electron donor system, and of electron transport (17-24). The bioenergetic systems and processes also evolved to a considerable degree (25-27).

On the basis of all that is known about this problem we may outline three main aspects or tendencies of evolution at different levels and the interrelations between them. Firstly, parallel with the evolution of the plant domain and of the plant species as a whole system, there is a very intensive and divergent molecular evolution. This concerns all important cell components, though to varying degrees. The slowly ascending development of the whole system is connected with a change in a number of microevolutionary states of the subordinated systems and this in rare cases leads to the appearance of molecular mutations.

As a second tendency, it may be noted that the molecular epigenesis at the DNA level and the consequent epigenesis at the level of proteins forms the basis of biochemical evolution (28). Therefore, it may be presumed that they form the basis of the evolution of the main mechanisms of self-regulation as well.

There is a third aspect, to which little attention has been paid so far. This is that mechanisms exist which integrate the specific feature of the primary structure of macromolecules with the whole system of self-regulation, and bring about an interaction with the other components of the plant metabolic system. To reveal these mechanisms is a significant task for those investigating self-regulation.

The dependence of enzymic activity on specific metabolic effectors is one of the possible routes by which the connection takes place between the enzyme systems and the specific components or metabolic pathways for a given organism. Comparative studies on the regulation mechanisms for a number of key enzymes in different types of organism indicate this. There have been studies which show that the properties of some allosteric enzymes depend on the type of metabolism and on the phylogenetic position of the organisms investigated (29-31). As Borris (32) notes, data are available to indicate how the control of three enzymes: ADP-glucoso-pyrophosphorylase, citrate synthetase, and glucoso-6-phosphate dehydrogenase is adapted to the type of metabolism of the different organisms by means of specific metabolic effectors, such as AMP, ATP, NADH, and others. These data have given Borris (32) ground to conclude that in the course of evolution there is a change in the sensitivity to a given control mechanism of enzymes that catalyze equivalent metabolic reactions. Thus, the enzyme action is adapted to the type of metabolism.

In our investigations (unpublished findings of Georgiev, Bakardjieva, and Georgiev), we have established some changes conditioned by evolution in the characteristic properties of peroxidase from species of algae, mosses, ferns, coniferous trees, and certain monocotyledons and dicotyledons (maize, barley, and peas). Parallel with the increase in the specific activity of the peroxidase from lower to higher organisms, some differences in the substrate specificity has been observed. For the peroxidase systems from the same plant groups, varying degrees of activation or inhibition have been established. This is caused by calcium or copper ions and by sodium azide, and depends on the phylogenetic position of the plant. We assume these results to be an example of a different sensitivity of the enzyme to the ionic effectors. It is well known that the kind and quantity of the ions is a specific feature of a given species, being connected with the type of metabolism (33).

It must be taken into account that not only the enzyme systems, but also a number of properties of other molecules (for example nucleic acids and nucleotides), are dependent on the metabolic features of the whole system through the respective metabolic effectors.

We shall now proceed to the question of the multiple molecular forms of the substances in the organism. According to Goldovsky

(34), the multiplicity in each group of substances is a general phenomenon. The reason for its existence may be seen in the great variety of biochemical reactions and in the existence of "sheaves" of similar reactions. The existence of isoenzyme components of many enzyme systems may be explained on this basis (34). The significance of multiplicity as an evolutionary acquisition in connection with the adaptation processes, and with a view to the evolution of the system, has been discussed by many authors (35-37). That mechanism which determines the more active inclusion of one or another form of substance in the metabolic processes will be of dominant importance for the self-regulation of the plant organism as a whole system. This allows divergent specialization of the biocatalysts to take place, resulting in the formation of optimal catalysts which are able to direct the biosynthetic and bioenergetic processes in one of the possible ways (38).

Another source for changing the systems of metabolic self-regulation occurs when substances assume a regulatory function which they have not previously had. This happens at a particular moment in evolution. A common origin of all living systems is a prerequisite for some unity in their chemical composition and in the basic biochemical reactions. But in the process of the morphological and biochemical improvement of a living system, some substances present in the organism may acquire a new function. All this increases the possibility for the system as a whole to achieve homeostasis. In this case, according to Goldovsky (39), a change, an expansion, an acceleration, or a limitation of the function may possibly take place. The phytohormones provide a typical example of this. Kefeli and co-authors (40) point out that substances functioning as phytohormones in plants are found in great quantities in many microorganisms, but in this case they do not possess any regulatory function. No doubt, in the process of formation of the complex multicellular organism with different tissues and organs the necessity has arisen for a coordination of the processes taking place in them. This function has been taken on by the substances of phytohormonal type.

At different times and with various qualitative changes in the organism it is probable that the polyfunctional enzyme systems have also acquired a new kind of activity. For example, it is well known that the peroxidase system often also manifests IAA-oxidase activity, but we know very little about the stage of evolution and the kind of organisms in which this activity appears. Recently Peive and others (41) have stated that the peroxidase system has also a nitrate reductase activity. This fact raises the same questions for investigation. The increase in the biochemical features by which living systems may differ also increases the degree of biochemical individuality (42). Thus new possibilities are revealed for the self-regulation of metabolism and for an evolution of control mechanisms.

The parameters in which control mechanisms may act are determined by the natural possibility that many systems may bring about an oscillatory course of the processes. This fact is closely related to the processes of self-regulation. Usually, one oscillating biochemical system is an interaction among different molecular structures. The oscillating processes are an expression of the integrative interaction of these molecules, the combination of which leads to an integration of activating and inhibiting processes (43). Such a course of the reactions brought about by the biocatalytic systems in a complex medium and in a more complex subsystem, such as a polyenzyme complex, is very characteristic. Biological endogenous rhythms provide an even more complex and integrative expression of oscillations of biochemical systems. In respect to self-regulation, these oscillating processes indicate the possibilities of the functional system for preserving the basic character of the function in varying conditions in and out of the biosystem. The amplitude of these oscillations could determine the evolutionary potential of a living system, its ability to adapt, and the biochemical limitation of the evolution of a given type of organism.

The self-regulation of the metabolic processes in plants may be associated with the activity of an integrative functional system which includes the interaction of the basic macromolecules--DNA, RNA, enzyme proteins, hormones, and energy-transferring compounds. All present-day data about the physiological and biochemical role of metal ions gives grounds for their inclusion in this integrated functional system of self-regulation (44).

As has already been pointed out, the evolution of this system may be considered from two aspects: the inclusion of new functional components (as in the case of phytohormones), and a change in the character of their interaction. The latter is the result of conformational changes in some components and of replacement in the process of phylogenesis of one component by another with an equal function. It is also the result of a complication in the structure of the components, connected with an increase in the number of functions (polyfunctionality) and results from a different dependence on the metabolic effectors.

It seems possible that the specific character of these interactions determines the respective type of metabolism in the different organisms. Tracing the common, typical relations among the components of the functional system, as well as the specific features of their interaction,is the main purpose of the studies on the self-regulation of the plant metabolism with a view to the evolution of living systems.

REFERENCES

1. Oparin, A. I., "Razvitie koncepcii strukturnih urovnei v biologii," pp. 235-247, "Nauka," Moskva, 1972.

2. Oparin, A. I., "Szizn--priroda, proizhoszdenie i razvitie," Izdat. Academii nauk SSSR, Moskva, 1960.

3. Anokhin, P. K., "Prinzipi sistemnoi organizacii funkcii," pp. 5-62, "Nauka," Moskva, 1973.

4. Soran, V., Prog. sti. 8, 168 (1972).

5. Ohta, T., and Kimura, M., Nature 233, 118 (1971).

6. Belozerski, A. N., and Mednikov, B. M., "Nukleinovie kisloti i sistematika organizmov," "Znanie," Moskva, 1972.

7. Jukes, T. H., Fourth International Conference Origin of Life, Barcelona, Abstracts, 1973.

8. Novak, V., and Liebl, V., Fourth International Conference Origin of Life, Barcelona, Abstracts, 1973.

9. Watts, R. L., "Chemical Evolution and the Origin of Life" (Buvet, R., and Ponnamperuma, C., eds.), pp. 14-43, North-Holland, Amsterdam, 1971.

10. Bakardjieva, N. T., Spisanie na BAN 5, 20 (1972).

11. Hall, D. O., Cammack, R., and Rao, K. K., Nature 233, 136 (1971).

12. Cammack, R., Chimia 26, 674 (1972).

13. Bayer, E., Jung, G., and Hagenmaier, H., "Prebiotic and Biochemical Evolution" (Kimball, A. P., and Oro, J., eds.), pp. 223-233, North-Holland, Amsterdam, 1971.

14. Yamanaka, T., J. Japan. Biochem. Soc. 43, 47 (1971).

15. Margoliash, E., Ninth International Congress of Biochemistry, Stockholm, 1973.

16. Webb, J. L., "Enzyme and Metabolic Inhibitors," Academic Press, New York, 1963.

17. Krasnovsky, A. A., Fourth International Conference Origin of Life, Barcelona, Abstracts, 1973.

18. Krasnovsky, A. A., "Prebiotic and Biochemical Evolution" (Kimball, A. P., and Oro, J., eds.), pp. 209-217, North-Holland, Amsterdam, 1971.

19. Goodwin, T. W., "Prebiotic and Biochemical Evolution" (Kimball, A. P., and Oro, J., eds.), pp. 200-209, North-Holland, Amsterdam, 1971.

20. Baltscheffsky, H., Fourth International Conference Origin of Life, Barcelona, Abstracts, 1973.

21. Evstigneev, V. B., Fourth International Conference Origin of Life, Barcelona, Abstracts, 1973.

22. Broda, E., Prog. Biophys. Molec. Biol. 21, 143 (1970).

23. Sironval, C., "Chemical Evolution and the Origin of Life" (Buvet, R., and Ponnamperuma, C., eds.), pp. 236-259, North-Holland, Amsterdam, 1971.

24. Metzner, H., International Symposium on the Origin of Life and Evolutionary Biochemistry, Varna, 1971.

25. Broda, E., "Chemical Evolution and the Origin of Life" (Buvet, R., and Ponnamperuma, C., eds.), pp. 224-236, North-Holland, Amsterdam, 1971.

26. Rubin, B. A., "Vazneishie problemi fotosinteza v rasteinievodstve," pp. 68-69, "Kolos," Moskva, 1970.

27. Egami, F., Fourth International Conference Origin of Life, Barcelona, Abstracts, 1973.

28. Florkin, M., "Chemical Evolution and the Origin of Life," (Buvet, R., and Ponnamperuma, C., eds.), pp. 366-381, North-Holland, Amsterdam, 1971.

29. Flechtner, V. R., and Hanson, R. S., Biochim. Biophys. Acta 222, 253 (1970).

30. Weitzman, P.D.J., and Jones, D., Nature 219, 270 (1968).

31. Furlong, C. E., and Preiss, J., J. Biol. Chem. 244, 2539 (1969).

32. Borris, R., Biol. Rundschau 10, 163 (1972).

33. Vernadsky, V. I., Dokl. Akad. nauk SSSR, 137 (1931).

34. Goldovsky, A. M., Zhur. evol. biochim. fiziol. 8, 353 (1972).

35. Shannon, L. M., Ann. Rev. Plant Physiol. 19, 187 (1968).

36. Iljin, V. S., Pleskov, V. M., and Razumovskaja, N. I., Zhur. evol. biochim. fiziol. 3, 519 (1967).

37. Jakovleva, V. I., Uspehi biol. chim. 9, 55 (1968).

38. Shnol, S. E., Zhur. obstei biologii 34, 331 (1973).

39. Goldovsky, A. M., "Zakonomernosti progressivnoi evoluzii," pp. 83-94, Leningrad, 1972.

40. Kefeli, V. I., Komizerko, E. I., Turezkaja, R. H., Koff, E.M., and Kutacek, M., "Immunitet i pokoi rastenii," pp. 200-212, "Nauka," Moskva, 1972.

41. Peive, J. V., Ivanova, N. N., and Drobisheva, N. T., Fiziol. rastenii 19, 340 (1972).

42. Goldovsky, A. M., "Organizacia i evolucia zivivo," pp. 160-163, "Nauka," Leningrad, 1972.

43. Hess, B., and Boiteux, A., Ann. Rev. Plant Biochem. 40, 237 (1971).

44. Bakardjieva, N. T., Filosofska misal 26, 65 (1970).

PHOTOCHEMICAL FORMATION OF ORGANIC COMPOUNDS FROM MIXTURES OF SIMPLE GASES SIMULATING THE PRIMITIVE ATMOSPHERE OF THE EARTH (1,2)

Wilhelm Groth

Institute for Physical Chemistry

The University of Bonn

Already in the 1938 English edition of his book "The Origin of Life" A. I. Oparin (3) gave strong reasons for the assumption that the primordial or primitive atmosphere of the Earth was a "reducing atmosphere," consisting mainly of hydrogen, methane, ammonia, water vapour, and noble gases. In addition he presented many facts supporting the assumption that carbon appeared first on the Earth's surface, not in the oxidized form of carbon dioxide, but in the reduced state in the form of hydrocarbons. He added that nitrogen first appeared, with a high degree of probability, like carbon, also in its reduced state, in the form of ammonia. These assumptions were later on supported by H. C. Urey (4), W. Kuhn (5), J. D. Bernal (6), and others.

In 1937, however, these ideas were not generally accepted.* Geologists and geophysicists concluded from the composition of the Earth's crust and from volcanic exhalations that the primitive atmosphere of the Earth did not contain any free oxygen and that water vapour and carbon dioxide could be supposed to be essential constituents of the primordial atmosphere of the Earth. In fact this view has been strengthened by many authors in recent years. First of all, the assumption was accepted that the Earth in its first stage was without an atmosphere because the primary gaseous constituents

*In recent years serious doubts have arisen concerning this assumption. Fox, S. W., and Dose, K. in their book: Molecular Evolution and the Origin of Life, Freeman, San Francisco (1972) cite several authors emphasizing that the content of a methane-ammonia-hydrogen-water atmosphere seems to be without a firm geologic and geophysical foundation.

escaped into outer space, and that the secondary atmosphere evolved from volcanic outgassings (7). Volcanic gases consist mainly of water vapour, CO_2, N_2, SO_2, H_2S, S, HCl, B_2O_3, small amounts of H_2, CH_4, CO, NH_3, and HF, and no oxygen. In any case CO_2 was, after water vapour, the main constituent.

The intention of our early experiments (7) was to demonstrate that the formation of free oxygen and of organic compounds at a time before the development of organic life could be made plausible by photochemical experiments in the laboratory.

Carbon dioxide and water vapour start to absorb ultraviolet light below 2000 Å. Radiation of these short wave lengths is, according to Planck's radiation law, emitted by the Sun. At present, such radiation does not penetrate to the Earth's surface in measurable quantities because the strong absorption by oxygen and ozone in the ultraviolet prevents it. In addition, there were astrophysical considerations (8) which made it probable that the absolute amount of the intensity of the Sun's radiation in the extreme ultraviolet, in any case from time to time and from special parts of the Sun's surface, is essentially larger than that of a black body of 5800° K.

Beginning in 1934, I had started the development of Harteck's xenon lamp (9) which emits the resonance wave lengths of xenon at 1469 Å and 1296 Å, and investigated the photochemical dissociation of carbon dioxide and water vapour (10). The following reactions occur:

$$(1) \quad CO_2 + h\nu \rightarrow CO + O$$

$$(2) \quad H_2O + h\nu \rightarrow OH + H$$

Later, it had been shown (11,12) that H atoms react with CO to give formaldehyde or glyoxal according to

$$(3) \quad H + CO\ (+ M) \rightarrow HCO\ (+M)$$

$$(4) \quad 2\ HCO \begin{matrix} \nearrow H_2CO \\ \searrow (HCO)_2 \end{matrix}$$

If mixtures of carbon dioxide and water vapour are illuminated by the xenon lamp, these reactions occur simultaneously. From the primarily formed CO and the dissociation products of water vapour, aldehydes are formed which can be detected by Schryver's reagent. Their condensation products precipitate as a layer on the window of the lamp.

This process is assumed to have played an important role in the Earth's atmosphere. Under the influence of the Sun's radiation, oxygen and carbon compounds can be formed from water and carbon dioxide. It would give an explanation for the first appearance of free oxygen in the primitive atmosphere and for the formation of certain carbon compounds that were probably the prerequisite for the evolution of organic life.

The hypotheses of A. I. Oparin, H. C. Urey, W. Kuhn, and J. D. Bernal, mentioned in the beginning, assumed that the Earth's primitive atmosphere contained hydrogen, methane, and water vapour. While hydrogen escaped into space, the remaining gases were slowly oxidized. In the course of this oxidation process, many simple organic compounds were possibly formed in the atmosphere as intermediate products.

The formation of organic compounds under conditions which would have been present in the reducing atmosphere was demonstrated by various laboratory experiments. S. L. Miller (13), as well as K. Heyns et al. (14), exposed a mixture of methane, ammonia, water vapour, and hydrogen to an electric spark discharge. From these experiments they obtained several amino acids, carboxylic acids, amines, and other organic compounds.

Electric discharges were probably not the strongest source of energy although they would form ions and radicals of high concentration near the Earth's surface. As emphasized by Urey (14) and Miller (15), the greater part of the energy would be contributed by the ultraviolet radiation of the Sun; the light energy below 2500 Å, which is now radiated to the Earth by the Sun, is in the order of 650 cal cm^{-2} y^{-1}, whereas the energy of electric discharges in the atmosphere is only about 4 cal cm^{-2} y^{-1}.

For this reason, the earlier photochemical experiments were resumed in 1957 (2) using mixtures of methane or ethane with ammonia and water vapour in mercury-sensitized as well as direct photolysis experiments. For irradiation the resonance wave lengths of mercury at 2537 Å and 1849 Å, of xenon at 1469 Å and 1296 Å, and of krypton at 1236 Å and 1165 Å were used.

The mercury-sensitized experiments were carried out in a circulating system containing a low pressure mercury discharge lamp of 1 m length with an efficiency of 3 x 10 quanta/sec, as shown in Fig. 1.

The solutions resulting from the experiments were analyzed by a procedure similar to that used by Miller. The amino acids were detected by a paper chromatographic test; quantitatively they were determined by ion exchange chromatography employing the method developed by S. Moore and W. H. Stein (16). The carboxylic acids were determined by means of the procedure described by W. A. Bulen et al. (17).

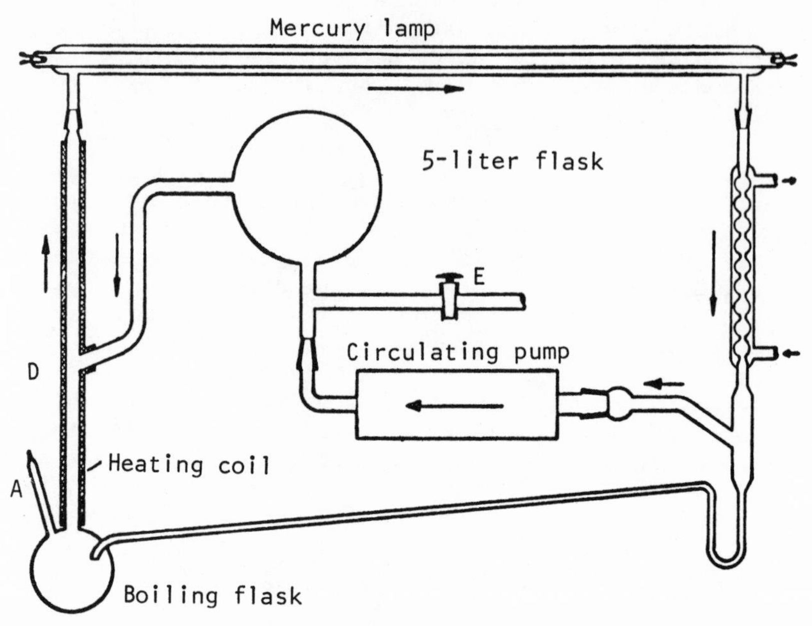

Figure 1. Apparatus used for the mercury-sensitized experiments.

The results of the mercury-sensitized experiments are compiled in Table I.

In all runs using mixtures of methane, ammonia, and water vapour, no reactions were observed even after an irradiation time of one week. In the runs using ethane, ammonia, and water vapour a chemical conversion became visible already after 6 hrs of irradiation. In the paper chromatographic test strong spots were determined to be due to glycine and α-alanine and, in addition; weaker spots appeared but were not identified. The results of the quanti-

Table I

Products of Mercury-Sensitized Irradiations

No.	Gas mixture (mm Hg)	Hg	Amino acids	Carboxylic acids	Hydro-carbons
1	CH_4(200), NH_3(200), H_2O	+	–	–	–
2	C_2H_6(400), NH_3(200), H_2O	+	+	+	+
3	C_2H_6(400), NH_3(200), H_2O	–	+	+	+
4	C_2H_6(600), H_2O, (NaOH)	+	–	+	+
5	C_2H_6(600), H_2O, (NaOH)	–	–	–	–

tative analyses of amino acids and fatty acids in μmoles are given in Table II.

For the direct photolysis with xenon and krypton resonance wave lengths, a thermal convection circulating system at a temperature of 55°C was used as is shown in Fig. 2.

Due to the low quantum efficiency of the xenon and the krypton lamp as compared to the mercury lamp, only a few micrograms of amino acids could be expected even under the most favorable conditions.

The chromatograms of the full runs of approximately 30 hrs, at a temperature of 55°C corresponding to a pressure of water vapour of about 120 torr with mixtures of 400 torr methane and 200 torr ammonia, showed several spots. One of them was definitely determined to be glycine; another spot could be attributed to α-alanine. The chromatograms of the blank tests did not show any spots.

That no reaction products are obtained in the mercury-sensitizing experiments with mixtures of methane, ammonia, and water vapour can be understood by comparing the cross sections of these gases for quenching the Hg resonance wave lengths. The values 3.0×10^{-16} cm^2 of ammonia and 1.0×10^{-16} cm^2 of water are 50 and 20 times greater, respectively, than the value 0.06×10^{-16} cm^2 of methane (18). Therefore the excited mercury atoms react nearly exclusively with the former two gases, and only H-atoms, OH-, NH_2-, and NH-radicals occur as photochemical primary products. On the other hand, methane is not attacked by radicals and H-atoms up to a temperature of 300°C (19).

Table II

Amounts of Amino and Fatty Acids (in μmoles)

Experiment No.	2	3	4
Glycine	32.0	24.5	--
α-Alanine	23.0	12.0	--
α-Aminobutyric acid	0.5	0.2	--
Total	55.5	36.7	
Formic acid	72	82	58
Acetic acid	203	234	136
Propionic acid	17	15	9
Total	292	331	203

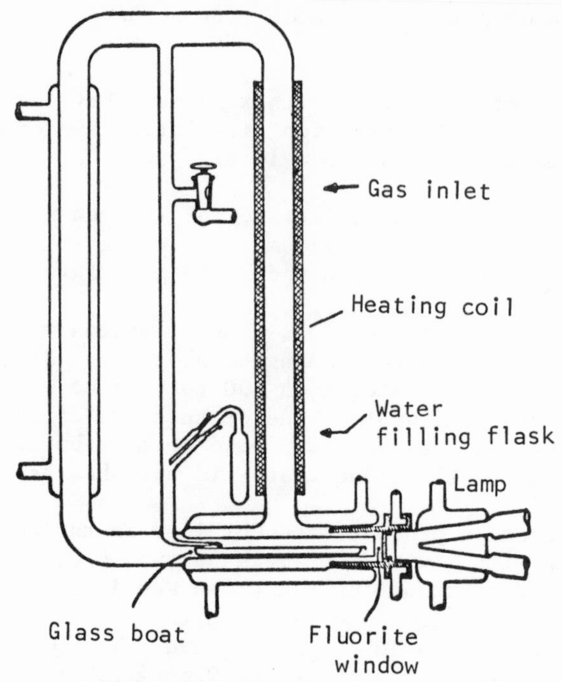

Figure 2. Circulating system operated with thermal convection.

The quenching cross section of ethane (0.11×10^{-16} cm^2) is not much greater than that of methane. Thus the quenching reaction with ethane yields only a small contribution to its dissociation. Ethyl and methyl radicals, however, are probably formed by the reaction of ethane with H atoms arising from the photolysis of ammonia and water vapour:

$$(5) \quad H_2H_6 + H \rightarrow C_2H_5 + H_2$$

$$(6) \quad C_2H_5 + H \rightarrow C_2H_6^*$$

$$(7) \quad C_2H_6 \rightarrow 2\ CH_3^*$$

This is confirmed by the non-sensitized runs using ethane, ammonia, and water vapour (runs nos. 3 and 5 of Table I), which yielded nearly the same results as the sensitized runs.

In non-sensitized experiments, ammonia is the only constituent which is dissociated by direct photolysis. The quantum energy available from the mercury lamp is not sufficient to dissociate ethane and water vapour directly (20,21). Therefore, reaction products were obtained from the runs using ethane and water vapour only if mercury sensitizer was present.

Under irradiation with xenon resonance wave lengths, the following primary reactions occur in mixtures of methane, ammonia, and water vapour:

$$(8) \quad CH_4 + h\nu \rightarrow CH_3 + H; \quad CH_4 + h\nu \rightarrow CH_2 + H_2$$

$$(9) \quad NH_3 + h\nu \rightarrow NH_2 + H; \quad NH_3 + h\nu \rightarrow NH + H_2$$

$$(10) \quad H_2O + h\nu \rightarrow OH + H$$

Considering the light intensities and the absorption coefficients at the wave lengths 1469 Å and 1296 Å, it is reasonable that the radicals and H-atoms are primarily formed at a comparable rate.

As a final product, glycine could be definitely determined, but even this in extremely small amounts, compared with the sensitized experiments. This may be explained mainly by the small quantum yield available from the xenon lamp.

These results were confirmed by Terenin (22) in 1959, who irradiated a mixture of methane, ammonia, and water vapour with a hydrogen lamp. In 1961, Dodonova and Siderova (23) irradiated a mixture of methane, ammonia, water vapour, and carbon monoxide with radiation wave length between 1450 Å and 1800 Å and detected the amino acids glycine, alanine, valine, and norleucine, as well as the amines methylamine, ethylamine, and some other compounds: hydrazine, urea, and formaldehyde. In 1966, Ponnamperuma and Flores (24) irradiated a mixture of methane and water vapour with the continuum of a helium lamp between 1000 Å and 2000 Å. After 48 hours of irradiation, only one-half percent of methane had been converted into organic compounds.

Therefore, more powerful sources for radiation in the extreme ultraviolet are necessary to produce experimentally higher effects on simulated primitive atmospheres and to determine quantum yields for the production of amino acids by ultraviolet light.

REFERENCES

1. Groth, W., and Suess, H., Naturwissenschaften 26, 77 (1938).

2. Groth, W., and Weyssenhoff, H. v., Naturwissenschaften 44, 510 (1957). Groth, W., and Weyssenhoff, H. v., Ann. Physik 4, 69 (1959). Groth, W., and Weyssenhoff, H. v., Planetary Space Sci. 2, 79 (1960).

3. Oparin, A. I., "The Origin of Life," Macmillan, London, 1938.

4. Urey, H. C., "The Planets," Oxford University Press, London, 1952. Urey, H. C., Proc. Nat. Acad. Sci. U.S. 31, 351 (1952).

5. Kuhn, W., Chem. Ber. 89, 303 (1956).

6. Bernal, J. D., "The Physical Basis of Life," Routledge and Kegan Paul, London, 1957.

7. Vinogradov, A. P., in "The Origin of Life on the Earth" (Oparin, A. I. et al., eds.), Pergamon, London, 1959. Berkner, L. V., and Marshall, L. C., Discussions Faraday Soc. 37, 122 (1964).

8. Unsold, A., "Physik d. Sternatmosphären," Berlin, 1937.

9. Groth, W., Z. Elektrochem. 42, 533 (1936).

10. Groth, W., Z. Phys. Chem. B37, 307 (1937).

11. Frankenburger, W., Z. Elektrochem. 36, 757 (1937).

12. Groth, W., Z. Phys. Chem. B37, 315 (1937).

13. Miller, S. L., Science 117, 528 (1953); ibid., J. Am. Chem. Soc. 77, 2351 (1955); ibid., Ann. N.Y. Acad. Sci. 69, 260 (1957); ibid., Biochim. Biophys. Acta 23, 480 (1957).

14. Heyns, K., Walter, W., and Meyer, E., Naturwissenschaften 44, 385 (1957). Heyns, K., and Pavel, K., Z. Naturforsch. 126. 97 (1954).

15. Miller, S. L., and Urey, H. C., Science 130, 245 (1959).

16. Moore, S., and Stein, W. H., J. Biol. Chem. 211, 893 (1954).

17. Bulen, W. A., Varner, J. E., and Burrel, R. C., Anal. Chem. 24, 187 (1952).

18. Bates, J. R., J. Am. Chem. Soc. 52, 3852 (1930).

19. Harteck, P., Bonhoeffer, K. F., and Geib, J. H., Z. Phys. Chem. A139, 64 (1928); ibid., A170, 1 (1934).

20. Darwent, B., J. Chem. Phys. 18, 1532 (1950).

21. Senftleben, H., and Rehren, J., Z. Phys. Chem. 34, 529 (1926).

22. Terenin, A. N., in "The Origin of Life on the Earth" (Oparin, A. I. *et al*., eds.), Pergamon, London, 1959.

23. Dodonova, N., and Siderova, A. L., *Biofizika* *6*, 149 (1961).

24. Ponnamperuma, C., and Flores, J., Abstr. 152nd Nat. Meeting Amer. Chem. Soc., New York (1966).

THE IRON-SULPHUR PROTEINS: EVOLUTION OF A UBIQUITOUS PROTEIN FROM THE ORIGIN OF LIFE TO HIGHER ORGANISMS

D. O. Hall, R. Cammack, and K. K. Rao

Department of Plant Sciences
University of London King's College
London SE24 9JF

Iron-sulphur (Fe-S) proteins are a group of metalloproteins containing one or more iron atoms liganded to cysteine sulphurs of the protein chain. They occur in all forms of life, from the most primitive bacteria to higher plants and animals in which their major function is as electron transfer agents in many biological reactions (1,2). Some of the properties of Fe-S proteins are listed in Table I. The ferredoxins are members of the group of Fe-S proteins which contain two or more non-haem iron atoms and an equivalent amount of labile sulphur per molecule. They are relatively small proteins with molecular weights ranging from 5500 to 12,500 and have been isolated from bacteria, algae, plants, and animals (3,4).

Geologists and palaeontologists are in general agreement that the Earth is about 4.6×10^9 years old (5). The atmosphere of the early Earth was probably "reducing" and would have consisted of gases and vapours such as CH_4, H_2O, H_2, NH_3, CO_2, HCN, H_2S, etc. Except for CH_4, these other molecules and their fragments have recently been reported as constituents of intergalactic clouds (6). The interaction of these molecules is thought then to have led to the formation of simple amino acids, purines, porphyrins, etc. The energy sources for the reactions would have been heat (volcanic), UV radiation (Sun), or electrical (lightning). The amino acids in turn condensed on the surface of the Earth forming peptide chains and proteins; iron and sulphide would have been abundant and available for combination with these polypeptides. Thus, there probably existed some complexes of iron (or even iron-sulphur) compounds with polypeptides which could carry out a variety of oxidation-reduction reactions long before the development of organisms with genetic replication machinery and the capacity to synthesize proteins on the

Table I

Properties of Iron-Sulphur Proteins

Type of Fe-S chromophore	Source	No. of Fe and S	MW	Approx. no. of amino acids	Redox potential mV	No. of electrons transferred	Type of reactions catalyzed
4[Fe-S]							
(a) Clostridial type (8 Fe) ferredoxin	Anaerobic bacteria	8 Fe, 8 S	6000	55	-390	2	Phosphoroclastic reaction, CO_2 reduction
	Green photosynthetic bacteria Chromatium	8 Fe, 8 S	9000	81	-490	2	N_2 fixation, photosynthetic CO_2 fixation
(b) 4 Fe ferredoxins	D. gigas	4 Fe, 4 S	6000	56	-	-	Sulphite reduction
	B. polymyxa	4 Fe, 4 S	9000	79	-390	1	N_2 fixation, NADP photoreduction
(c) HIPIP	Chromatium	4 Fe, 4 S	9600	86	+350	1	Photosynthetic electron transport
2[Fe-S]							
(a) Plant ferredoxin	Algae and plants	2 Fe, 2 S	10500	97	-430	1	NADP photoreduction, photosynthetic-phosphorylation, N_2 fixation (blue-green algae)
(b) Adrenodoxin	Mammalian mitochondria	2 Fe, 2 S	12500	114	-270	1	Hydroxylation of steroids
(c) Putidaredoxin	P. putida	2 Fe, 2 S	12500	114	-240	1	Hydroxylation of camphor
1[fe]							
Rubredoxin	Anaerobic bacteria	1 Fe	6000	53	-60	1	
Rubredoxin	P. oleovorans (aerobic)	1 or 2 Fe	19000	174	-60	1 per Fe	Oxidation of hydrocarbons
Complex Fe-S proteins							
Xanthine oxidase	Bacteria Mammals	8 Fe, 8 S 2 Mo, 2 FAD	275000				Oxidation of xanthines, aldehydes

ribosomes.

The "abiogenic" hypothesis of the origin of life enunciated by Oparin in 1924, and later developed by other authors, assumes the existence of biochemically active colloidal droplets or "coacervates" prior to the occurrence of the "primary organism." According to Oparin, "The formation of the coacervates was a most important event in the evolution of primary organic matter and in the process of the generation of life" (7).

We postulate that one of the molecules that functioned as a catalyst in the electron transfer processes in the primordial organism was a precursor of the present-day ferredoxin molecule. The protoorganism which originated in the early reducing atmosphere, about 3.4×10^9 years ago, would have been completely heterotrophic, growing and multiplying at the expense of abiogenically synthesized organic molecules. The metabolism of these early heterotrophs was probably similar to that of the present-day fermentative anaerobes like the clostridia, which obtain their entire energy requirements from fermentation of organic nutrients and can survive only in an oxygen-free atmosphere. Ferredoxin plays an essential role in a number of reactions catalyzed by these bacteria, such as the conversion of acetyl phosphate to pyruvate, reduction of molecules such as N_2, HCN, and C_2H_2 in the presence of nitrogenase, and metabolism of hydrogen in the presence of hydrogenase (8, 8a). Both the nitrogenase and hydrogenase enzymes contain iron sulphur prosthetic groups. Haem proteins are relatively uncommon in these obligate anaerobic bacteria. Lemberg and Legge (9) have suggested that the chemical bond energy of primitive chemosynthesis may well have been formed by the action of hydrogenase with molecular hydrogen: $H_2 \rightarrow 2H^+ + 2e^-$, a reaction requiring ferredoxin as a catalyst. The midpoint oxidation-reduction potential (E'_o) of ferredoxin is about -400 mV, a value close to that of hydrogen, so ferredoxin is a suitable electron carrier for a reducing environment. Clostridial ferredoxin can transfer one or two electrons at a time to suit the type of reaction it catalyzes (e.g.: carbon metabolism or photosynthetic electron transport; 10,11).

The ferredoxins of anaerobic fermentative bacteria are relatively small protein molecules of 55 amino acid residues and a molecular weight of about 5500 daltons. The prosthetic groups consist of two clusters of four Fe and four S atoms, the eight iron atoms being individually linked to the eight cysteine residues of the protein chain (12). The primary structure of ferredoxin from six of these species is known (13,14). The amino acids present in these ferredoxins are mostly the simple ones which are thermodynamically stable, as was pointed out by Eck and Dayhoff (15). Only nine of the twenty proteinaceous amino acids are common to all these ferredoxins, these being Gly, Ala, Pro, Val, Glu, Asp, Ser, Cys, and Ile. These common amino acids constitute 91% of the amino acid content of Clostridium butyricum ferredoxin (16). We regard these amino acids as simple since they are easily synthesized under simulated prebiotic conditions. Moreover, they are more abundant in the "amino acid pool" recently detected in meteorites and lunar soils (13). Thus these common amino acids could easily have been present in the prebiotic Earth. By condensation of these amino acids, a small polypeptide chain similar to the Clostridium apoferredoxin could possibly have formed. Subsequently iron and sulphur were incorporated into the polypeptide to give a catalytically active molecule; the reaction proceeds nonenzymatically in vitro in an oxygenless atmosphere (17). Thus it seems logical to assume that a "prototype" of the present-day eight-iron ferredoxin molecule was functioning in the period at which life originated on Earth.

The amino acid sequences of ferredoxins from six obligate anaerobes are shown in Fig. 1. There is a striking homology in the sequences. The amino terminus of all the ferredoxins is alanine and most of the amino acid substitutions are neutral in nature. Particularly interesting from a functional point of view is the invariance of the eight cysteine residues. As already mentioned, these cysteine residues are all linked to the iron atoms in the prosthetic group. Further, as shown in the figure, the molecule can be arranged into two almost identical halves, each containing four cysteines so that each half has a potential capacity to hold a (4Fe-4S) cluster. This repeating array of identical amino acids in the sequence of clostridial ferredoxin suggests that the protein might be evolved from a simpler molecule by gene duplication (15,18,19). In the present-day ferredoxin, each four-iron cluster is held by three cysteines from one group and one from the other (12). However, the separate halves of the molecule can be reconstituted to give ferredoxin-like molecules (see 2).

It is postulated that the next type of energy metabolism to develop, after that of the anaerobic fermenters, would be that of the photosynthetic bacteria of the green and red sulphur types (represented here by the Chlorobium and Chromatium species, respectively). These bacteria could use reduced sulphur compounds, e.g. H_2S, thiosulphate, etc., in conjunction with light energy to provide the necessary redox potential to fix carbon. Thus they would no longer be reliant on the dwindling resources of abiogenic organic carbon compounds for growth and reproduction. The eight-iron ferredoxin present in Chlorobium appears to be similar in composition and function to those in clostridia (20). Chromatium eight-iron ferredoxin is similar to that of Chlorobium in function, but its amino acid sequence has an extra 26 residues compared to those of the clostridia (21). However, the Chromatium ferredoxin molecule can be aligned to the clostridium ferredoxin (13,21) with the eight functional cysteine residues occupying invariant positions. These ferredoxins function as one-electron carriers in photosynthesis and two-electron carriers in carbon metabolism. In photosynthetic bacteria there

	1																											28
A	Ala	Phe	Val	Ile		Asn	Asp	Ser	Cys	Val	Ser	Cys	Gly	Ala	Cys	Ala	Gly	Glu	Cys	Pro	Val	Ser	Ala	Ile	Thr	Gln	Gly	Asp	Thr
B	Ala	Tyr	Lys	Ile		Ala	Asp	Ser	Cys	Val	Ser	Cys	Gly	Ala	Cys	Ala	Ser	Glu	Cys	Pro	Val	Asn	Ala	Ile	Ser	Gln	Gly	Asp	Ser
C	Ala	Tyr	Val	Ile		Asn	Glu	Ala	Cys	Ile	Ser	Cys	Gly	Ala	Cys	Asp	Pro	Glu	Cys	Pro	Val	Asp	Ala	Ile	Ser	Gln	Gly	Asp	Ser
D	Ala	Tyr	Val	Ile		Asn	Asp	Ser	Cys	Ile	Ala	Cys	Gly	Ala	Cys	Lys	Pro	Glu	Cys	Pro	Val	Asn		Ile	Gln	Gln	Gly		Ser
E	Ala	His	Ile	Ile		Thr	Asp	Glu	Cys	Ile	Ser	Cys	Gly	Ala	Cys	Ala	Ala	Glu	Cys	Pro	Val	Glu	Ala	Ile	His	Glu	Gly	Thr	Gly
F	Ala	His	Ile	Ile		Thr	Asp	Glu	Cys	Ile	Ser	Cys	Gly	Ala	Cys	Ala	Ala	Glu	Cys	Pro	Val	Glu	Ala	Ile	His	Glu	Gly	Thr	Gly

	29																										55	
A	Gln	Phe	Val	Ile	Asp	Ala	Asp	Thr	Cys	Ile	Asp	Cys	Gly	Asn	Cys	Ala	Asn	Val	Cys	Pro	Val	Gly	Ala	Pro	Asn	Gln	Glu	
B	Ile	Phe	Val	Ile	Asp	Ala	Asp	Thr	Cys	Ile	Asp	Cys	Gly	Asn	Cys	Ala	Asn	Val	Cys	Pro	Val	Gly	Ala	Pro	Val	Gln	Glu	
C	Arg	Tyr	Val	Ile	Asp	Ala	Asp	Thr	Cys	Ile	Asp	Cys	Gly	Ala	Cys	Ala	Gly	Val	Cys	Pro	Val	Asp	Ala	Pro	Val	Gln	Ala	
D	Ile	Tyr	Ala	Ile	Asp	Ala	Asp	Ser	Cys	Ile	Asp	Cys	Gly	Ser	Cys	Ala	Ser	Val	Cys	Pro	Val	Gly	Ala	Pro	Asn	Pro	Glu	Asp
E	Lys	Tyr	Gln	Val	Asp	Ala	Asp	Thr	Cys	Ile	Asp	Cys	Gly	Ala	Cys	Gln	Ala	Val	Cys	Pro	Thr	Gly	Ala	Val	Lys	Ala	Glu	
F	Lys	Tyr	Glu	Val	Asp	Ala	Asp	Thr	Cys	Ile	Asp	Cys	Gly	Ala	Cys	Glu	Ala	Val	Cys	Pro	Thr	Gly	Ala	Val	Lys	Ala	Glu	

A. *Clostridium butyricum*
B. *C. pasteurianum*
C. *C. acidiurici*
D. *Peptococcus aerogenes*
E. *C. tartarivorum*
F. *C. thermosaccharolyticum*

Figure 1. Amino acid sequences of bacterial eight iron ferredoxins (13,14).

seems to be more than one iron-sulphur protein involved in electron transfer processes (22,23).

The Athiorhodaceae, or non-sulphur purple bacteria of the Rhodospirillum type, which are facultative photoheterotrophs, may be intermediates between the Thiorhodaceae and the algae. When grown anaerobically in the light, R. rubrum produces eight-iron and two-iron ferredoxins; but only a two-iron ferredoxin is formed when the cells are grown aerobically (24). Both the sulphur and non-sulphur purple bacteria contain the same type of high-potential iron sulphur protein with a single 4 Fe - 4 S center (25,26). Some authors suggest that the athiorhodaceae may be an evolutionary "dead end" rather than the possible ancestor of any other group (27).

The exact period at which the sulphate-reducing bacteria of the desulfovibrio type originated on the Earth is still not clear (28, 28a). Improved techniques by which the sulphate deposits of old rocks could be dated may throw some light on this subject. The primary structure of a ferredoxin from one species of sulphate-reducer, Desulfovibrio gigas, is known. It contains one active 4 Fe - 4 S center in a molecule of 56 amino acid residues. The initial half of this ferredoxin molecule (Fig. 2) can be aligned with a fair degree of homology with the first half of a clostridial eight-iron ferredoxin while the latter half of the ferredoxin shows some homology with a segment of plant-type ferredoxin molecule (29). Though it is premature to form any conclusions about the evolution of the sulphate reducers from a single ferredoxin sequence, it is tempting to speculate that the sulphate reducers originated after the advent of the photosynthetic bacteria and before the appearance of the blue-green algae. It is possible that the nitrate-reducing bacteria also originated along with the sulphate reducers (see Refs. 30,31). Sulphate and nitrate could have acted as electron acceptors before the organisms which use oxygen as the final electron acceptor in their energy metabolism had evolved. The photosynthetic bacteria using reduced sulphur compounds and the sulphate reducers living on oxygenated sulphur compounds could have had a long period of parallel association; such symbiosis still occurs among present-day organisms.

The transition from an anaerobic to an aerobic atmosphere appears to have started about 2×10^9 years ago with the appearance of O_2 in the atmosphere as a consequence of the earlier evolution of the blue-green algae with their photosynthetic apparatus capable of liberating oxygen from water. Ferredoxin is an essential catalyst in photosynthesis. The blue-green algal and plant ferredoxins are of the same type (32,33); they contain about 97 amino acid residues (Fig. 3) and two atoms of Fe and S per molecule. The two-iron atoms are linked to cysteines 39, 44, 47, and 77 of the protein chain. Though a cysteine is present in position 18 of many plant-type ferredoxins also, in Equisetum (34) and Aphonthece ferredoxins a valine residue is found at position 18 (Matsubara, personal communication). Homologies are found

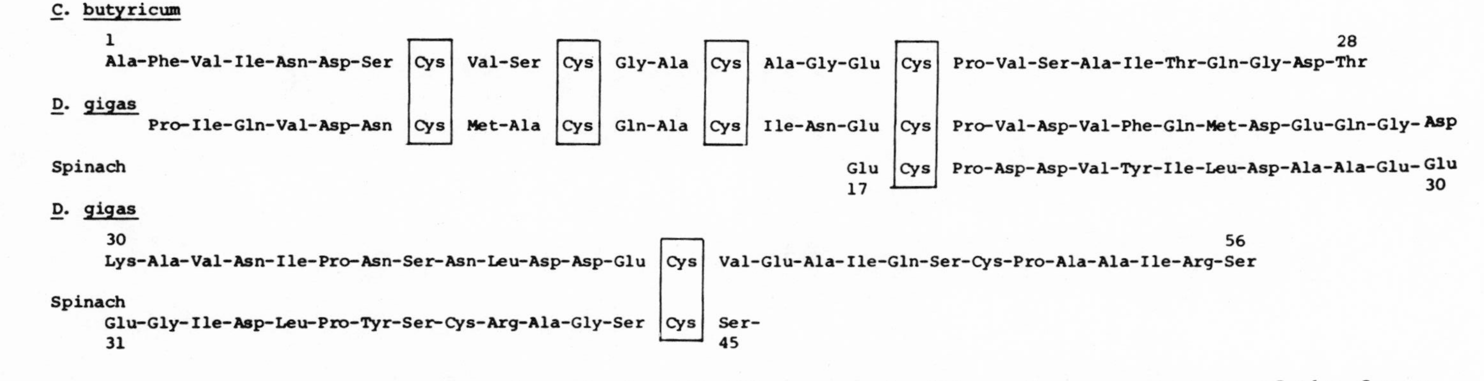

Fig. 2. Sequence of *D. gigas* ferredoxin (29,13) compared with analogous segments of the ferredoxins of *C. butyricum* and Spinach.

```
     1                                                                          20
(A)  Ala-Thr-Tyr-Lys-Val-Thr-Leu-Lys-Thr-Pro-Ser-Gly-Asp-Gln-Thr-Ile-Glu  Cys  Pro-Asp
(B)  Ala-Ala-Tyr-Lys-Val-Thr-Leu-Val-Thr-Pro-Thr-Gly-Asn-Val-Glu-Phe-Gln  Cys  Pro-Asp
(C)      Ala-Phe-Lys-Val-Lys-Leu-Leu-Thr-Pro-Asp-Gly-Pro-Lys-Glu-Phe-Glu  Cys  Pro-Asp
(D)  Ala-Thr-Tyr-Lys-Val-Lys-Leu-Val-Thr-Pro-Ser-Gly-Gln-Gln-Glu-Phe-Gln  Cys  Pro-Asp
(E)  Ala-Ser-Tyr-Lys-Val-Lys-Leu-Val-Thr-Pro-Glu-Gly-Thr-Gln-Glu-Phe-Glu  Cys  Pro-Asp

     21                                                                         40
(A)  Asp-Thr-Tyr-Ile-Leu-Asp-Ala-Ala-Glu-Glu-Ala-Gly-Leu-Asp-Leu-Pro-Tyr-Ser  Cys  Arg
(B)  Asp-Val-Tyr-Ile-Leu-Asp-Ala-Ala-Glu-Glu-Glu-Gly-Ile-Asp-Leu-Pro-Tyr-Ser  Cys  Arg
(C)  Asp-Val-Tyr-Ile-Leu-Asp-Gln-Ala-Glu-Glu-Leu-Gly-Ile-Asp-Leu-Pro-Tyr-Ser  Cys  Arg
(D)  Asp-Val-Tyr-Ile-Leu-Asp-Gln-Ala-Glu-Glu-Val-Gly-Ile-Asp-Leu-Pro-Tyr-Ser  Cys  Arg
(E)  Asp-Val-Tyr-Ile-Leu-Asp-His-Ala-Glu-Glu-Glu-Gly-Ile-Val-Leu-Pro-Tyr-Ser  Cys  Arg

     41                                                                                60
(A)  Ala-Gly-Ala  Cys  Ser-Ser  Cys  Ala-Gly-Lys-Val-Glu-Ala-Gly-Thr-Val-Asp-Gln-Ser-Asp
(B)  Ala-Gly-Ser  Cys  Ser-Ser  Cys  Ala-Gly-Lys-Leu-Lys-Thr-Gly-Ser-Leu-Asn-Gln-Asp-Asp
(C)  Ala-Gly-Ser  Cys  Ser-Ser  Cys  Ala-Gly-Lys-Leu-Val-Glu-Gly-Asp-Leu-Asp-Gln-Ser-Asp
(D)  Ala-Gly-Ser  Cys  Ser-Ser  Cys  Ala-Gly-Lys-Val-Lys-Val-Gly-Asp-Val-Asp-Gln-Ser-Asp
(E)  Ala-Gly-Ser  Cys  Ser-Ser  Cys  Ala-Gly-Lys-Val-Ala-Ala-Gly-Glu-Val-Asn-Gln-Ser-Asp

     61                                                                     80
(A)  Gln-Ser-Phe-Leu-Asp-Asp-Ser-Gln-Met-Asp-Gly-Gly-Phe-Val-Leu-Thr  Cys  Val-Ala-Tyr
(B)  Gln-Ser-Phe-Leu-Asp-Asp-Asp-Gln-Ile-Asp-Glu-Gly-Trp-Val-Leu-Thr  Cys  Ala-Ala-Tyr
(C)  Gln-Ser-Phe-Glu-Asp-Asp-Gln-Gln-Ile-Gln-Gln-Gly-Trp-Val-Leu-Thr  Cys  Ala-Ala-Tyr
(D)  Gly-Ser-Phe-Glu-Asp-Asp-Gln-Gln-Ile-Gly-Glu-Gly-Trp-Val-Leu-Thr  Cys  Val-Ala-Tyr
(E)  Gly-Ser-Phe-Glu-Asp-Asp-Asp-Gln-Ile-Glu-Glu-Gly-Trp-Val-Leu-Thr  Cys  Val-Ala-Tyr

     81                                                                 97
(A)  Pro-Thr-Ser-Asp-Cys-Thr-Ile-Ala-Thr-His-Lys-Glu-Glu-Asp-Leu-Phe
(B)  Pro-Val-Ser-Asp-Val-Thr-Ile-Glu-Thr-His-Lys-Glu-Glu-Glu-Leu-Thr-Ala
(C)  Pro-Arg-Ser-Asp-Val-Val-Ile-Glu-Thr-His-Lys-Glu-Glu-Glu-Leu-Thr-Ala
(D)  Pro-Val-Ser-Asp-Gly-Thr-Ile-Gln-Thr-His-Lys-Glu-Glu-Glu-Leu-Thr-Ala
(E)  Ala-Lys-Ser-Asp-Val-Thr-Ile-Glu-Thr-His-Lys-Glu-Gly-Glu-Leu-Thr-Ala
```

(A) *Scenedesmus*

(B) Spinach

(C) *Leuceana glauca*

(D) *Taro*

(E) Alfalfa

Fig. 3. Amino acid sequences of plant-type ferredoxins. Invariant amino acids are underlined (13).

between plant and clostridial ferredoxins if their sequences are compared (35). Ferredoxins of the two-iron type are found in both aerobic and anaerobic bacteria (4,8). However, we do not know the sequences of any of these ferredoxins other than that of the aerobe Pseudomonas putida--putidaredoxin-used in the hydroxylation of camphor (36). Unfortunately, this protein does not show any homology to other ferredoxins.

Adrenodoxin, a two-iron ferredoxin isolated from adrenal mitochondria, resembles the plant ferredoxins in molecular weight and nature of the active center (37,38). But the primary structure of adrenodoxin shows no homology with the sequence of any of the known iron-sulphur proteins (39).

MODELS OF THE IRON-SULPHUR PROTEINS

The unusual chemical properties of iron-sulphur proteins have stimulated a lot of work in synthesizing organometallic complexes with iron-sulphur centers that resemble those of the proteins. Two recently prepared compounds (see Fig. 4) appear to be close analogues of the 4 Fe + 4 S and 2 Fe + 2 S centers respectively (40,41). It is interesting that Mayerle et al. (41) propose that the two-iron structure may be regarded as a fragment of the four-iron structure and is formally derived from it by incorporation of a cysteinyl sulphur at each iron and rupture of four Fe-S bonds. These model compounds are not water-soluble but if they can be made so by the use of suitable side-groups, it will open up many possibilities of checking their properties in biological systems and seeing if these properties, e.g. association with peptides, can regulate the biochemical function. The role of the protein is obviously important. For example, the model compounds are extremely sensitive to oxygen; one function of the protein moiety of ferredoxins appears to be to protect the iron-sulphur group from autoxidation. It may be significant that eight-iron ferredoxins of the clostridial type, in which the iron-sulphur groups are close to the surface (12), are not found in aerobic organisms. As another example, we know that the redox potential of the ferredoxin-like Chromatium high-potential iron-sulphur protein can be changed from +350 to -500 mV by altering the protein conformation with dimethylsulphoxide (42).

Complex iron-sulphur proteins, such as xanthine oxidase and nitrogenase, contain flavins and other metals such as molybdenum in addition to iron-sulphur groups. The basic iron-sulphur active center seems to be a useful unit to incorporate into large proteins to form complex electron transfer sequences within a single molecule. The resulting complex iron-sulphur proteins can catalyze reactions such as N_2-fixation and mitochondrial electron transfer from succinate or reduced pyridine nucleotides. We do not know the exact structure of the iron-sulphur groups in any complex protein. Spectroscopic

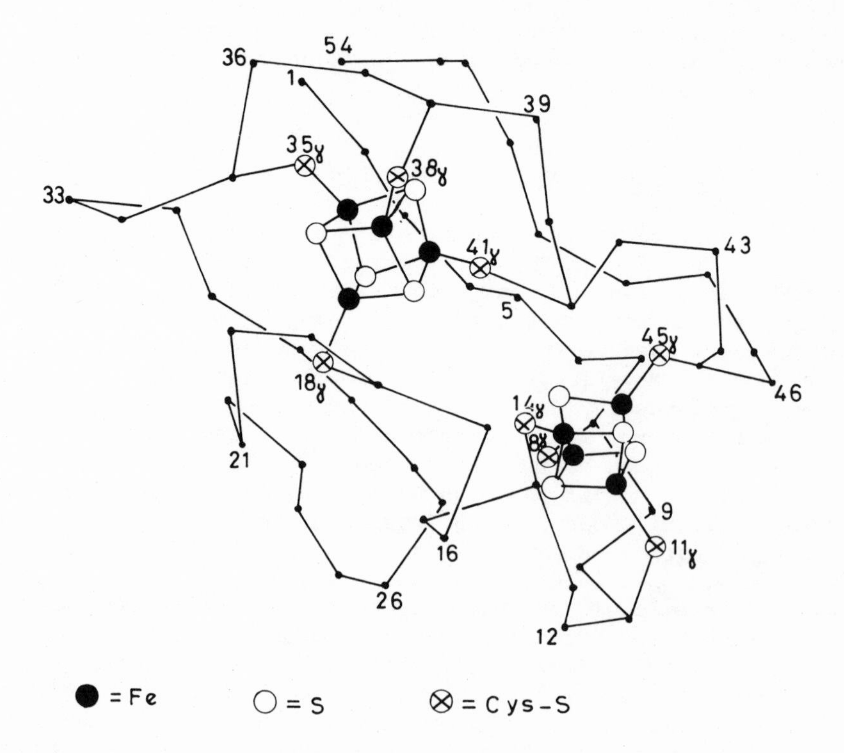

Figure 4a. Structure of bacterial eight iron ferredoxin. Redrawn from Ref. 12.

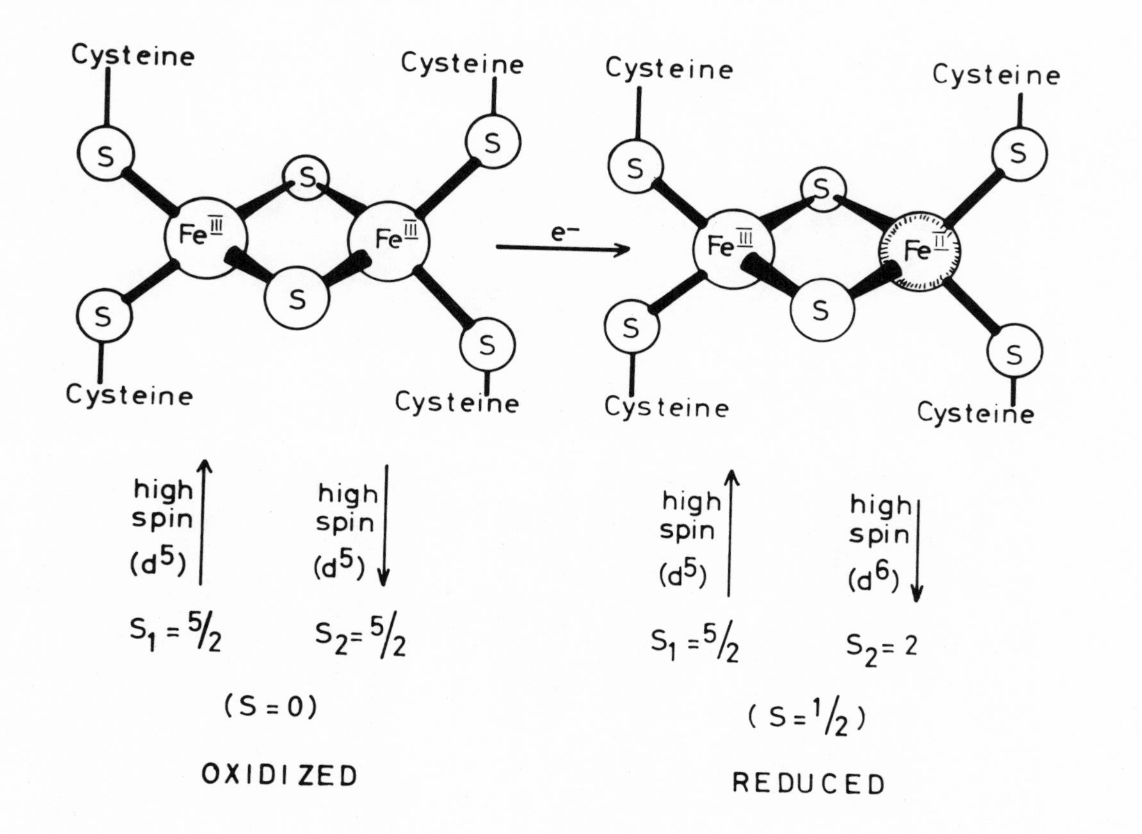

Figure 4b. Proposed model of the active center of two-iron ferredoxin and the change in the valence state of the iron atom during electron transfer (32,38).

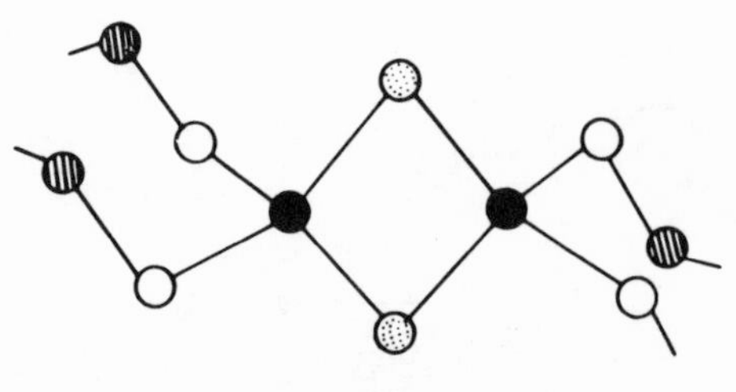

$(FeS(SCH_2)_2C_6H_4)_2^{2-}$

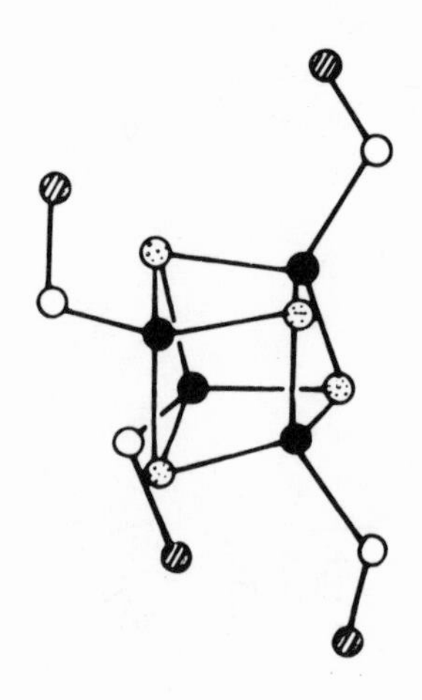

$(Fe_4S_4(SCH_2Ph_4)^{2-}$

● = Fe ⊙ = S (iron bonded) ○ = S ⊘ = dithiolene, C_5H_5,CH_3

Figure 4c. Synthetic analogues of the four-iron and two-iron clusters of ferredoxins (40,41).

evidence indicates that in many cases they are similar to the known two-iron or four-iron centers, but that some proteins, such as the iron-molybdenum protein of nitrogenase, may contain iron-sulphur centers of a novel type (see refs. 2 and 4).

CONCLUSION

The ferredoxins are involved in a diverse range of biochemical reactions from hydrogen, carbon, and nitrogen metabolism to photosynthesis, oxidative phosphorylation, and hydroxylation. They occur in all types of living organism and it is thus evident that they are ideal proteins for studies of biological evolution based on protein sequences. A tentative scheme of the evolution of ferredoxins, based on the arguments presented in this paper, is shown in Fig. 5. Using this type of scheme as a basis, further sequence and structural studies on the ferredoxins should provide further clues to the early development and evolution of living organisms.

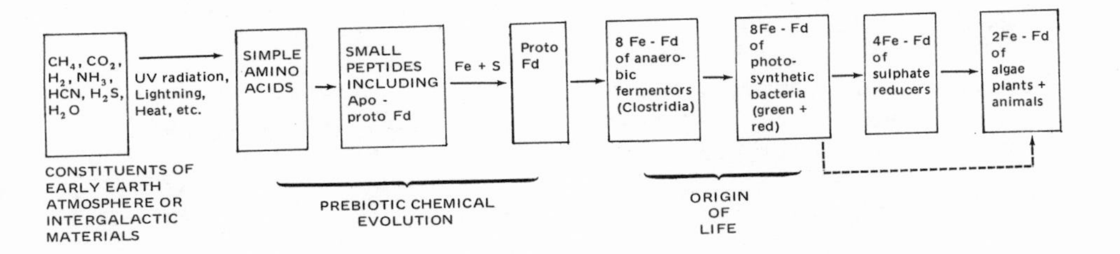

Figure 5. Scheme for the possible evolution of ferredoxins.

REFERENCES

1. Hall, D. O., and Evans, M.C.W., Nature 223, 1342 (1969).

2. Orme-Johnson, W. H., Ann. Rev. Biochem. 42, 159 (1973).

3. Buchanan, B. B., and Arnon, D. I., Advances Enzymol. 33, 119 (1970).

4. Hall, D. O., Cammack, R., and Rao, K. K., "Iron in Biochemistry and Medicine" (Jacobs, A., ed.), Chapter 8, Academic Press, New York, 1974.

5. Brooks, J., and Shaw, G., "Origin and Development of Living Systems," Academic Press, London, 1973.

6. Buhl, D., Nature 234, 332 (1971).

7. Oparin, A. I., "The Origin of Life," Dover Publications, Inc., New York, 1953.

8. Yoch, D. C., and Valentine, R. C., Ann. Rev. Microbiol. 26, 139 (1972).

8a. Silver, W. S., and Postgate, J. R., J. Theor. Biol. 40, 1 (1973).

9. Lemberg, R., and Legge, J. W., "Hematin Compounds and Bile Pigments," Interscience, New York, 1949.

10. Evans, M.C.W., Hall, D. O., Bothe, H., and Whatley, F. R., Biochem. J. 110, 485 (1968).

11. Eisenstein, K. K., and Wang, J. H., J. Biol. Chem. 244, 1720 (1969).

12. Adman, E. T., Sieker, L. C., and Jensen, L. H., J. Biol. Chem. 248, 3987 (1973).

13. Hall, D. O., Cammack, R., and Rao, K. K., "Theory and Experiment in Exobiology" (Schwartz, A. W., ed.), Vol. 2, p. 67, Wolters-Noordhoff, Groningen, 1972.

14. Tanaka, M., Haniu, M., Yasunobu, K. T., Himes, R. H., and Akagi, J. M., J. Biol. Chem. 248, 5215 (1973).

15. Eck, R. V., and Dayhoff, M. O., Science 152, 363 (1966).

16. Hall, D. O., Cammack, R., and Rao, K. K., Nature 233, 136 (1971).

17. Malkin, R., and Rabinowitz, J. C., Biochem. Biophys. Res. Comm. 23, 822 (1966).

18. Jukes, T. H., "Molecules and Evolution," p. 229, Columbia University Press, New York, 1966.

19. Dickerson, R. E., J. Mol. Biol. 57, 1 (1971).

20. Rao, K. K., Matsubara, H., Buchanan, B. B., and Evans, M.C.W., J. Bact. 100, 1411 (1969).

21. Matsubara, H., Sasaki, R. M., Tsuchiya, D. K., and Evans, M.C.W., J. Biol. Chem. 245, 2121 (1970).

22. Dutton, P. L., and Leigh, J. S., Biochim. Biophys. Acta 314, 178 (1973).

23. Evans, M.C.W., Telfer, A., and Reeves, S. G., Biochem. J. (in press).

24. Shanmugam, K. T., and Arnon, D. I., Biochim. Biophys. Acta 256, 489 (1972).

25. Bartsch, R. G., Methods Enzymol. 23, 644 (1971).

26. Carter, C. W., Kraut, J., Freer, S. T., Alden, R. A., Sieker, L. C., Adman, E., and Jensen, L. H., Proc. Nat. Acad. Sci. U.S. 69, 3526 (1972).

27. Klein, R. M., and Cronquist, A., Quart. Rev. Biol. 42, 105 (1969).

28. Broda, E., Progr. Biophys. Mol. Biol. 21, 143 (1970).

28a. LeGall, J., and Postgate, J. R., Advances Microbial Physiol. 10, 81 (1973).

29. Travis, J., Neuman, D. J., LeGall, J., and Peck, H. D., Jr., Biochem. Biophys. Res. Comm. 45, 452 (1971).

30. Yamanaka, T., Advances Biophys. 3, 229 (1972).

31. Schlegel, H. G., Intl. Symp. Origin of Life and Evol. Biochem. (Abstracts), Varna, Bulgaria (1971).

32. Rao, K. K., Cammack, R., Hall, D. O., and Johnson, C. E., Biochem. J. 122, 259 (1971).

33. Hall, D. O., Cammack, R., and Rao, K. K., Pure Appl. Chem. 34, 553 (1973).

34. Aggarwal, S. J., Rao, K. K., and Matsubara, H., J. Biochem. (Tokyo) 69, 601 (1971).

35. Matsubara, H., "Aspects of Cellular and Molecular Physiology" (Hamaguchi, K., ed.), p. 31, University of Tokyo Press, 1972.

36. Tsai, R. L., Gunsalus, I. C., and Dus, K., Biochem. Biophys. Res. Comm. 45, 1300 (1971).

37. Kimura, T., Structure and Bonding 5, 1 (1968).

38. Cammack, R., Rao, K. K., Hall, D. O., and Johnson, C. E., Biochem. J. 125, 849 (1971).

39. Tanaka, M., Hanui, M., Yasunobu, K. T., and Kimura, T., J. Biol. Chem. 248, 1141 (1973).

40. Herskowitz, T., Averill, B. A., Holm, R. H., Ibers, J. A., Phillips, W. D., and Weiher, J. F., Proc. Nat. Acad. Sci. U.S. 69, 2437 (1972).

41. Mayerle, J. J., Frankel, R. B., Holm, R. H., Ibers, J. A., Phillips, W. D., and Weiher, J. F., Proc. Nat. Acad. Sci. U.S. 70, 2429 (1973).

42. Cammack, R., Biochem. Biophys. Res. Comm. 54, 548 (1973).

EVOLUTION AND ECOLOGY OF PHOSPHORUS METABOLISM

M. Halmann

Isotope Department
Weizmann Institute of Science
Rehovot, Israel

INTRODUCTION

The particular role of phosphorus in living systems poses many riddles. Why does this element fulfill such important functions? Phosphate esters, acid anhydrides and amides serve as key intermediates in metabolic processes. The phosphate group is an essential link in the chain of nucleic acids - the storage molecules of hereditary information. Many important coenzymes and cofactors are phosphorus compounds. Attempts have been made to rationalize the role of phosphorus on the basis of the bonding characteristics of this element (1). In the present chapter, the evolution of phosphorus metabolism will be outlined, from the nucleosynthesis of the element in stars to its emergence as the limiting nutrient in the fertility of many natural waters. A most valuable guideline in our quest to understand the evolution of metabolism is Oparin's hypothesis that metabolic pathways appeared earlier than the enzymes which now catalyze these pathways:

"Consequently, in order to solve the problem of the initial genesis of an ordered protein structure suited to its purpose, it is first of all necessary to determine how a definite sequence of metabolic reactions could have been formed from the chaos of criss-crossing reactions; it is necessary to visualize clearly and to prove experimentally methods of formation in the 'primitive soup' of such initial systems in which, by an evolutionary process, a definite sequence of interactions with the external medium could be built up, gradually approaching the modern character of all living metabolism" (2).

Thus, primitive prebiological systems may be more comparable to ecosystems than to contemporary organisms (3).

As will be seen below, there are two major difficulties to be overcome in formulating a model of prebiological phosphorylation: the problem of concentrating phosphate from very dilute solutions, and the thermodynamic problem of activating orthophosphate to form organic phosphates.

COSMIC AND PLANETARY ORIGIN OF PHOSPHORUS COMPOUNDS

Phosphorus is one of the relatively minor elements in the universe. Its cosmic abundance is six orders of magnitude smaller than that of hydrogen and two to three orders of magnitude smaller than that of carbon, nitrogen and oxygen, the major elements of life. These data were obtained from the atomic emission lines of the Sun and of more distant stars. The origin of the element phosphorus, as that of the other elements of the periodic table, may be understood on the basis of the stepwise nucleosynthesis of atoms - starting with hydrogen as the primordial substance of newborn stars (4). The formation and decay of the only stable isotope of the element, ^{31}P, can be represented by a scheme in which phosphorus is produced by β^- decay from ^{31}Si and is consumed by neutron capture to form the radioactive isotope, ^{32}P (5):

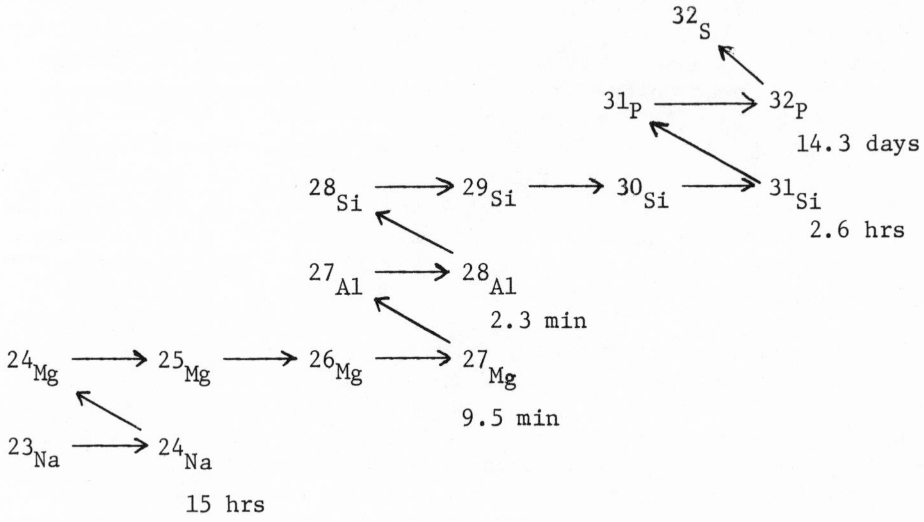

$$^{32}S$$
$$^{31}P \longrightarrow {}^{32}P \ (14.3\ \text{days}) \nwarrow {}^{32}S$$
$$^{28}Si \longrightarrow {}^{29}Si \longrightarrow {}^{30}Si \longrightarrow {}^{31}Si \ (2.6\ \text{hrs}) \nwarrow {}^{31}P$$
$$^{27}Al \longrightarrow {}^{28}Al \ (2.3\ \text{min}) \nwarrow {}^{28}Si$$
$$^{24}Mg \longrightarrow {}^{25}Mg \longrightarrow {}^{26}Mg \longrightarrow {}^{27}Mg \ (9.5\ \text{min}) \nwarrow {}^{27}Al$$
$$^{23}Na \longrightarrow {}^{24}Na \ (15\ \text{hrs}) \nwarrow {}^{24}Mg$$

β-decay is represented by diagonal arrows, while neutron capture is shown by horizontal arrows.

Various small molecules, containing the elements of carbon, hydrogen, oxygen, nitrogen, and sulfur, have been detected by radio-astronomy of the interstellar space. However, no phosphorus compounds have yet been detected among the interstellar molecules (6), which may be due to the low relative abundance of phosphorus.

It may have been hoped that some clue on the original forms of phosphorus in the primordial Earth could be obtained from the examination of meteorites that have fallen on the Earth and from the analysis of samples of the Moon's surface. In iron meteorites, the predominant minerals containing phosphorus were shown to be iron-nickel phosphides, such as schreibersite $(Fe,Ni)_3P$ or $(Fe,Co,Ni)_3P$ (7). On the other hand, in stony meteorites, phosphorus appears mainly in orthophosphate form, together with silicate phases, e.g. as the calcium phosphate mineral, merillite, $Na_2O{\cdot}Ca_3(PO_4)_2$. Thus, both reduced and oxidized forms of phosphorus occur in iron and stony meteorites, respectively. It is not immediately clear whether any conclusion can be drawn from these observations, as the meteorites may not represent primitive material.

Interest in the chemical composition of the lunar surface was derived at least in part from the belief that it may provide an indication of the primordial prebiological terrestrial soil. As a result of the Apollo tests, it is now realized that the surface of the Moon is not an average sample of the Solar System (8). Although the surface rocks of the Moon are igneous, they have been considerably modified by meteoric impact, thus producing a cover of lunar regolith. The age of the lunar rocks is at least as ancient as the oldest terrestrial rocks; for the basalts of the mare, the age was determined as 3.1-3.9 billion years, while for the highland rocks, the age must have been more than 4 billion years. In the lunar samples, the content of P_2O_5 was 0.08-0.13%, which is slightly less than the 0.13-0.42%, observed in terrestrial basalts (9). Reduced forms of phosphorus, such as schreibersite, as well as oxidized forms, such as apatite, $Ca_5(PO_4)_3(F,Cl)$, and whitlockite, $Ca_9MgH(PO_4)_7$, have been identified among the lunar minerals (8). In the rocks recovered in the Apollo 11 mission, one of the minerals was identified as fluorapatite (10).

The main conclusion which may be drawn from the study of lunar samples is that these, in contrast to terrestrial rocks, also contain reduced forms of phosphorus.

THE GEOCHEMICAL CYCLE OF PHOSPHORUS

In trying to create a model of the geochemical cycle of phosphorus on the primordial Earth, it is instructive to consider the geochemistry of this element on the more contemporary scene. In particular, it will be useful to search for the geochemical processes causing enrichment of phosphate.

In igneous rocks, the phosphorus is found only in small concentrations, usually about 0.1%. In the oceans, the concentrations are even very much smaller. Thus, most of the phosphate which is brought to the oceans by rivers and by rain is removed by precipitation or by absorption on clay minerals. Precipitation may be a direct inorganic process, such as formation of apatite minerals, or it may involve the biochemical cycle. In the present-day oceans, phosphate concentrations are usually low (0.1×10^{-6}M) in warm surface waters, but are progressively higher in colder deep waters. The main reason for the higher solubility of apatite minerals in deeper waters seems to be related to the higher concentrations of carbon dioxide which are possible in colder water and higher hydrostatic pressure (11). The increased carbon dioxide concentration results in lowered pH, enhancing solubilization of calcium phosphate. In surface waters, on the other hand, carbon dioxide is usually in equilibrium with atmospheric carbon dioxide.

The mineral of marine phosphorite (marine rock phosphate) is essentially fluorapatite, $Ca_5(PO_4)_3F$, with various substitution, such as by carbonate, sulfate and other ions. Important major deposits of phosphorite are found in the Ordovician (about 4.8×10^8 years old), e.g. the phosphate deposits of Tennessee and Kentucky, the Permian (about 2.6×10^8 years old), e.g. the phosphorites in the Rocky Mountain region of North America, and the Upper Cretaceous (about 0.8×10^7 years old) of Northern Africa and the Middle East. Such rich phosphate rocks contain 25 - 33% of P_2O_5.

In order to account for the formation of these and more recent phosphorite deposits, several theories have been proposed (11-15):

(a) Carbonate apatite is considered to have been formed by regular chemical precipitation, upon supersaturation of natural waters with respect to phosphate (11). Such deposits are formed by a cyclic process, involving "upwelling" of cold water rich in minerals (carbonate, nitrate and phosphate) in areas near the continental shelf, followed by warming up near the surface exposed to sunlight. Loss of carbon dioxide and increase in pH result in precipitation of both calcium carbonate and of carbonate apatite, which thus deposits in shallow basins and lagoons near the continental shelf (11).

(b) The above-described process of "upwelling" of mineral rich water into the trophogenic zone enables the growth of photosynthetic organisms, algae and photosynthetic bacteria, which are the primary producers in the food chain of the oceans. Blooms of such organisms ("red tides") are followed by massive growth of zooplankton, fish and other animals, and by precipitation of fecal pellets, containing high concentrations of phosphorus, mainly as ferric phosphate. Periodically, such blooms are followed by mass mortality (fish kills) and precipitation of organic matter. In the "near shore down-welling", part of this organic matter, reaching cold deep regions, becomes mineralized, releasing carbon dioxide and phosphate and thus causing an enrichment in these ions.

(c) Another important process for the formation of apatites is the replacement of calcium carbonate by phosphate (14,15). The transformation of calcite into carbonate apatite occurs at concentrations of both phosphate and calcium which are much lower than those required for apatite precipitation. A scheme for the replacement of calcite by carbonate apatite was proposed by Ames (15) on the basis of laboratory studies:

$$3Na_3PO_4 + 5CaCO_3 + NaOH \rightleftharpoons Ca_5(PO_4)_3OH + 5Na_2CO_3 \qquad (1)$$

Such a process would result in secondary enrichment of the deposit with phosphate.

(d) An interesting proposition is that phosphate precipitation may be associated with periods of volcanism, such as for the Ordovician phosphate rocks of Tennessee and Kentucky. Fluoride released by volcanic gases caused conversion of phosphate into the very insoluble fluorapatite (16). However, in most cases no stratigraphic correlation between phosphate deposits and volcanic phenomena could be ascertained.

Of the above four mechanisms for formation of marine phosphorites, only (b) requires biological processes. The other mechanisms may be purely inorganic.

In the interstitial waters of the ocean floor on the continental shelf, calcium phosphates reach oversaturation, resulting in dissolved phosphate concentrations of up to 3 mg/l P, or 10^{-4} M phosphate (13). These concentrations are 2 - 3 orders of <u>magnitude</u> higher than in surface waters. Baturin observed in these interstitial waters the precipitation of phosphorite nodules, consisting initially of gel-like calcium phosphate. These, by diagenetic processes of lithification, gradually changed into hard nodules of apatite, containing up to 33% of P_2O_5 (13). The high concentrations of phosphate in interstitial waters under reducing conditions were shown to be accompanied by very high concentrations of dissolved ferrous iron (17).

Under very strongly reducing conditions, phosphorus may presumably appear as phosphite or hypophosphite. Since the calcium salts of these phosphorus oxyanions are much more soluble than orthophosphate, it was proposed that these ions played a role as the prebiotic phosphorus compounds (18). However, these lower oxidation states are thermodynamically unstable, under any reasonable pressures of hydrogen gas (19).

Gulbrandson suggested that possibly seawater had evolved from a solution originally with a much lower Ca/P ratio than the high contemporary one (11). A lower calcium concentration in prebiotic seawater would result in a higher dissolved phosphate concentration.

As a conclusion; enrichment of phosphate on the prebiotic Earth may have occurred by cyclic processes, involving counter-current flow of cold water, rich in carbonate and various minerals, from deep regions near the continental shelf, followed by precipitation of calcite and apatite in shallow basins at higher temperature and alkalinity.

ECOLOGY OF PHOSPHATE TURNOVER BY CONTEMPORARY CELLS

A most remarkable observation is that the elementary composition of waters of the open ocean is equal to that of the plankton, mainly algae, growing in that water (20). The relative distribution of carbonate, nitrate and phosphate dissolved in the ocean is very close to 106:16:1 (ratio by number of atoms). This ratio is also that in which these atoms are consumed during photosynthesis, and by which they are released during respiration:

$$106CO_2 + 16NO_3^- + HPO_4^{--} + 122H_2O + 18H^+ \text{ (+ trace elements)}$$

$$\text{Photosynthesis} \downarrow\uparrow \text{Respiration}$$

$$138O_2 + C_{106}H_{263}N_{16}O_{110}P \text{ (empirical composition of plant protoplasm)}$$

The evolutionary significance of this striking coincidence in the composition of seawater and of the plankton is ambiguous. Either the primordial ocean already had the above elementary composition, and the primitive organisms adapted themselves to these conditions, or alternatively, the present day ocean water composition with respect to nitrogen and phosphorus is a secondary one, due to the modifying and controlling influence of the biological systems, just as the oxygen atmosphere is a product of plant photosynthesis.

In fresh water, the relative abundance of phosphate, by comparison with carbon dioxide and nitrate, is usually much lower than in sea water and does not correspond to the above equation for photosynthesis. Under these conditions, phosphate often becomes the limiting nutrient - determining the fertility of the water.

In a natural ecosystem, the main sources of soluble phosphorus are due to the weathering of rocks and airborne dust. In present-day unpolluted natural waters, the concentration of dissolved phosphate is controlled on the one hand by the input of the element (e.g. by the influx into a lake from its watershed, by leaching out from suspended minerals and from the sediment, and by excretion from aquatic organisms) - and on the other hand by the efflux of phosphorus (e.g. by the outflow from a lake, by adsorption on clay particles, by inorganic precipitation and sedimentation, and by assimilation into organisms). Some of these processes must have been important also on the prebiotic Earth. It is possible to estimate the extent of inorganic processes of phosphorus turnover occurring even now in natural waters with the help of ^{32}P-exchange experiments, while using specific inhibitors to eliminate the biological turnover (44).

In our recent experiments carried out in Lake Kinneret, ^{32}P-labelled orthophosphate in tracer concentrations (less than 10^{-7} M phosphate) was added to the lake water. The rate of exchange of the tracer with the suspended matter was then measured. As an inhibitor of both photophosphorylation and oxidative phosphorylation, m-chlorocarbonylcyanide phenylhydrazone was added,which is a potent uncoupler of these phosphorylation processes. The rates of phosphate turnover observed in the absence and in the presence of this inhibitor were 2.3×10^{-8} and 7×10^{-10} mole l^{-1} hr^{-1}, respectively. Thus tentatively we can conclude that 97% of the phosphate turnover in this water sample was due to biological processes, and probably the remaining exchange (about 3%) was due to inorganic processes. Presumably, in all contemporary natural waters most of the phosphate turnover is biological. On the other hand, the remaining very slow exchange could be a measure of the abiotic turnover which may have occurred even in the primordial ecosystem.

THE ROLE OF ACTIVE AND PASSIVE PHOSPHATE TRANSPORT

In the contemporary microorganisms, the rate-limiting step in the uptake of phosphate is often the transport of this ion through the cell membrane, which is followed by the more rapid energy-producing reactions leading to adenosine triphosphate.

For bacteria, such as *Escherichia coli*, it was shown that at small external phosphate concentrations, below 1.0×10^{-6} M, the uptake of phosphate is an active process. Using cold-shock techniques, a "phosphate-binding-protein" of molecular weight 42000 was isolated from the bacterial membrane (21). At higher external phosphate concentrations, the uptake was shown to be a passive diffusive transport process.

Studies on the rates of phosphate uptake as a function of internal phosphate concentration have been made for various groups of algae. In the case of Chlorella, at least two distinct phosphate absorption sites could be identified, with K_m values of about 3 x 10^{-6} and 3 x 10^{-4} M, respectively (22). The extent of active pumping of phosphate into the plant cell was directly measured in the giant marine alga, Nitella tranlucens, in which the phosphate concentration in the cell sap was found to be several hundred times higher than in the external medium (23).

THE EVOLUTION OF PHOSPHATE METABOLISM

The metabolic pathway most universally common to all contemporary organisms is glycolysis, which reversibly connects glucose, via hexose-and triose-phosphates, to the three carbon products, such as pyruvate and glycerol. Fermentation is probably the most primitive biological process creating energy. The production of adenosine triphosphate in fermentation, and in the glycolytic pathway in all organisms, is due to the mechanism of "substrate-level phosphorylation", in which energy-rich phosphorylated substrates, such as diphosphoglycerate and phosphoenolypyruvate serve as intermediates in the transfer of phosphoryl groups to adenosine diphosphate. Substrate-level phosphorylation must be very ancient in evolution, as it occurs even in an anoxygenic atmosphere in the dark, and as it does not require organelles (24).

While we do have laboratory models of the prebiological synthesis of sugar phosphates (see Section on Prebiological Phosphorylation, below), no plausible model of the synthesis of the required energy-rich phosphorylated substrates has yet been proposed. For the glycolytic degradation of sugars, a model pathway may be seen in the alkaline degradation of glucose 6-phosphate, which is converted via fructose 6-phosphate and glyceraldehyde 3-phosphate to lactic acid (25). However, this model is energetically "useless", as it does not produce the required energy-rich phosphorylated substrates.

The following sequence of pathways of metabolic evolution was proposed by Egami (26):

Fermentation → nitrate fermentation → nitrate respiration→ oxygen respiration.

Thus, oxidative phosphorylation involving the respiratory chain is considered to be a later development.

MODELS OF PREBIOLOGICAL PHOSPHORYLATION

The sequence of chemical reactions which, on the prebiotic Earth, led to the formation of the chemical building blocks of life, must also have included phosphorylation reactions. As mentioned in the Introduction, there are two major hurdles to be overcome in creating a model of prebiotic phosphorylation - the problem of energy source and the problem of concentration.

Phosphorylation, the conversion of orthophosphate into organic phosphates, is thermodynamically very unfavorable in the presence of an excess of water. It becomes possible after activation of orthophosphate with the help of activating agents, to form reactive phosphorus intermediates,

e.g. $$NCCN + HPO_4^{--} \longrightarrow NC - C(NH) - OPO_3^{--} \qquad (3)$$

Phosphorylation proceeds by interaction of the reactive phosphate with a nucleophilic reagent, an alcohol, sugar, nucleoside, nucleotide or amine, to produce a more stable phosphate ester or amide.

One important approach towards formation of reactive phosphates is by dehydration-condensation of acid salts of orthophosphoric acid. As examples, the well-known thermal condensation of the di- and mono-anions of orthophosphate to pyrophosphate and to trimetaphosphate, respectively may be presented:

$$2HPO_4^{2-} \xrightarrow{200^\circ} {}^{2-}O_3POPO_3^{2-} + H_2O \qquad (4)$$

and

$$3H_2PO_4^{-} \xrightarrow{500-600^\circ} P_3O_9^{3-} + 3H_2O \qquad (5)$$

The pyrophosphate bonds formed are the loci of reactivity of such linear or cyclic polyphosphates. Heating of acid salts of phosphoric acid has been used with success to phosphorylate nucleosides (27-31). These reactions may be carried out by evaporation of aqueous solutions of urea, orthophosphate and of the nucleoside to dryness, and heating to temperatures of e.g. 65° (31). The intermediate formation of a pyrophosphate bond is a probable assumption in such reactions, although it was not always definitely demonstrated.

Another approach to the activation of phosphate is with the help of an external activating agent. The most successful agents have been molecules containing nitrilic groups (32-36). As a model of the phosphorylation of sugars by orthophosphate in moderately concentrated solutions (0.1 - 0.01 M phosphate), the cyanogen-induced phosphorylation was investigated (33-36). The reaction was typically carried out by introducing cyanogen gas into an evacuated flask containing neutral, or slightly alkaline (pH 6.8 - 8.8), solutions of sugars and sodium phosphate. We found that the reaction occurs preferentially at the glycosidic hydroxyl group of the sugars, producing glycosyl phosphates. Thus, with the aldose sugars ribose, arabinose, xylose, glucose, mannose, galactose, glycosamine and N-acetylglucosamine, the main products were the corresponding aldose-1-phosphates. One case of a ketose sugar, fructose, was also studied (36). In this case the product was fructose-2-phosphate, which confirms the above conclusions that phosphorylation occurs mainly at the glycosidic position.

In the case of D-glucose, in addition to the main phosphorylation product, α-D-glucopyranose-1-phosphate, the reaction mixture also contained small yields of α-D-glucose-1,6-diphosphate (35). The synthesis of this important cofactor seems of considerable interest in the context of prebiological evolution.

The approach of heating-dehydration or of an added activating agent overcomes the thermodynamic problem. However, it seems difficult to conceive of such a process occurring on a wide geographical scale, rather than only in relatively isolated areas, such as on volcanic ashes, or on the bed of a dried-out lake.

The problem of concentration is even more serious than that of activation. As discussed above, the concentration of phosphorus in igneous rocks is only about 1.1%, and in ocean or freshwater it is even much lower, usually in the range of about 10^{-7} M. Up to now, attempts to create laboratory models of phosphorylation, using such low concentrations of phosphate, have not been successful. These failures may be due to the fact that the activating agents XY used (see equation 3) do not discriminate sufficiently in favor of reaction with phosphate, by comparison with their reaction with water (37). Present-day organisms are capable of such discrimination, and as discussed above (see section Role of Active and Passive Phosphate Transport), by the process of active transport are able to pump phosphate against a concentration gradient of several hundred fold.

If we assume that the prebiotic phosphorylation system lacked such an "active transport" concentration mechanism, we have to look for an alternative concentration process. One suggestion, proposed by Schwartz (38) is an evaporating pond model, in which calcium ions are held in solution by chelation with carboxylate ions, e.g. oxalate, thus preventing the precipitation of apatite. In a laboratory simulation, a concentration of 10^{-3} M phosphate was achieved, and under these conditions, high yields in the phosphorylation of nucleosides were achieved, using various condensing agents.

An alternative approach in finding prebiotic scenes of concentration of phosphate is to look under what conditions such concentration does occur even under contemporary conditions. Such conditions do exist even now in the interstitial waters of the sediments of lakes, estuaries and on the continental shelf (see Section 3, above). In these waters, under reducing conditions which keep iron dissolved as ferrous ions - the dissolved orthophosphate reaches 6×10^{-4} M concentration, which may be thousand fold that in the surface waters (17). The assumption of reducing conditions, with iron in the ferrous state - are in agreement with the general consent on the composition of the primitive Earth's atmosphere. In this respect it is irrelevant whether this atmosphere was strongly reducing, composed of hydrogen, methane and ammonia (39-41) or only weakly reducing, consisting of carbon dioxide and nitrogen (42,43). This model requires, however, that the supersaturation of phosphate in the interstitial waters occurs even in the absence of biological processes. Under such conditions, activation of phosphate, e.g. by nitrilic reagents such as cyanogen, may have led to phosphorylation reactions.

ACKNOWLEDGEMENTS

I wish to thank Dr. A. Nissenbaum and Dr. O.A. Christophersen for valuable advice and discussions on the origin of phosphate rock deposits.

REFERENCES

1. Wald, G., in "Horizons in Biochemistry" (Kasha, M., and B. Pullman, B., eds.), p.127, Academic Press, New York, 1962.

2. Oparin, A.I., "Genesis and Evolutionary Development of Life", p.95, Academic Press, New York, 1968.

3. Margalef, R., "Perspectives in Ecological Theory", p.100, University of Chicago Press, Chicago, 1968.

4. Burbidge, E.M., Burbidge, G.R., Fowler, W.A., and Hoyle, F., Rev. Mod. Phys. 29, 552 (1957).

5. Halmann, M., in "Analytical Chemistry of Phosphorus Compounds" (Halmann, M., ed.), p.1, Wiley-Interscience, New York, 1972.

6. Brown, R.D., Chem. in Britain 9, 450 (1973); Shimizu, M., Prog. Theor. Phys. 49, 153 (1973).

7. Buseck, P.R., Science 165, 169 (1969).

8. Mason, B., Chem. in Britain 9, 456 (1973).

9. Engel, A.E., and Engel, C.E., Science 167, 527 (1970); Maxwell, J.A., Abbey, S., and Champ, W.H., ibid., 530.

10. Fuchs, L.H., in "Proceedings of the Apollo 11 Lunar Science Conference", Vol. 1 (Levinson, E.A., ed.) Supplement 1 to Geochim. Cosmochim. Acta 34, 475 (1970).

11. Gulbrandsen, R.A., Econ. Geol. 64, 365 (1969); Kazakov, A.V., Akad. Nauk SSR, Trud. Inst. Geol. Nauk 114, Geol. Ser. No. 40, 1, (1950).

12. Bentor, Y.K., Internat. Geol. Cong. 19th, Algiers 1952, Compt. rend., sec. 11, pt. 11, pp. 93-101, 1953.

13. Baturin, G.N., Nature (Physical Science) 232, 61 (1971).

14. Degens, E.T., "Geochemistry of Sediments", p.141, Prentice-Hall, 1965.

15. Ames, L.L., Jr., Econ. Geol. 54, 829 (1959).

16. Mansfield, G.R., Amer. J. Sci. 238, 863 (1940).

17. Bray, J.T., Bricker, O.P., and Troup, B.N., Science 180, 1362 (1973).

18. Gulick, A.A., Amer. Scientist 43, 479 (1955).

19. Horowitz, N.H., and Miller, S.L., Fortsch. Chem. Org. Naturstoffe 20, 423 (1962).

20. Ryther, J.H., and Dunstan, W.M., Science 171, 1008 (1971).

21. Medveczky, N., and Rosenberg, H., Biochim. Biophys. Acta 211, 158, (1970).

22. Jeanjean, R., Blasco, F., and Gaudin, C., Compt. rend. Acad. Sc. Ser. D. 270, 2946, (1970).

23. Smith, F.A., Biochim. Biophys. Acta 126, 94 (1966).

24. Broda, E., in "Progress in Biophysics and Molecular Biology", Vol. 21 (Butler, J.A.V., and Noble, D., eds.), Vol. 21, p. 143, Pergamon Press, Oxford, 1970.

25. Degani, Ch., and Halmann, M., Nature 216, 1207 (1967).

26. Egami, F., Z. Allgem. Mikrobiol. 13, 177 (1973).

27. Ponnamperuma, C., and Mack, R., Science 148, 1221 (1965).

28. Waehneldt, T.V., and Fox, S.W., Biochim. Biophys. Acta 134, 1 (1967).

29. Rabinowitz, J., Sherwood, C., and Ponnamperuma, C., Nature 218, 442 (1968).

30. Osterberg, R., and Orgel, L.E., J. Mol. Evol. 1, 241 (1972); Osterberg, R., Orgel, L.E., and Lohrmann, R., ibid. 2, 231 (1973).

31. Bishop, M.J., Lohrmann, R., and Orgel, L.E., Nature 237, 162 (1972).

32. Steinman, G., Lemmon, R.M., and Calvin, M., Proc. Nat. Acad. Sci. U.S. 52, 27 (1964); Steinman, G., Kenyon, D.H., and Calvin, M., Nature 206, 707 (1965); Ferris, J.P. Science 161, 53 (1968).

33. Halmann, M., Sanchez R.A., and Orgel, L.E., J. Org. Chem. 34. 3702 (1969).

34. Degani, Ch., and Halmann, M., J. Chem. Soc. 1459 (1971).

35. Degani, Ch., and Halmann, M., Nature (New Biology) 235, 171 (1972)

36. Kawatsuji, M., "Cyanogen-Induced Phosphorylation of D-Fructose", M. Sc. Thesis, Feinberg Graduate School, Weizmann Institute of Science, Rehovot, 1973.

37. Lohrmann, R., and Orgel, L.E., Science 161, 64 (1968).

38. Schwartz, A.W., Biochim. Biophys. Acta 281, 477 (1972); Schwartz, A.W., van der Veen, M., Bisseling, T., and Chittenden G.J.F., BioSystems 5, 119 (1973).

39. Oparin, A.I., "The Origin of Life", (Transl. by Morgulis, S.), Macmillan, New York, 1938.

40. Urey, H.C., "The Planets", Yale University Press, New Haven, 1952.

41. Miller, S.L., and Urey, H.C., Science 130, 245 (1959).

42. Rubey, W. W., Geol. Soc. Amer. Special Papers 62, 631 (1955); Rubey, W. W., in "The Origin and Evolution of Atmospheres and Oceans" (Brancazio, P. J., and Cameron, A.G.W., eds.), Wiley, New York, 1964.

42. Revelle, R. J., J. Marine Res. 14, 446 (1965).

44. Halmann, M., to be published.

AMINO ACID SYNTHESIS BY GLOW DISCHARGE ELECTROLYSIS:

A POSSIBLE ROUTE FOR PREBIOTIC SYNTHESIS OF AMINO ACIDS

Kaoru Harada
Institute for Molecular and Cellular Evolution and
Department of Chemistry
University of Miami
Coral Gables, Florida 33134 USA

INTRODUCTION

A considerable number of studies on the formation of amino acids under inferred prebiotic conditions using various organic and inorganic compounds have been reported in the past twenty years. In these studies, several elegant analytical methods, developed since the early 1950s, have played a crucial role in amino acid analyses from these complex reaction products. Amino acids were synthesized from gas mixtures by applying various types of energy, such as electric discharge, ultraviolet rays, ionizing radiation, thermal energy, and other sources such as shock waves. Amino acids were also synthesized from chemically reactive compounds such as hydrogen cyanide, formaldehyde, and ammonia. These accumulated results indicate that the formation of amino acids occurs rather easily, in many ways, under various inferred prebiotic conditions.

Amino acids were synthesized by the use of both spark and silent electric discharges from hydrogen, methane, ammonia, and water (1,2). The mechanism of amino acid formation was proposed as the Strecker type reaction. Amino acid formation by oligomerization of hydrogen cyanide has been reported (3-5), and the pathway has been considered as one of typical prebiotic synthesis of amino acids. On the other hand, several radical type amino acid formations using ultraviolet rays, x-rays, and ionizing radiation have been reported since the 1950s. However, the yields of amino acids by these reactions are generally very low.

Hasselstrom *et al.* (6) identified glycine and aspartic acid formed by irradiation of an aqueous solution of ammonium acetate by β-rays. Dose and Ettre (7) x-irradiated the same aqueous solution and observed the formation of β-alanine in addition to glycine and aspartic acid. Dose and Risi (8) also x-irradiated an aqueous solution of ammonium salts of fatty acids and confirmed the formation of homologous amino acids. The formation of α-amino acids was not favored. They also found that amino acids were formed from amines and carbonate ions in aqueous solution when irradiated with x-rays. The latter reaction is carboxylation of an amine. Deschreider (9) applied UV light (2537 Å) to various aqueous mixtures of fatty acids and ammonium salts and reported the formation of amino acids. Glycine and alanine were formed from succinic acid, while glycine, alanine, and aspartic acid were formed from maleic acid. Propionic acid yielded only glycine, but not alanine. Cultera and Ferrari (10) have described a series of amino acid formations by similar photochemical reactions. Mehran and Pageau (11) observed the formation of aspartic acid from an aqueous solution of alanine irradiated with γ-rays from ^{60}Co. In this experiment, the existence of carbon dioxide in the solution enhanced the formation of aspartic acid.

A possible prebiotic pathway of the formation of amino acids (especially glycine, alanine, and aspartic acid) by radical reaction could be drawn as shown in Fig. 1 (12). The scheme indicates that carboxylic acids or their corresponding nitriles could be aminated or coupled together to form glycine, alanine, aspartic acid, and β-alanine.

GLOW DISCHARGE ELECTROLYSIS

Glow discharge electrolysis (GDE) (13,14) is a chemical change due to the glow discharge between a liquid containing ions and the electrode above (or in contact with) the liquid. In this system, the liquid phase which contains chemical substances is regarded as an electrode. The ions produced by the electric discharge are accelerated by the electric field; they react with substances in the liquid electrode. This system is very different from that employing electric discharge between metal electrodes in gases under reduced pressure.

The chemistry of GDE has been studied mainly on inorganic compounds such as water, ammonia, and metal ions in aqueous solutions (15). However, little work has been done on the chemistry of GDE using organic compounds. Klemenc (16) studied the gaseous product of GDE of methanol, ethanol, formic acid, and acetic acid in water. Recently, Gore and Hickling (17) studied the product of GDE of aqueous and anhydrous formic acid and acetic acid. They found oxalic acid formed from formic acid; they also found succinic acid, malic acid, tricarballylic acid, and oxalic acid formed from acetic acid. These results strongly indicate the ·COOH radical from formic acid

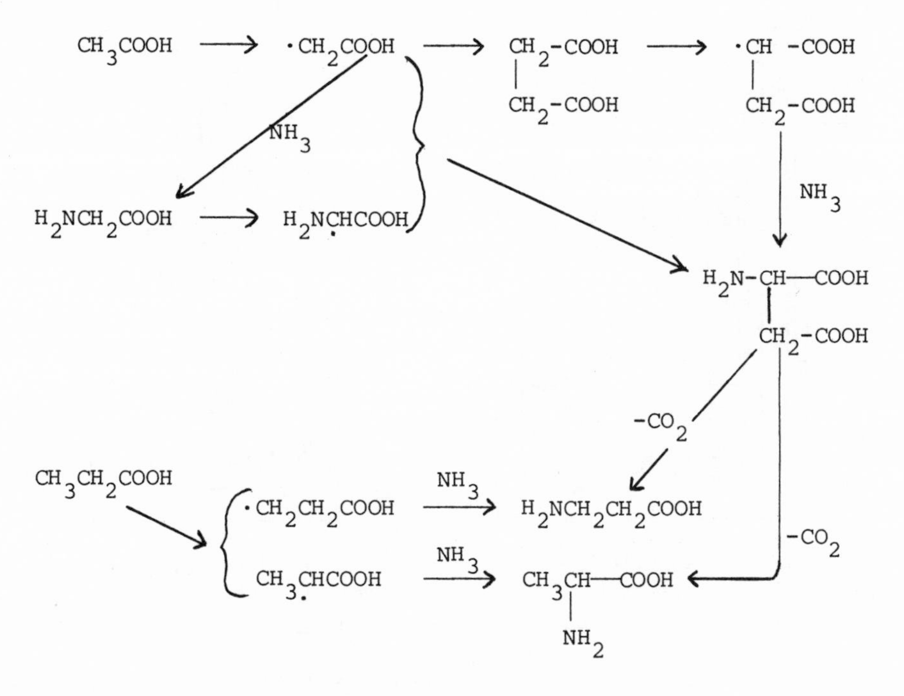

Figure 1. Reactions of GDE

and the $\cdot CH_2COOH$ radical from acetic acid to be key intermediates in formation of C_2 and C_4 compounds. These results are consistent with the scheme shown in Fig. 1. Similarly, formation of oxamide, $(CONH_2)_2$, from a solution of formamide by GDE was confirmed (18). Polymerization of acrylonitrile (19) and acrylamide (20) by GDE was reported.

The present paper reports representative results of amino acid formation by GDE from (a) carboxylic acids by amination with the $\cdot NH_2$ radical using ammonia and from (b) amines by carboxylation with the $\cdot COOH$ radical using formic acid. Other studies of amino acid formation from amines by carboxylation using carbon dioxide or cyanide, or by coupling reaction using GDE will be discussed elsewhere.

EXPERIMENTAL AND RESULTS

The process of contact glow discharge electrolysis (CGDE) (14, 21) was conducted in this study by contact of the platinum anode with an aqueous reaction mixture (Fig. 2). Two types of electrolysis tubes were used for GDE. One was a U-shaped tube (Fig. 2a); the other was a single tube (Fig. 2b). In the U-tube, the cathode

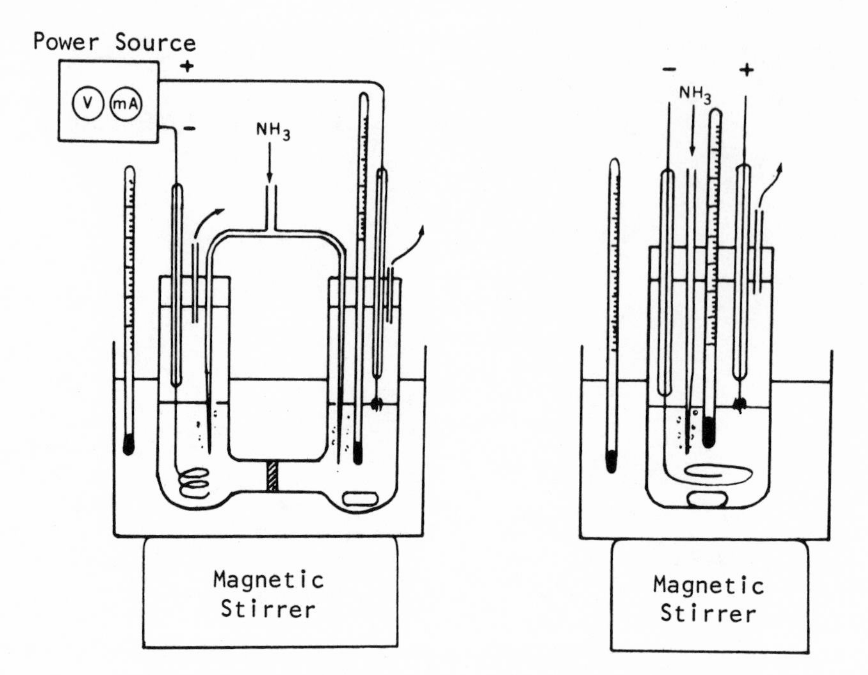

Figure 2. Apparatus for amination reaction.

and anode compartments were separated by a fine-porosity glass frit. The electrolysis was carried out in the anodic compartment. In the amination reaction of carboxylic acid, CGDE was carried out under saturation of ammonia gas while stirring. All reactions were carried out under one atmosphere. The reaction temperature was measured by thermometers in the reaction tube immersed in the methanol-dry ice bath (Fig. 2). The applied electricity (DC) was 450-600 V and 20-100 mA. The reaction time ranged from one-half hour to six hours. Some reaction mixtures were colored quickly by CGDE while others were colored slowly. After cessation of the reaction, the reaction products were evaporated to dryness under reduced pressure and the residues were diluted appropriately for amino acid analysis. Some of the reaction mixtures were hydrolyzed with 4-6 N hydrochloric acid; the amino acid contents were then analyzed. The reaction mixture was treated with 2,4-dinitrofluorobenzene (22,23); the resulting DNP-amino acids were separated both by column chromatography (24,25) and thin layer chromatography. The major amino acids were confirmed as DNP-derivatives.

The chemistry of GDE seems similar to that observed in radiation chemistry and to other kinds of discharge electrolysis. However, GDE is characterized by rather low individual particle energies and by a dose rate which is uniquely high. This is an advantageous point in laboratory study compared with conventional radiation chemistry. Therefore, the GDE experiments can be carried out on a small scale in a relatively short time (e.g. 30 min. to 6 hrs.). The approximate expenditure of energy for the formation of 10^{-6} mole of amino acids in the amination and carboxylation reactions is 0.1 to 1.0 Kcal. When the reaction time is less than 2 hrs, the expenditure of energy per 10^{-6} mole of amino acid(s) is 0.1∿0.3 Kcal. However, for the prolonged reaction, the energy is increased to 1 Kcal. Table I shows the energy for the synthesis of amino acids using CGDE as compared to various energies for the formation of glycine from constituents of reducing atmospheres using other techniques (26).

Figs. 3 to 9 represent the amino acids which formed from carboxylic acids with ammonia by CGDE. Figs. 10 to 15 show the amino acid analyses of the products from amines and amino acids using formic acid by CGDE. A coupling reaction (Fig. 16) and a glycine formation by simultaneous amination and carboxylation of methane by CGDE (Fig. 17) are also included. These amino acid formations by CGDE do not represent optimal conditions.

Table I

Approximate Energy for the Formation of Glycine (Amino Acids; 26)

Form of energy	Kcal/10^{-6} mole of glycine (amino acids)
CGDE	0.1-1.0
Spark discharge	2.4
Silent discharge	19
Ultraviolet light (1165-1470 Å)	220
X-rays	5
β-rays	1.0
Heat	2

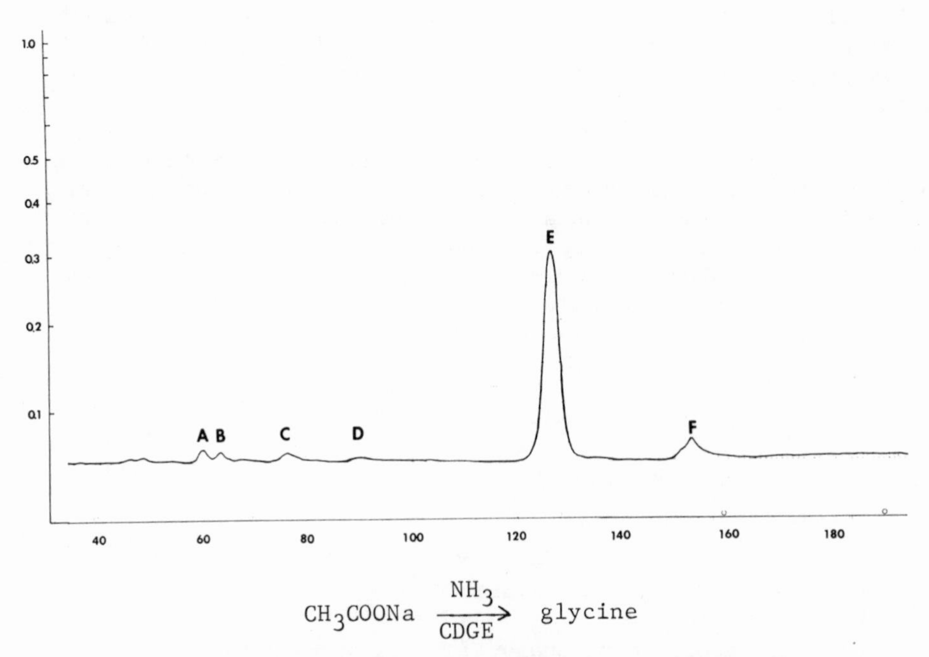

$$CH_3COONa \xrightarrow[CDGE]{NH_3} \text{glycine}$$

A,B,C,D, unknown; E, glycine; F, unknown + buffer change

Figure 3. Sodium acetate (0.0025 mole) was aminated in concentrated ammoniacal solution (10 ml) by CGDE. The reaction conditions were as follows: electric current, 65 mA, reaction temperature, 15°C; reaction time, 6 hrs. The yield of glycine (peak E) was 3.5% from sodium acetate.

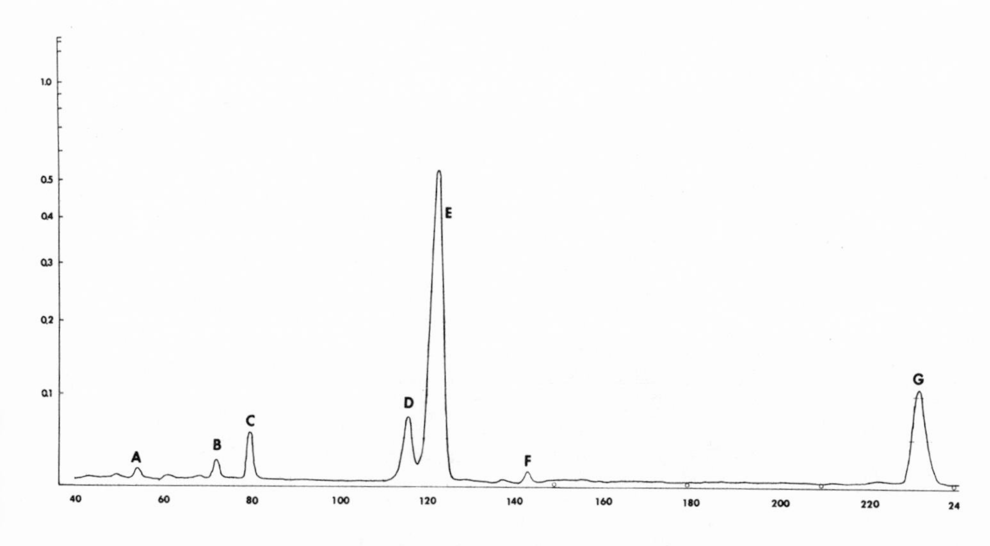

$$CH_3CH_2COOH \xrightarrow[CGDE]{NH_3} \text{alanine, } \beta\text{-alanine, glycine}$$

A,B,C, unknown; D, glycine; E, alanine; F, unknown; G, β-alanine

Figure 4. Propionic acid (0.0025 mole) was aminated in concentrated aqueous ammonia (10 ml) by CGDE. The reaction conditions were as follows: electric current, 75 mA; reaction temperature, 15°C; reaction time, 3 hrs. Alanine (peak E, yield 6.9%) was synthesized by amination of α-carbon, and β-alanine (peak G, yield 5.3%) was formed by amination of β-carbon of propionic acid. A small amount of glycine (peak D) was formed by the splitting of an α-β carbon linkage. The glycine formation by α-β cleavage of carboxylic acid is a general phenomenon in GDE experiments.

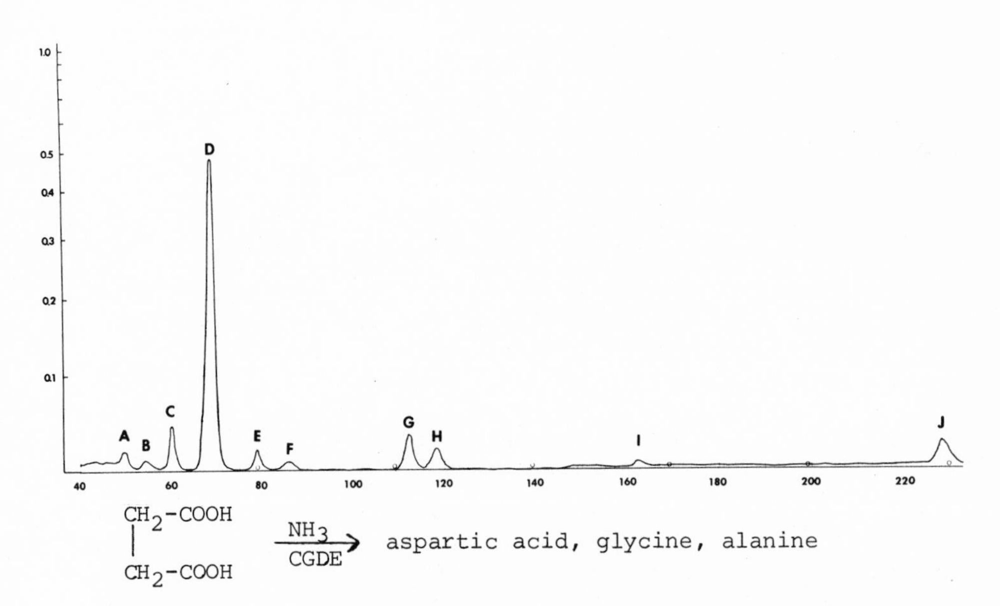

A,B,C, unknown; D, aspartic acid; E. F., unknown; G, glycine; H, alanine; I, unknown; J, β-alanine

Figure 5. Succinic acid (0.0025 mole) was aminated in concentrated aqueous ammonia (10 ml) by CGDE. The reaction conditions were as follows: electric current, 50-70 mA; reaction temperature, 15°C; reaction time, 6 hrs. The major amino acid product is aspartic acid (peak D, yield 9.1%). Glycine (peak G, yield 0.5%) was formed by α-β bond cleavage. Alanine (peak H, yield 0.3%) was formed by β-decarboxylation and β-alanine (peak J, yield 1.6%) was formed by α-decarboxylation of aspartic acid. In the several aspartic acid formations, a peak corresponding to β-hydroxyaspartic acid (peak C) was usually found with aspartic acid. However, the nature of the amino acid at peak C is not fully characterized yet. Similarly, peaks corresponding to threonine and serine (peaks E and F) were commonly found in the reaction products of GDE experiments. These are also not yet characterized.

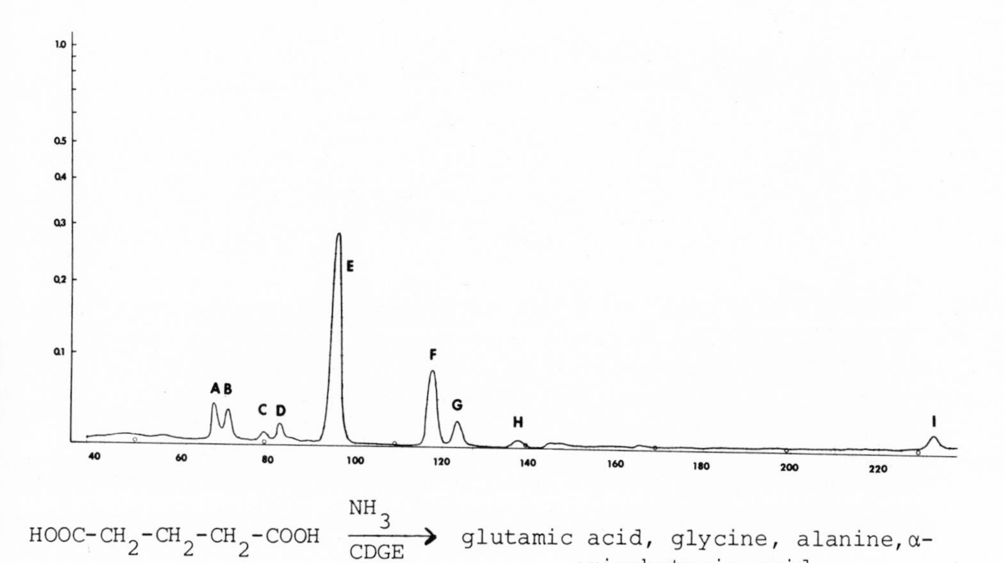

$$HOOC\text{-}CH_2\text{-}CH_2\text{-}CH_2\text{-}COOH \xrightarrow[CDGE]{NH_3} \text{glutamic acid, glycine, alanine, } \alpha\text{-aminobutyric acid}$$

A,B,C,D, unknown; E, glutamic acid; F, glycine; G, alanine; H, α-aminobutyric acid; I, unknown

Figure 6. Glutamic acid (0.0025 mole) was aminated in concentrated aqueous ammonia (10 ml) by CGDE. The reaction conditions were as follows: electric current, 75 mA; reaction temperature, 15°C; reaction time, 3 hrs. The main product was glutamic acid (peak E, yield 3.1%). Glycine (peak F, yield 0.8%) and alanine (peak G, yield 0.2%) were also formed. The peak of an expected product, β-aminoglutaric acid, was not identified.

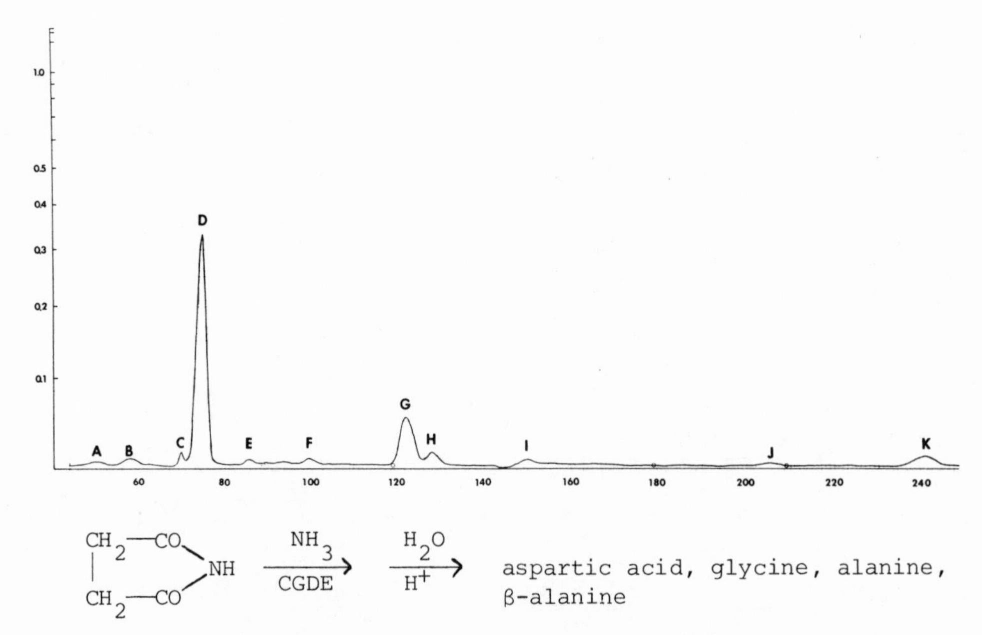

A,B,C, unknown; D, aspartic acid; E,F, unknown; G, glycine; H, alanine; I,J, unknown; K, β-alanine

Figure 7. Succinimide (0.0025 mole) was aminated in concentrated aqueous ammonia (10 ml) by CGDE. The reaction conditions were as follows: electric current, 75 mA; reaction temperature, 5°C; reaction time, 3 hrs. After the reaction was over, the ammoniacal solution was evaporated under reduced pressure, and the residue was hydrolyzed with 6 N hydrochloric acid for 6 hrs. The main product was aspartic acid (peak D, yield 3.2%). Glycine (peak G, yield 0.8%) and alanine (peak H) were also found in the hydrolyzate. The peak K corresponds to that of β-alanine (yield 0.7%).

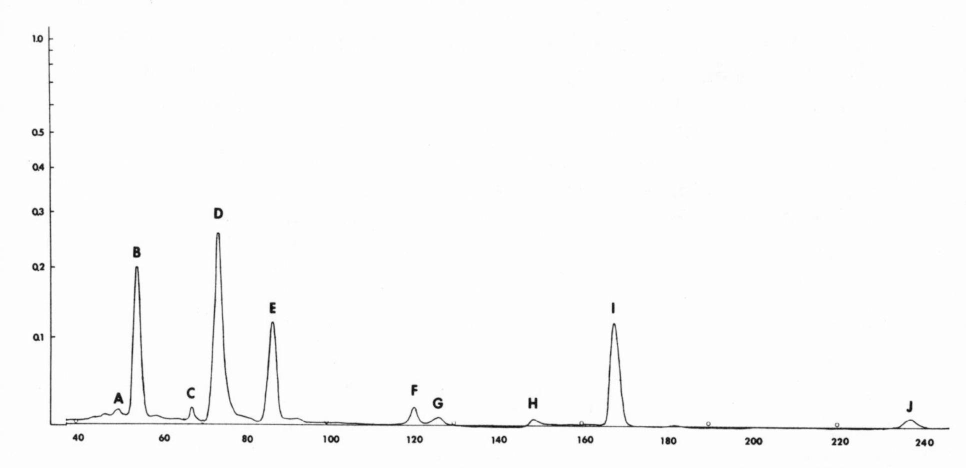

a) A,B,C, unknown; D, aspartic acid; F, glycine; G, alanine, H, buffer change; I, isoasparagine; J, β-alanine

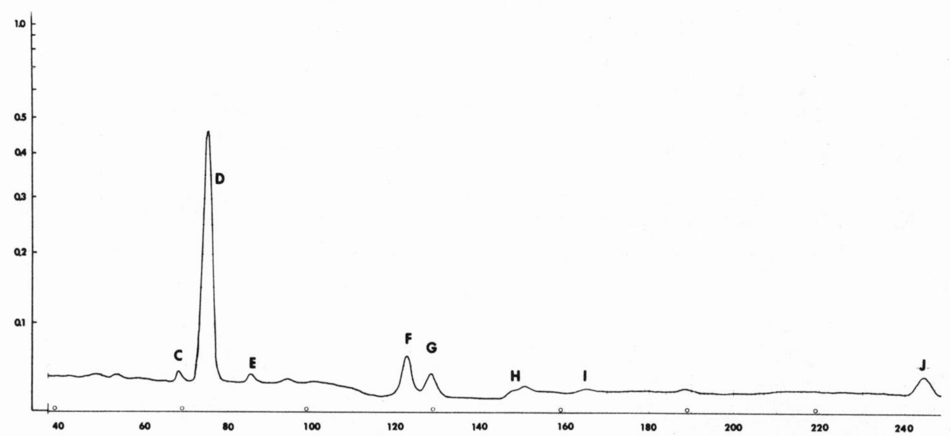

b) C, unknown; D, aspartic acid; E, unknown; F, glycine; G, alanine; H, buffer change; I, trace of isoasparagine; J, β-alanine

$$\begin{array}{l} CH_2\text{—}CO \\ | \qquad\qquad \searrow O \\ CH_2\text{—}CO \nearrow \end{array} \xrightarrow{NH_3} \begin{array}{l} CH_2\text{-}CONH_2 \\ | \\ CH_2\text{-}COO^-NH_4^+ \end{array} \xrightarrow[CGDE]{NH_3} \text{(Fig. 8a)} \xrightarrow[H^+]{H_2O} \text{(Fig. 8b)}$$

Figure 8. Succinic anhydride (0.0025 mole) was dissolved in 10 ml of concentrated aqueous ammonia and the solution was subjected to CGDE. The reaction conditions were as follows: electric current, 55 mA; reaction temperature, 5°C; reaction time, 3 hrs. The amino acids in the reaction mixture were analyzed without hydrolysis (Fig. 8a). Asparagine (peak E, yield 1.1%) and isoasparagine (peak I) were formed by CGDE. However, a larger aspartic acid peak

(peak D, yield 2.6%) was found. This suggests that the CGDE process would hydrolyze the amide bond of asparagine and isoasparagine formed. Glycine (peak F, 0.1%), alanine (peak G), and β-alanine (peak J) were also observed. The nature of the large peak B is not known; however, the peak B disappeared after acid hydrolysis (Fig. 8b). The yields of aspartic acid, glycine, alanine, and β-alanine after hydrolysis were 4.5, 0.5, 0.3, and 1.2%, respectively.

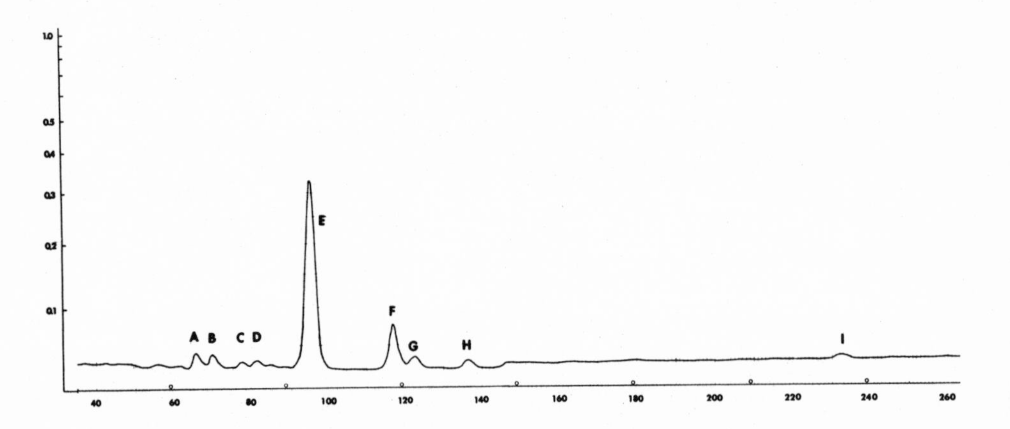

a) A,B,C, unknown; D, glutamine; E, glutamic acid; F, glycine; G, alanine; H, α-aminobutyric acid; I, unknown; J, isoglutamine

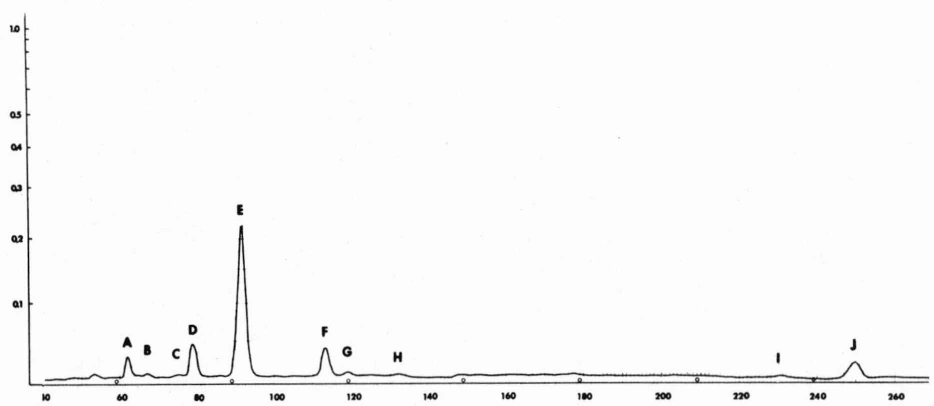

b) A,B,C, unknown; D, trace of glutamine; E, glutamic acid; F, glycine; G, alanine; H, α-aminobutyric acid; I, unknown

$$CH_2\begin{cases}CH_2\text{—}CO\\ \quad\quad\quad\;\; O\\ CH_2\text{—}CO\end{cases} \xrightarrow{NH_3} CH_2\begin{cases}CH_2\text{-}CONH_2\\ CH_2\text{-}COO^-NH_4^+\end{cases} \xrightarrow[CGDE]{NH_3} \text{(Fig. 9a)} \xrightarrow[H^+]{H_2O} \text{(Fig. 9b)}$$

Figure 9. Glutaric anhydride (0.0025 mole) was dissolved in 10 ml of concentrated aqueous ammonia. The solution was then treated by CGDE. The reaction conditions were as follows: electric current, 50-65 mA; reaction temperature, 5°C; reaction time, 3 hrs. Glutamine (Fig. 9a, peak D) and isoglutamine (peak J) were found in the reaction mixture. However, the largest peak is that of glutamic acid (peak E, yield 2.0%). This fact also indicates that the amide bond of glutamine and isoglutamine was hydrolyzed during the CDGE process. Glycine (peak F) and alanine (peak G) were also found in the reaction mixture. After hydrolysis of the reaction mixture (Fig. 9b), glutamine (peak D) and isoglutamine (peak J) disappeared and the peak of glutamic acid (peak E) became larger. The yield of glutamic acid after hydrolysis was 3.8%. The peak of β-aminoglutaric acid, which is an expected product of the amination reaction, was not identified.

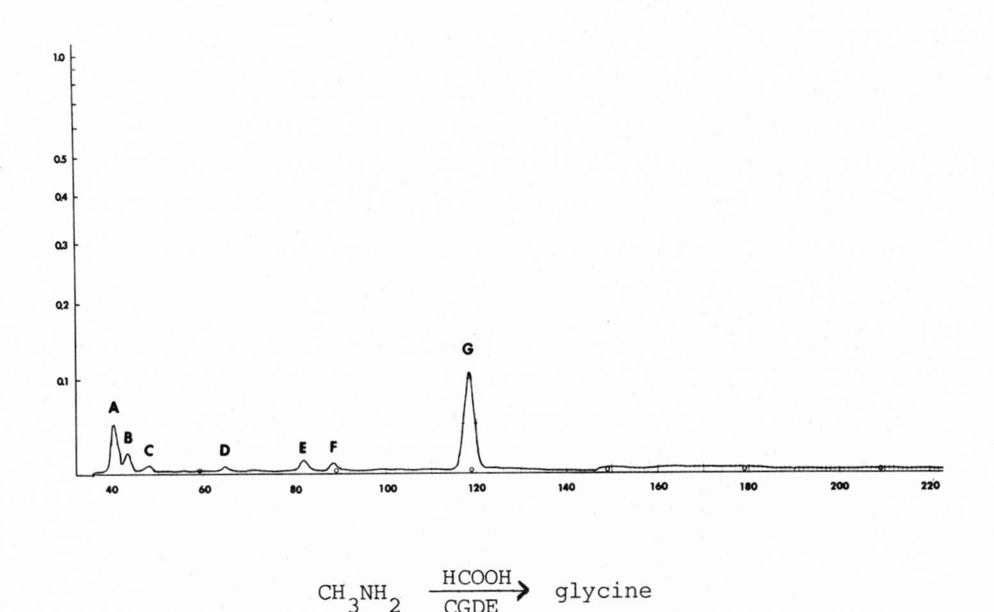

$$CH_3NH_2 \xrightarrow[\text{CGDE}]{\text{HCOOH}} \text{glycine}$$

A,B,C,D,E,F, unknown; G, glycine

Figure 10. Methylamine (0.0025 mole) in 10 ml of 30% formic acid was subjected to CGDE. The reaction conditions were as follows: electric current, 50 mA; reaction temperature, -10 to 5°C; reaction time, 2 hrs. Glycine (peak G, yield 1.2%) is the major amino acid product. Other experiments indicate that it is not necessary to use so much formic acid to carboxylate various amines.

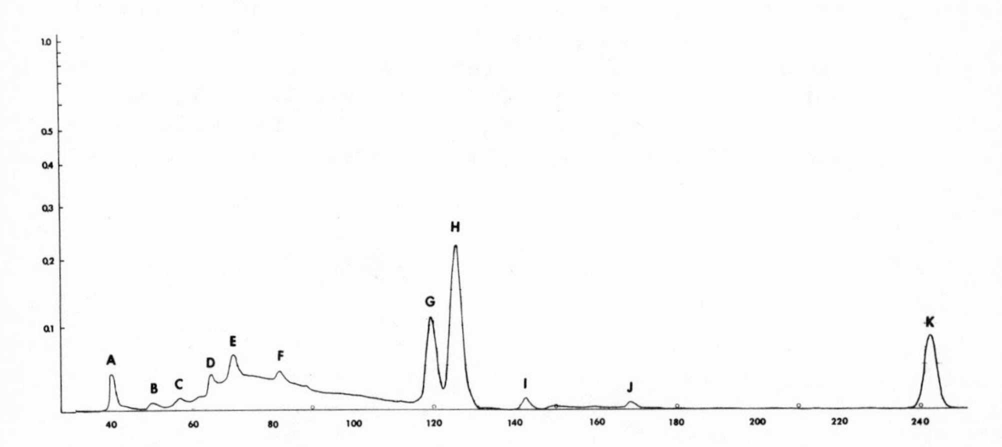

$$CH_3CH_2NH_2 \xrightarrow[\text{CGDE}]{\text{HCOOH}} \text{glycine, alanine, } \beta\text{-alanine, aspartic acid}$$

A,B,C,D, unknown; E, aspartic acid; F, unknown; G, glycine; H, alanine; I,J, unknown; K, β-alanine

Figure 11. Ethylamine (0.0025 mole) in 10 ml of 30% formic acid was subjected to CGDE. The reaction conditions were as follows: electric current, 50 mA; reaction temperature, 10°C; reaction time, 2 hrs. Major amino acid products are glycine (peak G, yield 1.3%), alanine (peak H, yield 2.6%), and β-alanine (peak K, yield 4.3%). Peak E is probably aspartic acid.

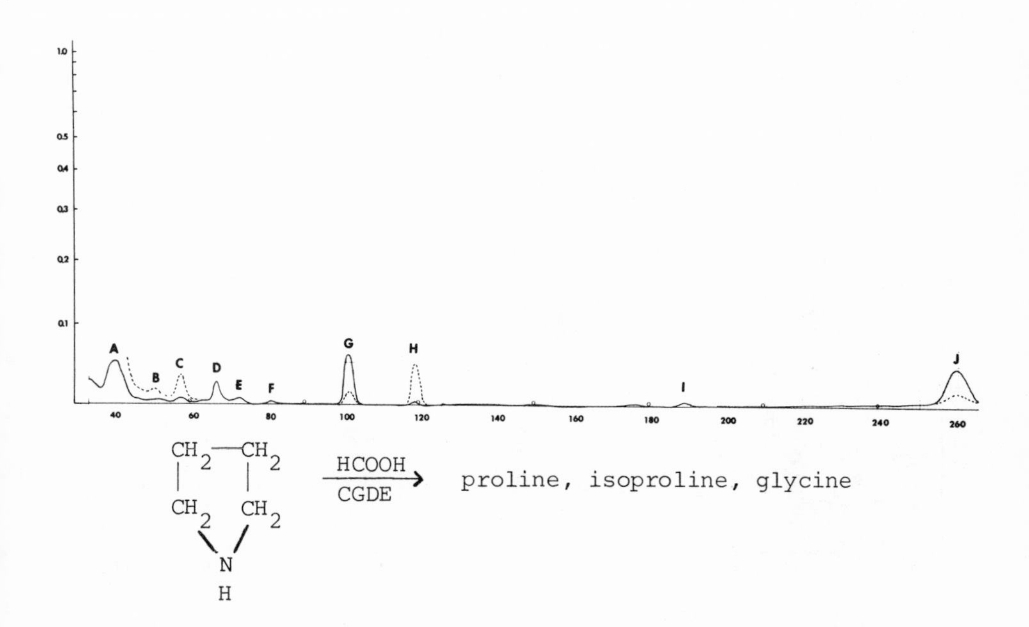

A,B,C,D,E,F, unknown; G, proline; H, glycine; I, unknown; J, isoproline (?)

Figure 12. Pyrrolidine (0.0025 mole) in 10 ml of 30% formic acid was subjected to CGDE. The reaction conditions are as follows: electric current, 65-70 mA; reaction temperature, 10°C; reaction time, 1 hr. Fig. 12 shows the elution pattern measured at 440 nm, whereas the dotted line shows the pattern measured at 570 nm. Proline (peak G, yield 3.5%) is one of the major products of amino acid. Peak J is probably β-carboxypyrrolidine (isoproline). The yield of this amino acid is 4.9% when the amino acid is calculated on a proline basis. Formation of glycine (peak H) was observed. Peaks D, E, or F, which are sensitive to the light of 440 nm, are probably carboxylated pyrrolidine. The peak D corresponds to that of hydroxyproline.

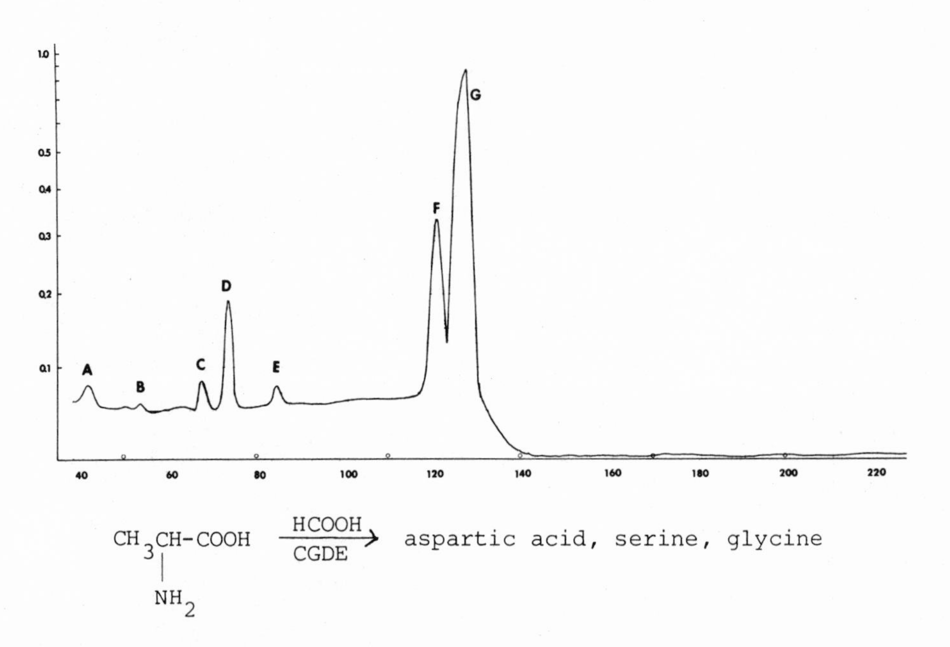

A,B,C, unknown; D, aspartic acid; E, serine; F, glycine; G, alanine

Figure 13. Alanine (0.0025 mole) in 10 ml of 30% formic acid was subjected to CGDE. The reaction conditions were as follows: electric current, 60 mA; reaction temperature, 15°C; reaction time, 2 hrs. Major amino acid products were aspartic acid (peak D, yield 1.2%) and glycine (peak F, yield 3.2%). The amount of alanine recovered after CGDE was 13.6%. Peak C is probably β-hydroxyaspartic acid and peak E might be a hydroxyamino acid.

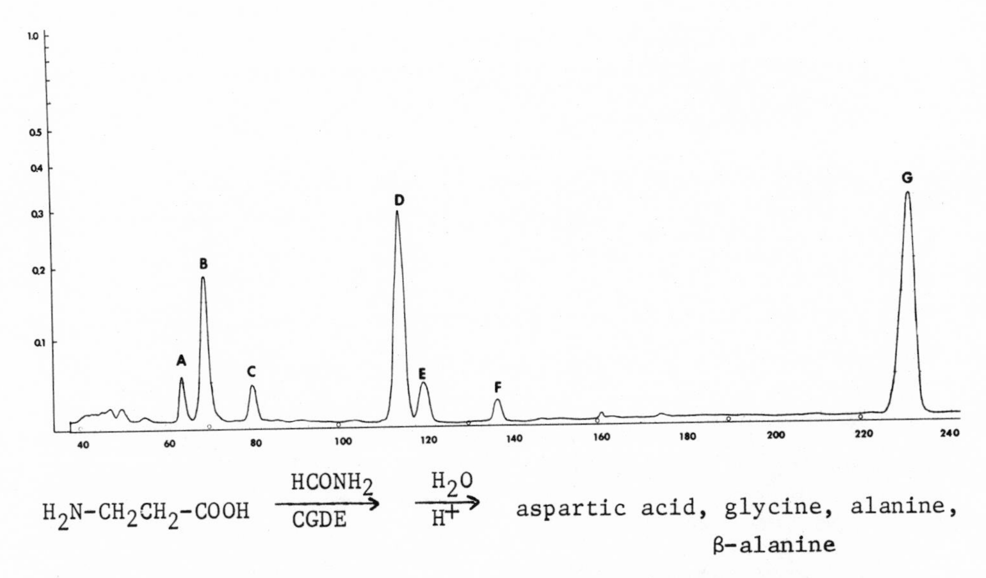

$$H_2N\text{-}CH_2CH_2\text{-}COOH \xrightarrow[\text{CGDE}]{HCONH_2} \xrightarrow[H^+]{H_2O} \text{aspartic acid, glycine, alanine, } \beta\text{-alanine}$$

A, unknown; B, aspartic acid; C, serine (?); D, glycine; E, alanine; F, unknown; G, β-alanine

Figure 14. β-Alanine (0.0025 mole) and formamide (0.005 mole) were dissolved in 10 ml of water which contained 0.0025 moles of sodium hydroxide. The solution was subjected to CGDE. The reaction conditions were as follows: electric current, 75 mA; reaction temperature, 38°C; reaction time, 45 min. After the reaction was over, the reaction product was hydrolyzed with 4 N hydrochloric acid for 8 hrs. After the reaction, 14.7% of β-alanine was recovered. Formation of aspartic acid (peak B, yield 1.6%) and glycine (peak D, yield 2.6%) were observed. Peaks A and C are probably β-hydroxyaspartic acid and a hydroxy amino acid, respectively. A small amount of alanine (peak E, yield 0.5%) was formed. The formation of a large amount of glycine indicates that the β-carbon of β-alanine was carboxylated first (aspartic acid formation); cleavage of α-β linkage then took place under the conditions employed.

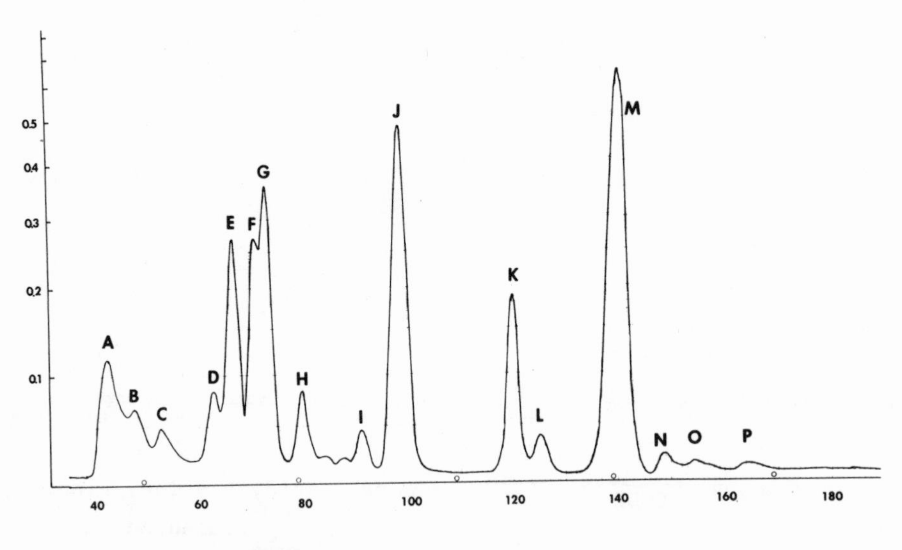

$$CH_3CH_2(NH_2)-COOH \xrightarrow[CGDE]{HCOOH} \beta\text{-methylaspartic acid, glutamic acid, glycine, alanine}$$

A,B,C,D,E,F,G, unknown; one of E,F, or G, β-methylaspartic acid; H, threonine (?); I, homoserine (?); J, glutamic acid; K, glycine; L, alanine; M, α-aminobutyric acid; N,O,P, unknown.

Figure 15. α-Aminobutyric acid (0.0025 mole) in 10 ml of 30% formic acid was subjected to CGDE. The reaction conditions were as follows: electric current, 60 mA; reaction temperature, 15°C; reaction time, 2 hrs. After the reaction, 8.7% of α-aminobutyric acid was recovered. The formation of glutamic acid (peak J, yield 5.9%), glycine (peak K, yield 2.1%), and alanine (peak L, yield 0.2%) were observed. The analytical result of acidic amino acids was complex (peaks A to G). The formation of β-methylaspartic acid was expected. One of the peaks (E, F, or G) could be β-methylaspartic acid.

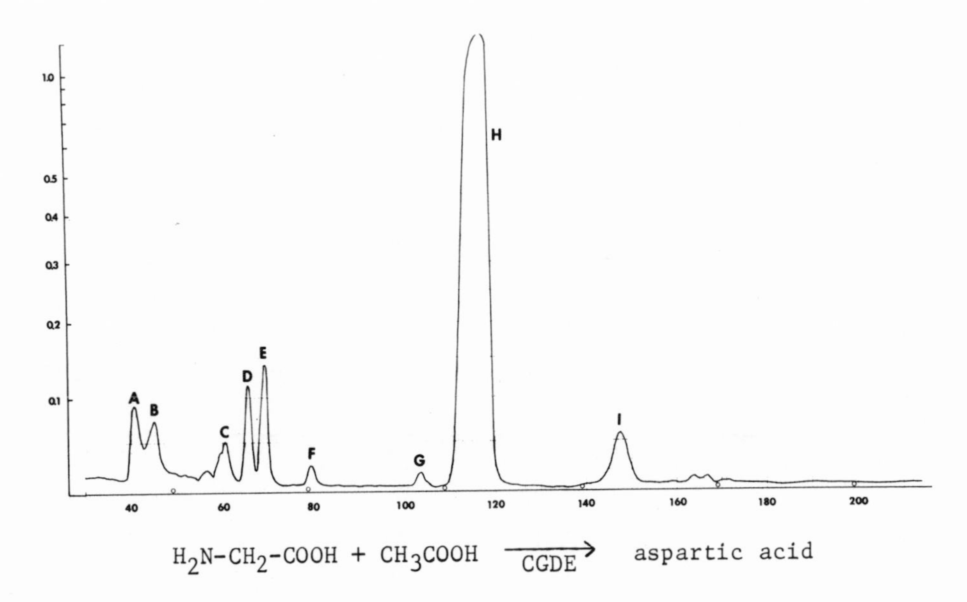

$$H_2N\text{-}CH_2\text{-}COOH + CH_3COOH \xrightarrow[CGDE]{} \text{aspartic acid}$$

A,B,C,D, unknown; E, aspartic acid; F,serine (?); G, unknown; H, glycine; I, unknown

Figure 16. Glycine (0.005 mole) and acetic acid (0.005 mole) were dissolved in 10 ml of water containing 1 ml of concentrated aqueous ammonia. The solution was subjected to CGDE. The reaction conditions were as follows: electric current, 100 mA; reaction temperature, 20-30°C; reaction time, 1 hr. 45 min. The amino acid analysis showed 19.4% recovery of glycine (peak H) after the reaction. The formation of aspartic acid (peak E, yield 0.6%) was observed. Peak D might be β-hydroxyaspartic acid. The formation of aspartic acid was also confirmed by column chromatography and thin layer chromatography after dinitrophenylation of the reaction product.

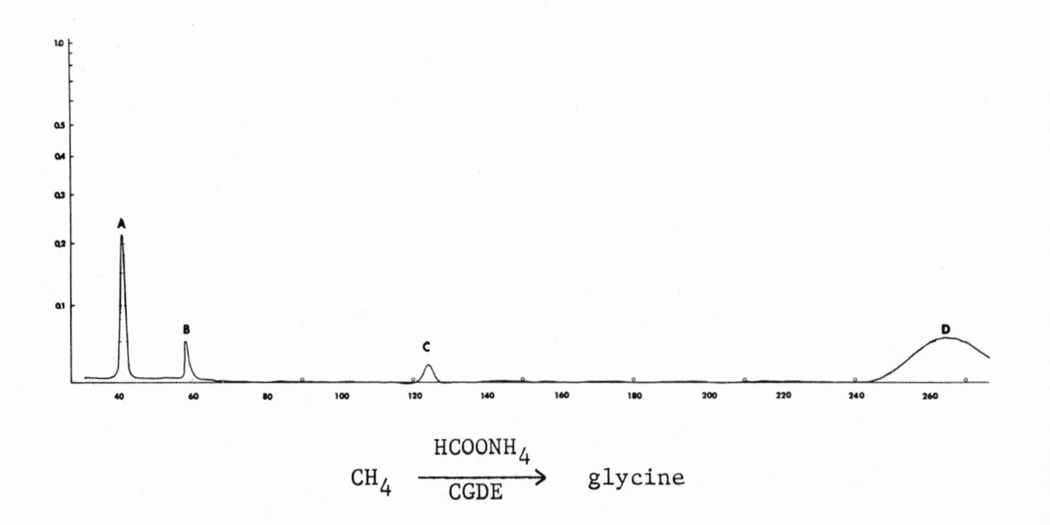

$$CH_4 \xrightarrow[\text{CGDE}]{HCOONH_4} \text{glycine}$$

A,B, unknown; C, glycine; D, unknown

Figure 17. Ammonium formate (0.01 mole) was dissolved in 10 ml of 10% aqueous ammonia. To this solution, methane was introduced slowly and CGDE was carried out on the surface of the aqueous layer. The reaction conditions were as follows: electric current, 50-80 mA; reaction temperature, 28°C; reaction time, 2 hrs. 30 min. Amino acid analysis of the product showed four peaks. Two peaks were very acidic ninhydrin-positive materials (peaks A and B). Formation of glycine (yield 0.01%) and a broad peak (peak D) in a rather basic area were observed.

SUMMARY

1. Amino acids were synthesized from aliphatic carboxylic acids by direct amination using CGDE in aqueous ammoniacal solution.

2. Amino acids were synthesized from aliphatic amines by direct carboxylation using CGDE in aqueous formic acid solution.

3. Hydrocarbons were also aminated and carboxylated simultaneously in aqueous solutions of ammonium formate using CGDE.

4. The yield of total amino acids is up to 13%. The yield could be higher if optimum conditions were used.

5. The energy expenditure for the formation of amino acids is much lower than that used in other prebiotic syntheses of amino acids (Table I).

6. Fragmentations of amino acids (or their precursors) during CGDE were observed. These are: (a) hydrolysis of the amide bond, (b) decarboxylation, (c) cleavage of the α-β bond of amino acids, (d) some cleavage of the β-α bond of amino acids, (e) possible hydroxylation of carbon compounds.

7. The amino acid products by CGDE are rather clean in composition and most of the major products could be explained by the regularity of fragmentation mentioned in 6.

8. The experimental results suggest that similar radical-type amination, carboxylation (or cyanide formation), and coupling reactions all take place in electric discharge experiments other than GDE.

9. The application of GDE to organic compounds is a new area of chemistry. GDE is also interesting as a possible method of synthesis of bioorganic compounds under prebiotic conditions.

ACKNOWLEDGMENTS

This work was supported by Grant no. NGR-10-007-052 of the National Aeronautics and Space Administration. The author wishes to express his thanks to Mr. Charles R. Windsor for amino acid analyses and to Mr. Tameo Iwasaki for valuable discussion. Contribution no. 268 of the Institute for Molecular and Cellular Evolution, University of Miami.

REFERENCES

1. Miller, S. L., Science 117, 528 (1953).
2. Miller, S. L., J. Am. Chem. Soc. 77, 2351 (1955).
3. Oró, J., and Kamat, S. S., Nature 190, 442 (1961).
4. Lowe, C. U., Rees, M. W., and Markham, R., Nature 199, 219 (1963).
5. Harada, K., Nature 214, 479 (1967).
6. Hasselstrom, T., Henry, M. C., and Murr, B., Science 125, 350 (1957).
7. Dose, K., and Ettre, K., Z. Naturforsch. 13b, 784 (1958).
8. Dose, K., and Risi, S., Z. Naturforsch. 23b, 581 (1968).
9. Deschreider, A. R., Nature 182, 528 (1958).
10. Cultera, R., and Ferrari, G., Ann. Chimica 47, 1321 (1957); 47, 1331 (1957); 48, 1410 (1958).
11. Mehran, A. R., and Pageau, R., Can. J. Biochem. 43, 1359 (1965).
12. Harada, K., Protein, Nucleic Acid and Enzyme 6, 65 (1961).
13. Hickling, A., and Ingram, M. D., J. Electroanal. Chem. 8, 65 (1964).
14. Hickling, A., in: "Modern Aspects of Electrochemistry"(Bockris, J. O., and Conway, B. E., eds.), Vol. 6, p. 329, Plenum Press, New York, 1971.
15. See literature cited, ref. 14.
16. Klemenc, A., Z. Elektrochem. 56, 694 (1953).
17. Gore, G. H., and Hickling, A., ref. 14, p. 352.
18. Brown, E. H., Wilhide, W. D., and Elmore, K. L., J. Org. Chem. 29, 5698 (1962).
19. Woodman, J. F., U.S. Patent 2,632,729 (1953).

20. Denaro, A. R., and Hough, K. D., Electrochim. Acta 18, 863 (1973).

21. Hickling, A., and Ingram, M. D., J. Chem. Soc. 783 (1964).

22. Sanger, F., Biochem. J. 39, 507 (1945).

23. Rao, K. R., and Sober, H. A., J. Am. Chem. Soc. 76, 1328 (1954).

24. Perrone, J. C., Nature 167, 513 (1951).

25. Court, A., Biochem. J. 58, 70 (1954).

26. Fox, S. W., and Dose, K., "Molecular Evolution and the Origin of Life," p. 90, W. H. Freeman and Co., San Francisco, 1972.

PREFIGURED ORDERING AND PROTOSELECTION IN THE ORIGIN OF LIFE

Dean H. Kenyon

Department of Cell and Molecular Biology
California State University
San Francisco, California 94132

INTRODUCTION

We are all indebted to Professor Oparin for reopening the origin of life problem in a scientific context (1,2). His writings were the original inspiration of what has become an impressive body of experimental research, especially in the last two decades. The immense value of Oparin's general conceptions on biogenesis is that they provide the overall theoretical framework in which specific experiments are conducted and interpreted. In recent years the view that life is an inevitable outcome of the properties of matter and energy, a view long held by Professor Oparin, has gained increasing support. The alternative view that life was the result of a lucky random combination of chemical substances was popular when the experimental data on origins were still scanty. However, this erroneous view has persisted in uninformed discussion of the subject in many biology textbooks and in the recent criticism of the chemical theory of origins by the new creationists (3). The new data, taken *in toto*, indicate that the origin was in some sense, which must be carefully spelled out, foreordained from the beginning (4).

Although considerable progress has been made in recent years, a large gap remains between the most complex protocells so far produced in the laboratory and the simplest living cells, i.e., the mycoplasma. Proteinoid microsystems (5), coacervate droplets (6), and ammonium thiocyanate microspheres (7,8) exhibit only the most rudimentary analogs of some of the characteristic properties of living protoplasm. Much additional work will be required in order to narrow the gap appreciably. We know much more about the possible earliest phases of cosmic evolution than we do about the transition from protocells to minimally alive units of self-replicating protoplasm. For the purposes of this discussion we assume that the threshold to the living state was crossed when the properties of evolving protocells reached the level of molecular complexity exhibited by present day free-living mycoplasma. On this view we are still a long way from the artificial synthesis of a living cell, assuming that this achievement is possible in principle, which has not been demonstrated.

In this paper we shall review the highlights of the experimental data which indicate the inherent tendency of cosmically abundant matter and energy to move toward the carbon-based living state that we know here on Earth. We shall discuss the use of the Darwinian concept of natural selection in attempting to account for the final phases of the origin of life and show that this concept in its classical form cannot be applied to the prebiological era without important qualifications. Implications of these ideas for a number of specific problems are then discussed.

PREFIGURED ORDERING IN CHEMICAL EVOLUTION

On the basis of a large body of investigations conducted mostly since Oparin's original writings we can depict the highlights of cosmic evolution by the diagram shown in Fig. 1. The diagram begins with the ultimate creation, that is, the bringing into being of just those categories of matter and energy and the universal laws describing their interactions that we find today, rather than conceivable alternatives. For this original endowment we can construct no empirically testable explanation. This is the ultimate cosmological problem and represents the purely historical dimension of the theological doctrine of _creatio ex nihilo_.

The striking parallels between the relative cosmic abundances of reactive elements (especially H, C, O, and N) and the elemental composition of living matter have been pointed out by many authors (5,9,10). The valences of carbon, oxygen, and nitrogen and the marked tendency of molecular hydrogen to escape from the surface of a condensing protoplanet easily account for the relative deficiency of hydrogen in living matter. If it can be shown that these most abundant reactive elements are uniquely suited for the living state, as Henderson (11), Wald (12), and Needham (9) have argued, then movement toward carbon-based life is discernible in the earliest stage of cosmic evolution as a favored direction.

The next stage of chemical evolution, the formation of the primitive gases (CH_4, NH_3, H_2O, CO, N_2, CO, etc.) can be regarded as a probable outcome of the combining properties of the cosmically abundant reactive elements. Many of these molecules as well as other

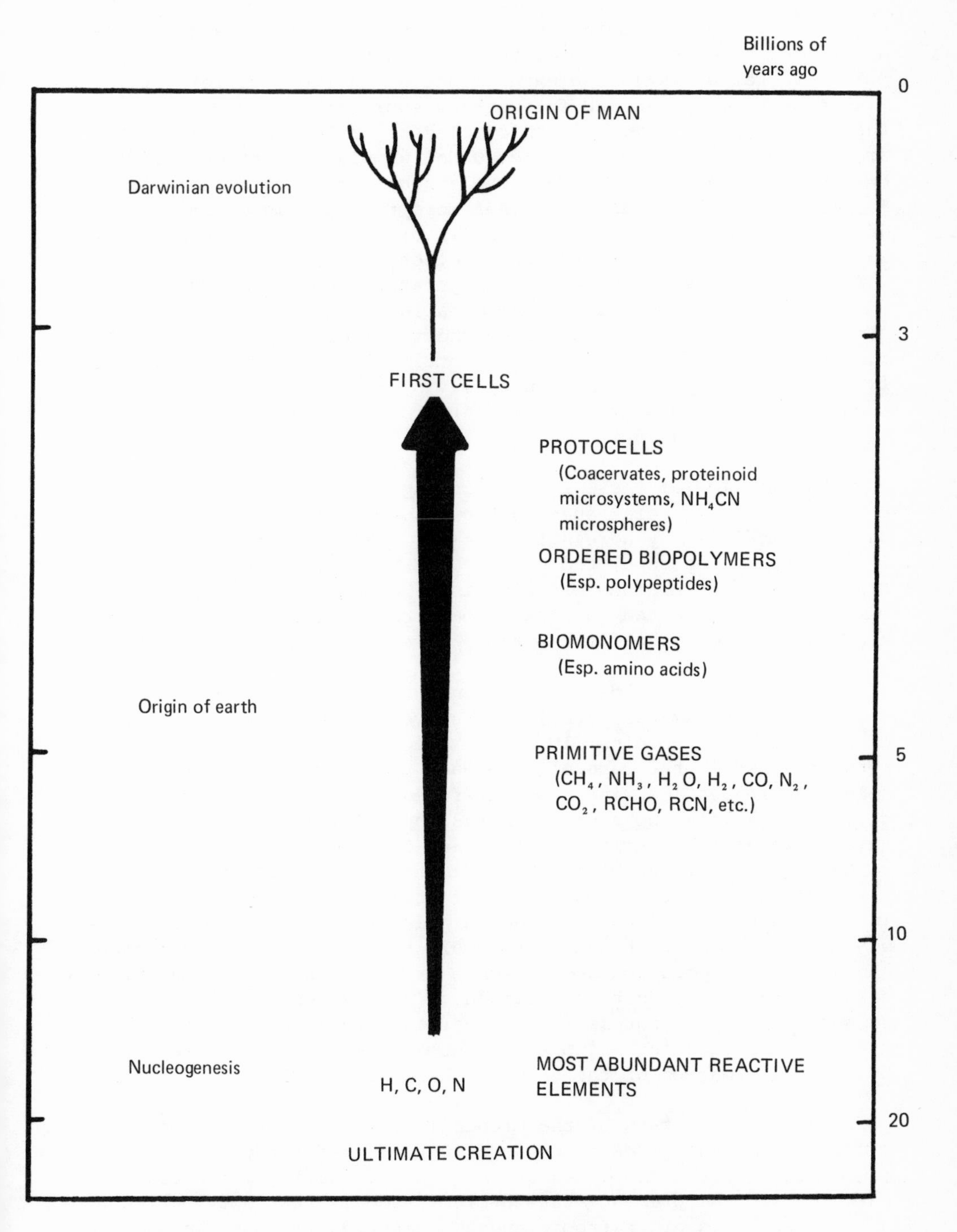

Figure 1. Major phases of cosmic evolution.

simple compounds including HCN, HCHO, cyanoacetylene, and methylacetylene, have been detected in deep space (13,14). Scores of primitive Earth simulation experiments employing a variety of initial gaseous mixtures and free energy sources have demonstrated that the types of compounds which play key roles in living matter are formed in appreciable yields under simple conditions (4,5,15). Most notable here are the relatively high yields of amino acids. For example, Miller has shown that more than 200 mg of amino acids are produced when a primitive gas mixture containing about 1 gm of initial methane is subjected to an electric discharge (16). Bar-Nun et al. found that in a high temperature shock tube 30% of the initial NH_3 is converted into amino acid product (17). Of course not all of the amino acids formed under simulated primitive Earth conditions occur in contemporary proteins (18). Nevertheless, the four most abundant amino acids of living matter, glycine, alanine, aspartic acid, and glutamic acid, are consistently formed in appreciable yields (4,5,15).

The meaning of these results is highlighted by the following consideration: several hundred thousand organic compounds have been characterized. Only a relatively small subset of these occur in life (several thousand) and yet many of them are formed preferentially in the simulation experiments. In spite of the non-biological compounds detected in these experiments a trend toward the living state long before the first life appeared is clearly indicated.

The relative yields of different amino acids in some of the simulation experiments parallels the average amino acid composition of contemporary proteins (19,20). Moreover, evidence has recently been obtained in the case of one model system indicating that the relative frequencies with which different free amino acids combine closely parallels the average nearest neighbor frequencies of amino acid residues in a series of proteins (21). Although more work needs to be done on this problem the preliminary data suggest that before life originated spontaneous sequence ordering of amino acid residues in growing polypeptides resulted in polymers having a general resemblance, at the nearest neighbor level, to proteins. The evidence for nonrandom incorporation and sequence ordering in the thermal condensation of amino acids to proteinoid is now extensive (5). These sequence ordering processes are governed largely by the stereochemistry of the reacting amino acids themselves and to a lesser degree by the conformation of the growing chain (5,21-23).

The above picture of the prebiotic origin of ordered polypeptides assumes that free amino acids were formed first and later combined via dehydration condensation. An alternative which cannot be ruled out in the present state of knowledge is that polypeptide precursor backbones arose from cyanogen compounds without the intermediate formation of free amino acids (24,25). These structures would then gradually be converted to polypeptides upon hydrolysis, possibly within ammonium cyanide microspheres (26). This alternative route to biopolymers has been largely neglected because of the great complexity of cyanogen chemistry but merits detailed investigation because of the presumed key roles of nitriles at several levels of prebiotic chemical evolution (4). Nevertheless, by both routes the primordial appearance of ordered biopolymers would appear to have been highly probable.

The next presumed major phase of biogenesis involved the spontaneous formation of morphological units of microscopic dimensions. The ease with which such protocell units arise under possible primitive Earth conditions has been abundantly documented, especially in the elegant experiments of Sidney Fox and his collaborators on the proteinoid microsystems. The remarkable array of properties of these model protocells has been detailed elsewhere (5). For our purposes it is sufficient to note that preformed primitive polypeptides (proteinoids) have properties enabling them to aggregate spontaneously to form remarkably uniform spherical units of bacterial dimensions which contain complex internal morphology including a double wall, exchange materials with the ambient medium, grow, cleave in two, fuse, exhibit weak catalytic activity, and move when ATP is added to the medium. Protocells containing both proteinoid and polynucleotide have even been shown to carry on a primitive kind of protocoding activity (27,28). The proteinoid microsystem is a compelling laboratory model for the high-probability prebiotic origin of discrete individual units of evolving organic matter which could conceivably compete with one another and thus provide the basis for a primitive selection process.

Alternative models for the actual historical precursors of cells include Oparin's coacervates(2,6) and the microscopic structures which arise spontaneously in aqueous solutions of ammonium cyanide (26) and ammonium thiocyanate (7,8). These latter two largely neglected systems deserve extensive chemical study. A simple transformation between proteinoid microsystems and coacervate-like structures has been demonstrated,thus blunting the criticism that Oparin's coacervates are not suitable protocell models since they are formed from highly evolved proteins (29). Whichever model or combination of models best depicts the protocell stage the conclusion that such units arose with high probability before life is unassailable. In fact it has been argued that they were an inevitable outcome of the properties of materials abundant on the surface of the primitive Earth (30, 31).

We have now nearly reached the limits of the available laboratory information of the origin of life. The tantalizing conclusion suggested by the above survey of experimental data is that in every phase of cosmic evolution from the origin of the elements to the appearance of protocells there is a discernible preferential movement toward the carbon-based living state. Several writers have argued on theoretical grounds that trends in chemical evolution may be rooted in one or another of the chemical and physical properties of carbon compounds. For example, Pullman suggests that unsaturated compounds because of their high resonance energies and highly mobile π-electron clouds would be expected to arise and persist before the appearance of life and would also be uniquely suited to their present major roles in biochemistry (23). Black has presented the interesting argument that thermodynamic factors governing the prebiotic synthesis of hydrophobic and hydrophilic portions of polypeptides were the principal driving forces of biogenesis (32,33).

Henderson argued early in this century that water and carbon dioxide are maximally and uniquely suited for the living state in virtually every one of their chemical and physical properties (11). The values of many of the physical properties of water are higher than those of any other known substance. Since water is formed from the two most abundant reactive elements in the cosmos it must have been present in vast quantities long before the first life appeared. These and similar considerations led Henderson to conclude that the universe is biocentric, by which he meant that "The properties of matter and the course of cosmic evolution are ... intimately related to the structure of the living being and to its activities ..." (11, p.312). Updated, more detailed, but essentially similar conclusions were reached by Wald (12) and Needham (9). These arguments are certainly consistent with, and in fact are greatly strengthened by the new experimental data on origins. Of course one could argue that emerging life simply made use of substances that were available. Since water was abundant at the surface of prebiotic Earth it is not surprising that this substance should serve as both the external and internal milieux of cells as well as a key reactant in biochemistry. But this begs the question. Water is not only common in the cosmos. It is also the best of all known substances for supporting the living state. That is the enigma.

It is perhaps permissible to introduce two new terms at this stage. To denote the concept that the evolution of matter at every stage "anticipates" the final outcome, i.e., carbon-based biochemistry, we propose the term lookingaheadness. In spite of its obvious anthropomorphic overtones this term accurately epitomizes our interpretation of the empirical data of biogenesis. Secondly, the living state exhibits a deep rootedness in the cosmos extending all the way back to the ultimate creation of matter and energy.

PROTOSELECTION IN THE DEVELOPMENT OF CELLS FROM PROTOCELLS

Oparin has described in detail a plausible means by which a prebiological version of natural selection operating in a collection of competing protocells could have resulted in the origin of the first self-replicating cells (30). Other investigators have utilized this general mode of thought in discussing the difficult areas of biogenesis (13,34-37). While these arguments have provided some valuable insights, they are not convincing. Some essential ingredient is missing. The problem is not that a strict interpretation of any biological definition of natural selection, such as "the differential reproduction of genotypes in a population of living organisms," would not allow the application of the concept to biogenesis. Obviously, in prebiological assemblages of organic matter including protocells there were neither genotypes nor reproduction. While there were populations of similar looking units, there were no living organisms.

In the protocell stage of biogenesis some form of selection undoubtedly took place. We can reasonably assume variability in chemical composition, structure, and reactions from one protocell to the next (keeping in mind the remarkable uniformity of proteinoid microsystems). Competition among these units for raw materials can readily be visualized. Protocells which acquired the ability to take in large quantities of materials from the ambient hydrosphere and transform them into their own substance would have increasingly dominated the total mass of protocell units and in this way were selected (30). Thus we have a crude prebiological analog to natural selection which we can term protoselection. The difficulties inherent in this concept of selection can be seen by the following considerations. It is assumed that each of the properties necessary for movement of evolving protocells toward the living state arose by competition among essentially randomly varying individual units. But what guarantee do we have that any protocell could in fact have acquired some degree of enhancement of the properties necessary for life? It is assumed that this must have occurred in order for biogenesis to take place, but convincing experimental demonstrations are lacking.

This problem is crucial in what is perhaps the most difficult aspect of biogenesis, i.e., the origin of self-duplication. Professor Oparin writes, "Thus, when the original systems were extending quickly and massively, the only ones selected for further evolution were those in which the reactions of the networks were coordinated in such a way that there developed constantly repeated chains . . ." (30, p. 87). We are accustomed to regarding natural selection as a process which ensures the survival of more highly fit variants from a collection of essentially randomly produced variants. The random character of mutations, the long-term source of variations, is emphasized in spite of the qualifications that some genes are more mutable than others and that any given purine or pyrimidine has only a limited number of possible mutant forms. (We do assert that there is no correlation between the probability of a given mutation and the need for that mutation to enhance fitness in a given natural environment.) When the idea is transferred to the prebiological period it is assumed that variations among individual protocells will range over a spectrum sufficiently wide to guarantee that among the variants there will always be some with enhanced life-like properties (given the long spans of time available). Protocells bearing these properties will become increasingly dominant in the primitive scenario precisely because the properties of life by their nature ensure survival and propagation. Again in spite of the interesting features observed in the model protocells so far studied, we do not yet have convincing evidence that such properties can undergo spontaneously the very considerable enrichment required for development into living protoplasm.

In the earliest phases of prebiotic chemical evolution, "selection" is nothing other than the expressions of tendencies which inhere in the starting materials themselves interacting with energy and the general environment, itself determined by general laws. The concept of selection is really not applicable here at all. *Prefigured ordering* is a better term. As cosmic evolution moved closer to biogenesis, selection in an increasingly biological sense must have matured. The protocell stage is crucial here since variability among individual units allowed for some form of protocompetition. However, internal constraints rooted in the properties of the primitive polymers as well as those emerging at the level of whole protocells must have been at least as important as protoselection in the transition from protocells to cells.

There is no reason to believe that the operation of internal directing constraints which played so prominent a role in the early phases of chemical evolution should have ceased to exert their influence in the later stages of biogenesis. I contend that these constraints must have sharply limited the range of possible variations in composition, structure and reactions in protocells, but that among the allowed (and possibly favored) possibilities were those on the main road to life. That is, protocells which contained analogs of many of the essential aspects of cells emerged with high probability from probable precursor states. The "search" was not random over a wide spectrum of possibilities but confined within a narrow envelope. Protoselection at this stage essentially refined the broad outlines preset by the nature of the most abundant reactive elements in the cosmos. If this analysis is correct, there is no reason why further research should not readily narrow the gap between protocells and living organisms.

IMPLICATIONS

The arguments developed in this paper have important implications for a number of biological and philosophical problems. A detailed analysis is inappropriate here but some comments are in order.

1. Monophyletic versus polyphyletic origins: it is widely believed that all present and fossil forms of life descended from a single primordial population of microorganisms although the idea is not universally accepted (38). According to this view the origin of cells from protocells occurred only once, or if several times, one line quickly gained the ascendency and gave rise to all subsequent forms of life. A corollary view is that once the primordial population was established life did not arise again *de novo*. Stated another way, neobiogenesis is impossible under conditions prevailing on the surface of the Earth after the origin.

The evidence for the monophyletic theory of biogenesis is impressive. It is based on the Darwinian picture of evolution with all its supporting lines of evidence, especially the unity of biochemistry. How could multiple independent origins of life result each time in the same basic patterns of biochemistry? It is difficult to answer this question in the context of a simple extrapolation of Darwinian natural selection into the prebiotic era. But if the general thesis of this paper is correct, namely, that the most abundant reactive matter in the cosmos moves spontaneously toward the living state in a way which involves the favored production of substances uniquely suited for life, i.e., there is a universal prebiological chemistry, then the possibilities of polyphyletic origins and of neobiogenesis must be seriously reconsidered (4, 38-40).

However, two serious difficulties remain: (a) the genetic code is universal (41). Many argue that it is also arbitrary in the sense that there are no discernible physical or chemical reasons for the assignment of particular codons (or anticodons) to the corresponding amino acids (13,41,42). If the code is in fact arbitrary in this sense it is very difficult to see how the same code could have arisen on several occasions independently. Others argue that there is a systematic relationship between codons and amino acids based on chemical properties (5,27,28,41-46). In this connection the recent work of Fox *et al.* on coding interactions in proteinoid microsystems is especially interesting and supports the possibility of polyphyletic origins (27,28). The second difficulty (b) involves the origin of the optical isomer preferences in present life. In spite of many attempts to discover a lawful basis for these preferences (e.g., L-amino acids), no such basis has been demonstrated convincingly (47). If the isomer preferences are arbitrary, then polyphyletic origins of Earth life would appear to be ruled out.

2. Extraterrestrial life: the data on biogenesis strongly support the view that life is widespread in the cosmos and that wherever it occurs it is based on essentially the same biochemistry we know on Earth, i.e., there is a universal cosmic biochemistry (12). This is a working assumption in the design of devices to detect signs of extraterrestrial life (48). But until unequivocal signs of an exobiology are found we should withhold judgment on this matter especially since the problem of biogenesis on Earth is far from solved. Although the weight of our preliminary evidence is against it, it may be the case that ours is the only biosphere in the cosmos, in which case we would be faced with an extraordinary problem: how was the strong drift toward life prevented everywhere else in the cosmos except on Earth?

3. Neo-Darwinism: we have seen that the concept of natural selection must be employed cautiously in interpreting the emergence of cells from protocells and modified to take into adequate account the major role of prefigured ordering. But there is no reason to suspect that after the origin internal guiding constraints did not continue to play a role in determining the main directions of evolutionary change. The role of internal limiting factors in evolution has been pointed out by several authors (49,50). In Henderson's thesis natural selection is not a sufficient principle for explaining the course of evolution. Equally significant are the limitations imposed by the chemical and physical properties of living matter (11). I suspect that the new data on origins will provide the basis for major revisions in the foundations of Neo-Darwinism. If the evolution of Earth life were to be rerun, a somewhat different phylogeny would probably result but the outlines would not be much different. Whether or not there would be any analog to the looking-aheadness discernible in the origin would depend on how closely we wish to specify the details of the emerging species.

4. Teleology revisited: teleologic thought has long been considered unacceptable in biology although some authors have argued convincingly that this proscription is more apparent than real in the actual conduct of research (51,52). It is not my goal here to analyze the use of teleological concepts in biological research but rather to point out that the new data on origins are consistent with the idea of a goal-oriented process of cosmic evolution. A classical example of a process that was considered to be teleological is embryogenesis. The end result of the process, the newborn organism, was considered to be the "first cause" of all the intervening events starting with the fertilized egg. Each step was thought to point toward the final result. We would say today that the newborn organism inheres in the chemical properties of the starting material,

namely the hereditary substance. This is similar to the conception that life inheres in the properties of the starting materials, the most abundant reactive elements in the cosmos, i.e., that the universe is biocentric. Why then do we hesitate to say that the goal of cosmic evolution is carbon-based life? Part of the reason, I submit, is that full acceptance of this conclusion would bring us face to face with an impenetrable mystery, namely the original source of the perdurable categories of matter and energy and the universal principles governing their interactions, which themselves are inexplicable in strictly scientific terms. They are the ultimate givens and they seem to have an arbitrary, nondeducible character.

CONCLUSIONS

The recent lines of laboratory and theoretical investigation inspired by Professor Oparin's brilliant reopening of the scientific problem of biogenesis have resulted in a body of new data and conceptions with far-reaching implications for the whole structure of biology. Perhaps the most significant of these results concerns the attitude and research strategy of many of the investigators working on the origin of life problem. The long reductionist phase of biological investigation which began with the chemical studies of living matter in the early nineteenth century and had gained momentum ever since culminating in the recent elucidation of molecular events accompanying hereditary phenomena, now has a powerful corrective counterbalance. This counterbalance is the "constructionist" research strategy in which laboratory events move spontaneously in the same direction as cosmic evolution itself toward the formation of functioning wholes of increasing intricacy and beauty (53). The spirit of this mode of inquiry has much in common with the dawning global ecological consciousness, which very likely is the most advanced terrestrial phase of cosmic evolution.

Some may contend that the argument of the present paper goes inexcusably beyond the available laboratory information. But it is precisely the extraordinary way in which the new data on origins point irresistibly beyond themselves toward the partial resolution of long-standing problems of both scientific and philosophic interest that makes the biogenesis problem especially fascinating to this investigator.

It was Professor Oparin's profound insight that even an exhaustive description of the molecular events which occur in living cells would not be sufficient for understanding the material basis of life (2,30). It is essential also to consider the historical origin and development of living matter. It now seems apparent that even an exhaustive molecular description of the origin of life from the earliest phases of cosmic evolution, while it would certainly deepen our insights into living nature, would not be thoroughly satisfying. It would not account for the fundamental categories of matter and energy and the perdurable principles governing their interactions.

REFERENCES

1. Oparin, A.I., "Proischogdenie zhizni," Moscovsky Robotchii, Moscow, 1924.

2. Oparin, A.I., "The Origin of Life", Macmillan, New York, 1938.

3. Transcripts of the public hearing of the California State Board of Education, November 9, 1972, Sacramento, California.

4. Kenyon, D.H., and Steinman, G., "Biochemical Predestination", McGraw-Hill, New York, 1969.

5. Fox, S.W., and Dose, K., "Molecular Evolution and the Origin of Life", W.H. Freeman, San Francisco, 1972.

6. Oparin, A.I., in "The Origins of Prebiological Systems", (Fox, S.W., ed.), p. 331, Academic Press, New York, 1965.

7. Steinman, G., Smith, A., and Silver, J., Science 159, 1108 (1968).

8. Smith, A.E., Steinman, G., and Galand, C., Experientia 25, 255 (1969).

9. Needham, A.E., "The Uniqueness of Biological Materials", Pergamon Press, Oxford, 1965.

10. Oro, J., Ann. N.Y. Acad. Sci. 108, 464 (1963)

11. Henderson, L.J., "The Fitness of the Environment", Macmillan, New York, 1913.

12. Wald, G., in "Horizons in Biochemistry" (Kasha, M., and Pullman, B., eds.), p.127, Academic Press, New York, 1962.

13. Miller, S.L., and Orgel, L.E., "The Origins of Life on the Earth", Prentice-Hall, Englewood Cliffs, 1974.

14. Snyder, L.E., and Buhl, D., Sky and Telescope 45, 156 (1973).

15. Lemmon, R.M., Chem. Revs. 70, 95 (1970).

16. Miller, S.L., J. Am. Chem. Soc. 77, 2351 (1955)

17. Bar-Nun, A., Bar-Nun, S., Bauer, S., and Sagan, C., Science 168, 470 (1970)

18. Lawless, J.G., and Boynton, C.D., Nature 343, 405 (1973).

19. Harada, K., and Fox, S.W., Nature 201, 335 (1964).

20. Vegotsky, A., and Fox, S.W., in "Comparative Biochemistry", Vol. IV (Florkin, M., and Mason, H.S., eds.), p. 185, Academic Press, New York, 1962.

21. Steinman, G., and Cole, M.N., Proc. Nat. Acad. Sci. U.S. 58, 735 (1967).

22. Pattee, H.H., in "The Origins of Prebiological Systems", (Fox, S.W., ed.), p.385, Academic Press, New York, 1965.

23. Pullman, B., in "Exobiology" (Ponnamperuma, C., ed.), p. 136, North-Holland, Amsterdam, 1972.

24. Matthews, C.N., and Moser, R.E., Nature 215, 1230 (1967).

25. Matthews, C.N., in "Chemical Evolution and the Origin of Life" (Buvet, R., and Ponnamperuma, C., eds.), p. 231, North Holland, Amsterdam, 1971.

26. Labadie, M., Cohere, G., and Brechenmacher, C., Bull. Soc. Chim. Biol. 49, 46 (1967).

27. Fox, S.W., Yuki, A., Waehneldt, T.V., and Lacey, Jr., J.C., in "Chemical Evolution and the Origin of Life" (Buvet, R., and Ponnamperuma, C., eds.), p. 252, North-Holland, Amsterdam, 1971.

28. Nakashima, T., and Fox, S.W., Proc. Nat. Acad. Sci. U.S. 69, 106, (1972).

29. Smith, A.E., and Bellware, F.T., Science 152, 362 (1966).

30. Oparin, A.I., "Life: Its Nature, Origin and Development", Academic Press, New York, 1961.

31. Smith, A.E., Galand, C., and Bahadur, K., Spaceflight 11, 325 (1969).

32. Black, S., Biochem. Biophys. Res. Commun. 43, 267 (1971).

33. Black, S., Adv. Enzymol. 38, 193 (1973).

34. Hein, H.S., "On the Nature and Origin of Life", McGraw-Hill, New York, 1971.

35. Bernal, J.D., "The Origin of Life", World, Cleveland, 1967.

36. Orgel, L.E., "The Originsof Life", Wiley, New York, 1973.

37. Ehrlich, P.R., and Holm, R.W., "The Process of Evolution", McGraw-Hill, New York, 1963.

38. Kerkut, G., "Implications of Evolution", Pergamon, Oxford,1960.

39. Keosian, J., "The Origin of Life", 2nd Ed., Reinhold, New York, 1968.

40. Smith, A.E., and Kenyon, D.H., Persp. Biol. Med. 15, 529 (1972).

41. Ycas, M., "The Biological Code", North-Holland, Amsterdam, 1969.

42. Crick, F.H.C., J. Mol. Biol. 38, 367 (1968).

43. Woese, C., "The Genetic Code", Harper and Row, New York, 1967.

44. Woese, C., Proc. Nat. Acad. Sci. U.S. 59, 110 (1968).

45. Lacey, J.C., Jr., and Pruitt, K.M., Nature 223, 799 (1969).

46. Gavaudau, P., in "Chemical Evolution and the Origin of Life" (Buvet, R., and Ponnamperuma, C., eds.), p. 432, North-Holland, Amsterdam, 1971.

47. Bonner, W.A., in "Exobiology" (Ponnamperuma, C., ed.), p. 170, North-Holland, Amsterdam, 1972.

48. Ponnamperuma, C., and Klein, H.P., Quart. Revs. Biol. 45, 235 (1970).

49. von Bertalanffy, L., "Problems of Life", Wiley, New York, 1952.

50. Whyte, L.L., "Internal Factors in Evolution", Braziller, New York, 1965.

51. Polanyi, M., "Personal Knowledge", University of Chicago, Chicago, 1958.

52. Selye, H., "The Stress of Life", McGraw-Hill, New York, 1956.

53. Fox, S.W., Naturwissenschaften 60, 359 (1973).

LIFE'S BEGINNINGS - ORIGIN OR EVOLUTION?

John Keosian

Marine Biological Laboratory

Woods Hole, Massachusetts 02543, USA

A fundamental question is whether inanimate matter gave rise to life more-or-less suddenly, or whether life gradually evolved through stages from inanimate matter. In the first case, a precise definition of life is of paramount importance for it becomes both the experimental goal and the identifying criterion for recognizing life produced experimentally. In the second case, neither definitions nor criteria of life are of particular importance for there is no longer any single goal. Rather, the objective should be a quest, first, for spontaneous formation of multimolecular systems, and secondly, a study of their properties, and finally, the determination of the physical and chemical mechanisms involved in driving each successive stage to a higher level of organization. In such a progression, it becomes meaningless to draw a line between two levels of organization and to designate all systems below that line as inanimate and all systems above as living. The meaninglessness of the terms life and living on other grounds was long ago pointed out by Pirie (23).

Today, the search for origins has narrowed to two main lines of approach. In one, the gene is held to be the basis of life and its origin. In the other, microsystems variously termed microdroplets, coacervates, microspheres, or prebiological systems are considered to be the precursors of the first primitive cells. Each approach, however, assumes that the first living things, genes or cells, arose more-or-less suddenly from their environment. The background and shortcomings of each approach are discussed below.

The gene theory was inspired by Troland's "living enzyme" theory (26-28). His ideas were remarkable for those times for Troland, a Harvard physicist, took a strong stand against his contemporary biologists who had embraced a vitalistic (supernatural) outlook on the subject of life. Troland wrote in part:

"It is the purpose of the present paper to combat the thesis of the new vitalism by showing how a single physico-chemical conception may be employed in the rational explanation of the very life phenomena which the neo-vitalists regard as inexplicable on any but mystical grounds."

Troland put the "single physico-chemical conception" in these words:

"Let us suppose that at a certain moment in earth history, when the oceans are yet warm, there suddenly appears at a definite point in the oceanic body a small amount of a certain catalyzer or enzyme... The original enzyme was the outcome of a chemical reaction, that is to say, it must have depended on the collision and combination of separate atoms or molecules, and it is a fact well known among physicists and chemists that the occurrence and specific nature of such collisions can be predicted only by the use of the so-called laws of chance... Consequently we are forced to say that the production of the original life enzyme was a chance event... The striking fact that the enzymic theory of life's origin, as we have outlined it, necessitates the production of only a single molecule of the original catalyst, renders the objection of improbability almost absurd... and when one of these enzymes first appeared, bare of all body, in the aboriginal seas it followed as a consequence of its characteristic regulative nature that the phenomenon of life came too."

Troland considered the normal function of enzymes to be heterocatalysis (metabolic regulation). For this hypothetical "life enzyme" he proposed the additional property of autocatalysis (self-replication). He further proposed that these two properties together constituted the minimum criteria of life. Anything embodying these two properties could be considered alive. He looked upon the "life enzyme", a protein, as a living macromolecule, eventually known as a "moleculobiont."

Troland's theory has serious flaws in common with the derivative gene theory. The first detailed statement of the gene theory of life, obviously based on Troland's enzyme theory, was made by Muller (14). Muller substituted the "naked gene" for Troland's "enzyme bare of all body" and to Troland's two criteria of life, autocatalysis and heterocatalysis, he added mutability. The gene theory, like Troland's enzyme theory, is untestable for the reason that a meaningful experiment cannot be devised to test an event that depends on remote accidents. But an accidental combination of all of the necessary atoms or molecules to form a life enzyme, or of all of the necessary nucleotides to form a gene, is not a probable event because the synthesis of only a single molecule is postulated. It is an event of vanishing probability. Moreover, a life enzyme or a gene is alive only because the authors so argue.

By granting, for the sake of argument, the all-at-once appearance of a gene on the primitive Earth, what are the possibilities of its serving as the ancestor of all life? Ignoring all limitations, it would serve as a template for the synthesis of more of itself at an exponential rate; it would mutate and mutant genes would similarly increase in number and would themselves mutate; the original gene and each mutant variety would be responsible for the synthesis of a particular protein, and, finally, each protein would be specific for some structural or catalytic role. Trying to stay within the confines of these events, it is difficult to see what else would result beside the synthesis, in a short time, of an oceanful of the original gene, lesser quantities of various mutants, and amounts of the corresponding proteins. All of this assumes that the first gene had properties identical with those of modern genes, and overlooks the requirement for specific enzymes in the replication of DNA, as well as an apparatus, a variety of RNAs, and enzymes for the synthesis of proteins. Not insignificantly, the experimental evidence so far for the formation of polynucleotides under prebiological conditions is weak, molecular weights and yields are small, and properties in terms of the goal are lacking.

As long as very little was known about biochemistry, and nothing at all about prebiotic chemical evolution, one could fill the void with broad and logical sounding assertions. It is no longer permissible to dismiss the problem of how a single molecule, whether enzyme, hypothetical gene or coded DNA, can serve as the ancestor of all life.

The microsystems approach is the one taken by Oparin (15-17) and coworkers, Fox and coworkers (3,4), and others doing experiments on models of prebiological systems. These models are claimed to have many morphological and functional characteristics in common with contemporary cells, including primitive forms of multiplication (3). A serious drawback to considering such models as truly representative of prebiological systems is that they are formed and made to operate under present-day conditions and sometimes with biochemicals of contemporary origin. Also, once microsystems are constructed, they are subjected to various biochemicals and conditions calculated to induce activities known to be essential in some contemporary "primitive" cell. In such endeavors, modern cellular physiology and biochemistry strongly influence the investigators' experimentation. It is an open question whether identical models will form spontaneously and function similarly in solutions exposed to methane, ammonia, hydrogen cyanide, hydrogen sulfide and other gases of the primordial atmosphere with its high flux of short-wave ultraviolet rays and ionizing radiation. One can indeed argue that the present experimentation on microsystems is as much an attempt to demonstrate the possibility of neobiogenesis (continuing reorigin of life) as it is of biopoesis (the first origin of life).

The belief in an early and sudden appearance of life under prebiological conditions seems to be inspired more by an overriding desire to rush a living thing onto the scene than by the dictates of the conditions of the primordial Earth. Nevertheless, it is assumed that the first living things (presumed to be similar to present-day "primitive" heterotrophs) would be destroyed by those conditions, especially the high flux of short-wave ultraviolet rays. This difficulty is belittled by proposing that the first living things survived in caves or under depths of water not reached by the lethal rays, there to remain until more suitable conditions came about (millions or tens of millions of years?).

The need for such assumptions is strong evidence, it would seem, that what did originate under the harsh conditions were not living things, but a succession of microscopic physico-chemical systems of increasing complexity. The early members in that long line of systems did not need the full complement of biochemicals postulated for the "hot thin soup." The same reactants, reactions and forces which together produced organic compounds in the atmosphere and waters, did so, also, in the microsystems. Their chemistry was, in that sense, autotrophic, not heterotrophic. Further, each stage in that progression must be considered to have been viable under the conditions prevailing at the time of its appearance. Changes in the chemical and physical conditions of the environment impressed corresponding changes in the microsystems due to the relatively simple structure and chemistry of those early systems. Homeostasis is a property of complex systems possessing a multiplicity of alternate pathways, a condition not yet characteristic of the early microsystems. Adaptation of those systems to changing environmental conditions was the result of direct, non-Darwinian, interaction.

The chemistry of a microsystem confined by a semipermeable membrane would soon take on a character different from that in the vast environment that tends toward homogeneity. For one, products of reactions can accumulate beyond their solubility products and can precipitate out as solid phases,whereas this may never occur in the outside medium. Likewise, reactants confined in small spaces can soon reach their reaction levels, whereas their dissipation in the outside medium would prevent it. The formation of solid phases within the microsystems would favor the coupling of reactions especially important in the case of exergonic and endergonic reactions. The latter possibility is significant in the eventual changeover from physical energy outside the microsystem, to chemical energy inside it, for the synthesis of compounds. Under such circumstances, the further coupling of reactions into pathways might be favored.

The synthesis of biochemical compounds, and the development of coupled reactions, and of biochemical pathways, must be considered to have been a property of microsystems and not of the structureless medium.

Considering the primitive method of multiplication of microsystems, the retention of an increasingly complex structure and chemistry by successive generations must have been precarious. This would, in turn, have required a long period of nongenetic evolution of microsystems. Microfossils said to be two-and-a-half to three billion years old, may represent various stages in the evolution of microsystems. Structurally, the oldest microfossils have a closer resemblance to clusters of some of Fox's microspheres than to primitive algae. The nature of the strata in which those microfossils are found are said to be a characteristic of once-living algae. However, it may be maintained, as well, that such effects were produced by the chemical activities of the more complex microsystems of that era.

Mono-, di-, and oligo-nucleotides would have a ready role in the chemistry of increasingly complex microsystems. How, when, and in what role polynucleotides and nucleic acids became incorporated into that chemistry is hazardous to guess. But it would seem more logical to assume that nucleic acids were fashioned in the chemistry of microspheres as part of their evolving complexity, than to assume that nucleic acids arose independently and spontaneously, "destined" to take over the activities of microsystems. Further, being the result of the chemical activities of microsystems, nucleic acids would be expected to reflect that chemistry. As for coded nucleic acids, the sudden great proliferation of species, beginning about a billion years ago, is better explained by the appearance, at that time, of coded nucleic acids in already chemically complex microsystems, than by the hypothesis that it was due to the acquisition of oxidative metabolism by already well established heterotrophic cells.

The foregoing observations include reinterpretations of some prevalent views concerning life and its beginnings on Earth. In the following, some other views are examined and reinterpreted.

VITALISM AND THE SPONTANEOUS GENERATION QUESTION

It was once almost universally believed that there could never be a naturalistic explanation of the origin of life. That belief was based on two firmly held convictions; first, that only living things could synthesize organic compounds, and secondly, that the theory of spontaneous generation had been disproved beginning with the experiments of Redi (24) and ending with those of Pasteur (18-21).

The first conviction was shattered by the experiments of Miller (11,12) inspired by Oparin's hypothesis (15). The second conviction is still held by many, although experimenters on spontaneous generation had been few and their experiments were either very limited or inconclusive. The belief in spontaneous generation was nevertheless rendered untenable by advances in the biological sciences. The credit lies not with the few experimenters on spontaneous generation with their limited, inconclusive, and sometimes controversial results. Rather, it belongs to the numerous naturalists, anatomists, and physiologists of the seventeenth, eighteenth and nineteenth centuries, whose cumulative work on the anatomy, taxonomy, habitat, reproduction, life cycles, embryology and physiology of species after species of animals and plants shed light on the nature of life and its continuity and dispelled the more mystical beliefs. Thus was the concept of spontaneous generation abandoned, not disproved.

And, finally, Pasteur's conclusion regarding the meaning of his experiments: "Never will the doctrine of spontaneous generation recover from the mortal blow of this simple experiment" (29) was utterly unjustified given the limited scope and number of his experiments. Bastian (1) and Schäfer (25) pointed this out (evidently to closed minds) and Moore (13) succinctly observed:

"Life probably arose as a result of the operation of causes which may still be at work today causing life to arise afresh. Although Pasteur has conclusively proven that life did not originate in certain ways, that does not exclude the view that it arose in other ways. The problem is one that demands thought and experimental work, and is not an exploded chimera. Therein lies the value of Schäfer's contribution to the question, and it is a most refreshing and valuable one."

In more recent times, the problem was reviewed by Keosian (7-10) in the light of the present era of experimentation on the origin of life. The neobiogenesis controversy was examined in greatest detail in Keosian's 1964 reference (8, p.98). The point to be raised concerning this episode is not so much the question of the merits or demerits of the spontaneous generation and neobiogenesis hypotheses, but the possible fallibility of even universally held scientific views. This oft-mentioned but rarely observed caution is particularly pertinent today when, with the rapid accumulation of data, views all too soon attain the stagnating status of dogmas.

THE QUESTIONS OF UNIQUENESS

The conditions for the abiotic formation of organic compounds were special and occurred on Earth early in its history and never again. The conditions for the appearance of life are not as special. Primarily necessary is the presence of organic compounds including macromolecules, a condition which has existed even since those early times. But one must not make the mistake of thinking that each reorigin of life must be a replica of the events of the first instance. On the contrary, each reorigin would reflect the kind of chemical substances present and the nature of the physical conditions existing at the time of its origin.

CONCEPTS OF LIFE

The terms life and living were at first lay terms and in that context had uncertain boundaries. For centuries, science has been attempting to sharpen the lines; to this day attempts continue unsuccessfully. The reason is that any "universal" definition of life conceptually strips any organism to which it is applied, of functions - and therefore the structures carrying out those functions - not essential for the definition. Thus all living things are reduced by the limits of the definition from what they recognizably are, to the same idealized entity from which the definition is derived. This is indeed meaningless.

Matter driven by energy in an open system can go on to higher and higher levels of organization. The thing to bear in mind is that each level of organization has its own properties by which alone it can best be recognized. Also, each higher level, although incorporating structures and processes evolved at the lower levels, has new properties not predictable from the properties of the lower level. This is true of the whole progression from elementary particles through atoms and molecules to man. Each stage in that progression incorporates structures and processes of the lower level but emerges as a new stage with new properties and the propensity for arriving at a higher level of organization. Where does "life" fit in this progression? Nowhere. It makes little sense to attempt to squeeze into this hierarchy of stages a nebulous indefinable something called "life".

It would appear to be more realistic to approach the problem of life's beginnings not as an attempt to discover the precise point at which lifeless matter gave rise to the "first living thing," but rather as an examination of the physical and chemical mechanisms operating in the transitions of matter into higher and higher levels of organization. Then the first level of organization that can be considered "alive" will still be a matter of personal preference, but at least we will all be talking about the same things.

THE HOT DILUTE SOUP

Haldane (5) believed that conditions existed on the primitive Earth that converted the primordial oceans into a "hot dilute soup". Oparin (15) proposed a similar hypothesis which would account, he believed, for the formation of an abundance of organic compounds in prebiological times. Laboratory experiments testing that hypothesis do indeed yield a large variety of organic compounds and biochemicals. Two objections may be cited against accepting the hot dilute soup as representative of prebiological waters. For one, experiments are conducted in the confined space of laboratory vessels. The course of chemical reactions under such conditions, as within microsystems discussed previously, may differ greatly from that in the relatively limitless expanse of oceans. The term "simulated prebiological conditions" is not truly descriptive of such laboratory conditions. Pattee (22) suggested a setting measured in hundreds of cubic meters with sand and simulated tides as more in conformity with prebiological conditions.

Secondly, there seems to be no general agreement on what the term "prebiological conditions" signifies. The great variety of compounds claimed to have been synthesized under primitive Earth conditions can be accepted only if we ignore the fact that the list is an accumulation of data from experiments employing a variety of reactants under a variety of conditions, some being mutually exclusive. For example, an experiment designed to produce nucleic acids produces little else. On the other hand, experiments of the Miller type (11,12), while producing many organic compounds, produce only traces of bases and no nucleic acids.

THE HETEROTROPH HYPOTHESIS

The thinking and experimentation on the origin of life has been, and still is, strongly influenced, consciously or unconsciously, by the heterotroph hypothesis which requires the pre-existence of all of the organic and inorganic compounds necessary for the structure, metabolism and nourishment of the first living things. This hypothesis, developed in detail by Horowitz (6), claims that the first living things had to be completely heterotrophic, i.e., they depended entirely on the environment for all of the organic raw materials required by their metabolism. But the simplest heterotrophic cell is an intricate structural and metabolic unit of harmoniously coordinated parts and chemical pathways. Its spontaneous assembly out of the environment, granting the unlikely simultaneous presence together of all of the parts, is not a believable possibility. On the other hand, within prebiological microsystems, syntheses could conceivably be driven by ultraviolet rays acting upon the same reactants present in the outside medium. These systems were, in that sense, autotrophic. They were not only responsible for the synthesis of more and more complex organic compounds and "biochemicals," but on disintegration, they enriched the medium with these compounds. As mentioned earlier, the formation of complex organic compounds, of specific biochemicals, and of biochemical pathways was brought about within these autotrophic systems rather than in the structureless outside medium.

THE PROBABILITY OF LIFE ARISING FROM INANIMATE MATTER

In this connection the question that seems to make the least sense is this: "What is the probability *of the appearance of life* in a context in which its origin seems only remotely likely? That is the approach taken by Troland, Muller and most geneticists. But the question that makes more sense is this: "What is the probability *of the occurrence of a context* in which the appearance of life is highly to be expected?" It is proposed that an answer to the first question is academic because, from the accumulating evidence, the primitive Earth represented an answer to the second question. It was an environment in which the events leading to the formation of living things had a high probability of occurring.

THE EVOLUTION OF BIOCHEMISTRY

There is only one evolution - the evolution of matter from elementary particles through atoms and molecules to systems of higher and higher levels of organization. Each new level ushers in new properties which could not be predicted from the properties of the lower level. Bernal (2) put this in the following terms:

"In general, the pattern I propose is one of stages of increasing inner complexity, following one another in order of time, each including in itself structures and processes evolved at the lower levels. The division into stages is not in my opinion an arbitrary one. Although the evolution of life was continuous, for no stage could have been completely static, it cannot have been uniform. Discontinuities which occurred at later stages of evolution, such as the emergence of airbreathing forms, are likely to have been paralleled at the earlier biochemical stages at such jumps as the genesis of sugars, nucleic acids and fats. One of our major problems is to establish the correct order of the steps inferred from existing metabolism *as well as the postulating and checking of other steps which have been subsequently effaced by the success of more efficient biochemical mechanisms*."(Italics added).

The portion in italics cannot be emphasized too strongly. An emerging concept on the origin of life is that it is a part of the evolution of matter which takes place throughout the universe. The assumption that an almost complete biochemistry evolved, even to the level of biochemical pathways in the absence of living things, is inconsistent with this view. For if that were so, we would be saying that, while evolution brought about enormous morphological, functional and psychological developments, biochemistry--the chemistry of living things--changed but little because its fundamental patterns were set before life appeared.

REFERENCES

1. Bastian, H.C., "The Beginnings of Life", D. Appleton and Co., New York, 1872.
2. Bernal, J.D., "The Origin of Life on Earth" (Oparin, A.I. *et al.*, eds.), p. 38, Pergamon Press, New York, 1959.
3. Fox, S.W., *Pure Appl. Chem.* *34*, 641 (1973).
4. Fox, S.W., and Dose, K., "Molecular Evolution and the Origin of Life", p. 196, W.H. Freeman and Co., San Francisco, 1972.
5. Haldane, J.B.S., *Rationalist Annual*, 1928, reprinted in "Science and Human Life", Harper Brothers, New York, 1933.
6. Horowitz, N.H., *Proc. Nat. Acad. Sci. U.S.* *31*, 153 (1945).
7. Keosian, J., *Science* *131*, 479 (1960).
8. Keosian, J., "The Origin of Life", p. 98, Reinhold, New York, 1964.
9. Keosian, J., "The Origin of Life", 2nd ed., Van Nostrand Reinhold Company, New York, 1968.
10. Keosian, J., in "Molecular Evolution: Prebiological and Biological" (Rohlfing, D.L., and Oparin, A.I., eds.), p. 9, Plenum Press, New York, 1972.
11. Miller, S.L., *Science* *117*, 528 (1953).
12. Miller, S.L., *J. Am. Chem. Soc.* *77*, 2351 (1955).
13. Moore, B., "The Origin and Nature of Life", Henry Holt & Co., New York, 1912.
14. Muller, H.J., *Proc. Intern. Cong. Plant Physiol.* *1*, 897 (1929).

15. Oparin, A.I., "Vozniknovenie zhizni na zemle", Izd. AN ISSR, Moscow, 1936, English translation by Margulis, S., Macmillan Co., New York, 1938, reprinted, Dover Publications, Inc., New York, 1953.

16. Oparin, A.I., "The Origin of Life on Earth", Academic Press, New York, 1957.

17. Oparin, A.I., "The Chemical Origin of Life", Charles C Thomas, Springfield, Illinois, 1964.

18. Pasteur, L., Compt. rend. 50, 303 (1860).

19. Pasteur, L., Compt. rend. 50, 849 (1860).

20. Pasteur, L., Compt. rend. 51, 348 (1860).

21. Pasteur, L., Ann. Chim. Phys. 3, serie 64, 1 (1862).

22. Pattee, H.H., in "Molecular Evolution", Vol. 1 (Buvet, R., and Ponnamperuma, C., eds.), p. 42, North-Holland, Amsterdam (1971).

23. Pirie, N.W., in "Perspectives in Biochemistry" (Needham, J., and Green, D., eds.), The University Press, Cambridge, England, 1937.

24. Redi, F., "Esperienze Intorno alla Generatione degl' Insetti," Firenze, Italy, 1668.

25. Schäfer, E.A., Rept. Brit. Assoc. Adv. Sci. 3 (1912).

26. Troland, L.T., The Monist 24, 92 (1914).

27. Troland, L.T., Cleveland Med. J. 15, 377 (1916).

28. Troland, L.T., Am. Naturalist 51, 321 (1917).

29. Vallery-Radot, R., "The Life of Pasteur", Dover Publications, New York, 1960.

CHEMICAL EVOLUTION OF PHOTOSYNTHESIS:

MODELS AND HYPOTHESES

A. A. Krasnovsky

A. N. Bakh Institute of Biochemistry

Academy of Sciences of the USSR, Moscow

The life on our planet depends on photosynthesis of plants providing organic matter and oxygen to all the organisms living on the Earth. Now the question arises: does the origin of life on the Earth depend on photosynthesis too?

A. I. Oparin presented convincing arguments that photosynthesis was developed after the primary heterotrophs originated, their metabolism having been based on the use of organic matter of abiogenic origin (1).

The study of the pathways of biological evolution is based on the data of comparative biology of contemporary organisms and on paleontological information on the development of ancient species. Blue-green algae are usually regarded as the most ancient photosynthetic organisms.

A plausible hypothesis is that the blue-green alga and photosynthetic bacteria had a common precursor, a primitive autotrophic organism, which has not been found yet on the Earth or in the ancient rocks.

To trace the pathways of prebiotic evolution, experimental data derived from different sources of information are used (2,3).

Chemical analysis of ancient rocks and meteorites can provide data on the nature of primary carbonaceous matter. With success in space exploration, the samples from the planets will be available for chemical analysis. The spectroscopy of stars and interstellar space provides us with valuable information on the nature and transformations of organic and inorganic substances in the Universe.

Impressive data were obtained by simulating in the laboratory the reactions in models of the primary reducing atmosphere under the influence of ultraviolet, visible, and corpuscular radiation, radioactivity and electric discharges [see reviews (4,5)].

The model experiments thus sustained the hypothesis of the accumulation of highly active organic and inorganic substances in a primary ocean.

Diverse organic substances of abiogenic origin in the primary soup could chemically interact one with another in aqueous solution, being adsorbed by minerals or in structures of protobionts. These reactions could be accelerated in general by catalysis or photochemical activation. I wish to present here some model experiments which may be relevant to constructing a hypothetic picture of photochemical evolution.

CATALYSIS OR PHOTOCHEMISTRY?

In the case of an abundance of chemically active substances in the primary soup, the redox reactions would probably have determined the pattern of primary energetic metabolism.

As catalysts of these reactions, various inorganic components of the Earth's crust may have been involved, the catalytic action having been enhanced by complexing with organic ligands of abiogenic origin such as simple organic bases.

Along with components having catalytic properties, there were on the primary Earth inorganic and organic substances, such as abiogenic tetrapyrrole compounds, possessing photosensitizing activity.

At the first International Symposium on the Origin of Life (6) we stated the assumption that the photochemical activation of catalysts similar to contemporary coenzymes might be an intermediary metabolic step in light energy utilisation. The photochemical activity of flavin coenzymes was known at that time. Later we studied the photochemical activation of reduced pyridine nucleotides and analogs excited by near ultraviolet radiation (365 nm) (cf. 7). Photochemical activation enhanced NAD·H oxidation by electron acceptor molecules including viologens which have a more negative redox potential than pyridine nucleotides (8). In the latter case it was possible to store some light energy in the reaction products:

$$\text{NAD·H + Viologen} \longrightarrow \text{NAD + reduced Viologen}$$

We recently revealed catalytic action of quinones in cytochrome *c* autoreduction. Blue light absorbed by quinones greatly enhanced this reaction (9).

The superposition of catalytic and photosensitizing activity is pronounced in this case.

$$\text{electron donor} \xrightarrow{e} \overset{\text{blue light}}{\overset{\downarrow\ h\nu}{\text{quinones}}} \xrightarrow{e} \text{cytochrome } Fe^{+++}$$

The photochemical activity of cytochromes and other Fe-porphyrins proper is negligible as compared to the strong activity of magnesium complexes, the latter being able to effect photosensitization (to the red part of the spectrum) for the redox transformations of cytochromes.

The examples of a possible transition from a heterotrophic to an autotrophic mode of metabolism may be found in an extensive literature on the action of light on the metabolism of contemporary heterotrophs, for instance fungi (10).

These chlorophyll-deficient organisms usually contain photochemically active pigments--flavins and porphyrins--as intermediates in heme biosynthesis. In our laboratory were revealed Zn-porphyrins in various types of organisms using phosphorescence action spectra measurements in the cells (11). Finally, photochemically active reduced NAD and NADP are universal cell components.

We proposed (6) that in the course of evolution the photosensitizing and biocatalytic functions became specialized. A good example is to compare the properties of iron and magnesium porphyrins, the latter being effective photosensitizers inactive in the dark and extremely active when excited. On the contrary, the biocatalysts--iron porphyrin complexes--being extremely active in the dark, are practically inactive in light.

The significance of photosensitized substrate activation became more important in the course of exhaustion of active substances in the primary ocean.

It is generally considered that on the primary Earth there was no ozone layer in the upper parts of the atmosphere to absorb the shortwave ultraviolet radiation of the Sun. But the water layer absorbs shortwave ultraviolet too. So, in the case of reactions occurring in the ocean the primary photochemical reactions required sensitization to the longer wavelengths of the solar spectrum. In general, the primary photoreceptors-photosensitizers could have been either of inorganic or organic nature.

INORGANIC PHOTOSENSITIZERS

In the literature on chemical evolution, the use of inorganic photoreceptors has not yet been considered. But some inorganic substances that are components of the Earth's crust possess photosensitizing activity. They could play the role of primary photosensitizers. It is surprising that the oxides of titanium, zinc and tungsten possess high photosensitizing activity in redox reactions comparable with the activity of porphyrins and chlorophylls. These compounds are able to sensitize reactions with light energy storage in terminal stable products (12,13).

It was shown in our laboratory that titanium, zinc, and tungsten oxides under the action of ultraviolet radiation (365 nm) are able, in water media, to photosensitize oxygen evolution coupled with exogenous electron acceptor reduction (ferric ions, ferricyanide ions, p-benzoquinone were used).

The reaction follows the stoichiometry:

$$2\ Fe^{+++} + H_2O \longrightarrow 2\ Fe^{++} + 2\ H^+ + 1/2\ O_2 \quad ^*$$

The origin of oxygen from the water molecules was recently confirmed experimentally with heavy water labelled by oxygen: $H_2{}^{18}O$ (14). The quantum yield of the reactions studied (Table I) ran up to 1%.

Table I

Photosensitization of Redox Reactions by TiO_2, ZnO and WO_3 (365 nm)

Electron donor	Electron acceptor	Reaction products
H_2O	Fe^{+++}, $Fe(CN)_6{}^{+++}$	Oxygen, Fe^{++}, $Fe(CN)_6{}^{++}$
H_2O	p-Benzoquinone	Oxygen, hydroquinone
H_2O	Viologens	Reduced viologens,
Glycine	Redox-dyes	Dyes
Hydrazine		Oxidized electron
Thiourea, etc.		donor

*I doubt that oxygen evolved on the primitive Earth according to this equation; these experiments demonstrate only that inorganic photosensitizers may do the same work as porphyrins.

These experiments reproduce the Hill reaction or the photosystem II of photosynthesis. When oxygen is used as a final electron acceptor it is possible to reproduce the chloroplast Mehler reaction accompanied by hydrogen peroxide formation. Finally, inorganic photocatalysts are able to photosensitize viologens' reduction running up to the redox potential of the hydrogen electrode (15). So the action of photosystem I can also be simulated by using inorganic photocatalysts. The mechanism of photocatalytic reactions may be the following.

As a result of light quantum absorption by the crystalline lattice of the photocatalyst-semiconductor the electron transition from valence band to conduction band proceeds.

At the phase boundary, photocatalytic particles/aqueous solution molecules-electron donors and acceptors are absorbed on the electron-accepting and electron-donating active centers of the photocatalytic surface. The primary redox reaction is realized by interaction of electron donor with a hole and acceptor with electron.

The dismutation of primary radicals on the catalytic surface takes place, in the case of oxygen evolution--recombination of OH radicals leading to oxygen evolution (see scheme).

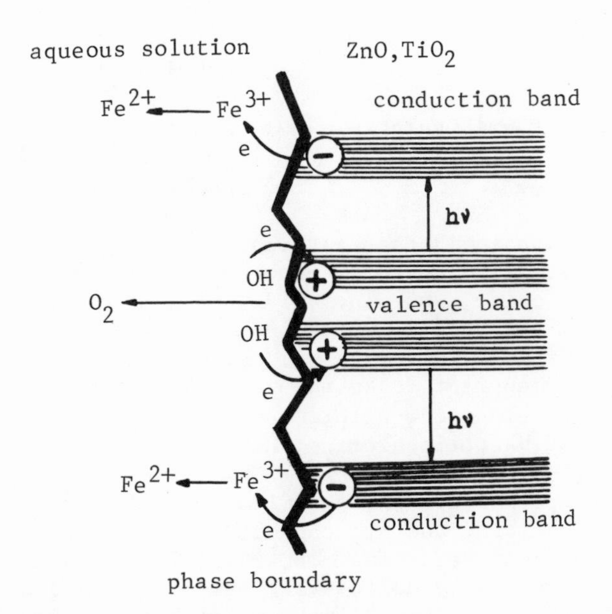

Figure 1. Visualized scheme of photocatalytic oxygen evolution on the photocatalyst surface.

These observations point to a feasible mode of involvement of inorganic photosensitizers into the structure of portobionts. Nevertheless, none of the existing organisms includes inorganic sensitizers in the patterns of their metabolism; those were probably "blind alleys" in the growth of the evolutionary tree.

In the course of evolution, most effective catalysts and photosensitizers had been developed by bonding inorganic ions to organic ligands, the porphyrin cycle being most widespread.

ABIOGENIC SYNTHESIS OF PORPHYRINS

Rothemund in 1936 (16) described the synthesis of porphin from pyrrole and formaldehyde. This reaction is thermodynamically spontaneous, proceeding slowly at room temperature in the dark in the presence of some catalysts. The presence of oxidants (oxygen, quinones) enhances the synthesis of porphin since the condensation of pyrrole and formaldehyde proceeds with abstraction of hydrogen.

In the solution of pyrrole and formaldehyde not only porphin but also more reduced compounds are formed: chlorin and bacteriochlorin (17). The synthesis and analysis of the most reduced substances requires anaerobiosis as these pigments are spontaneously oxidized in the presence of the oxygen of air. Chlorin may be formed from porphin not only by condensation of pyrrole and formaldehyde but via photoreduction of porphin too. The following scheme summarizes the processes listed above:

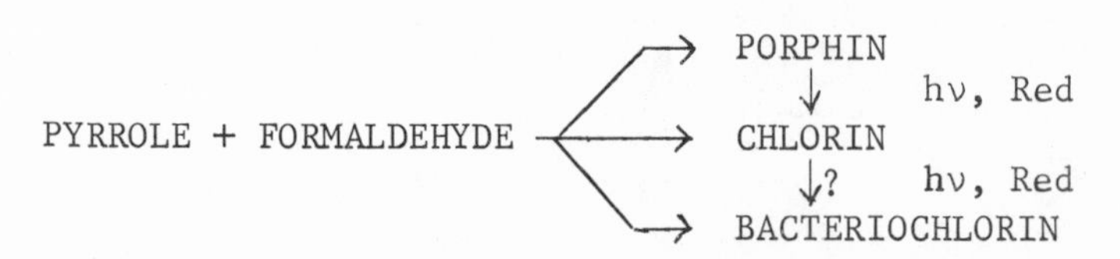

Our studies have shown that interaction between pyrrole and formaldehyde is catalyzed by some components of the Earth's crust, silica, titanium dioxide, and zinc oxide and some organic substances among them tryptophan, which is most active (17).

When analyzing ancient rocks and samples from planets it is necessary to search alongside with porphin, the more reduced pigments--chlorins and bacteriochlorins.

All these compounds are photochemically active; they possess photosensitizing activity and are capable of undergoing reversible photooxidation and photoreduction. However, in organisms free porphyrins are found mostly as intermediates of biosynthesis; among photosynthetic organisms magnesium complexes of porphyrins dominate and among catalysts, iron complexes dominate. The possibility of choosing the appropriate central metal atom in the porphyrin cycle in the process of evolution is illustrated by the following suggestion. The most widespread elements in the Earth's crust are Si (27.6%), Al (8.5%), Fe (5%), Ca (3.5%), Na (2.6%), K (2.5%), Mg (2%), Ti (0.6%).

The compounds of silicon and titanium are practically insoluble in water; this insulubility probably prevented the incorporation of these metals into porphyrin complexes. Complexes of Na, K and Ca with porphyrins are hydrolyzed in aqueous media. Only complexes of magnesium and iron are sufficiently stable to win in the competition. These complexes, having auxiliary coordination vacancies perpendicular to the plane of the porphyrin ring, are also capable of bounding reacting molecules. In the course of evolution, they made it possible to bind specific proteins.

The vanadyl and nickel porphyrins have been found in petroleum; they are probably formed as a result of chlorophyll degradation and central magnesium atom substitution.

The principal difference of magnesium porphyrin complexes from the ferric ones is that the first are active only after excitation by light quanta being converted into singlet or triplet excited states.

The studies in our laboratory have shown that excited porphyrins and their magnesium and zinc complexes are capable of reversible photochemical reduction or oxidation, that is, of accepting or donating an electron to the partner molecule (see scheme and review of ref. 18).

In the case of most longwave photosynthetic pigments, bacteriochlorophyll of Rhodopseudomonas viridis, the ability to reversible photooxidation is highly pronounced (19).

Reversible photoreduction: $P^* + Red \rightarrow {\cdot}P^- + {\cdot}Red^+$

Reversible photooxidation: $P^* + Ox \rightarrow {\cdot}P^+ + {\cdot}Ox^-$

In the triple system composed of molecules: electron donor (Red), electron acceptors (Ox) and pigment-excited molecules (P*) the photosensitization of the electron transport from Red molecules to Ox molecules proceeds at the expense of light energy absorbed by pigment molecules.

$$\text{Red} \xrightarrow{e} \text{excited pigment} \xrightarrow{e} \text{Ox}$$

Porphyrins can be incorporated into proteinoid structures as demonstrated in the experiments of Dose (20).

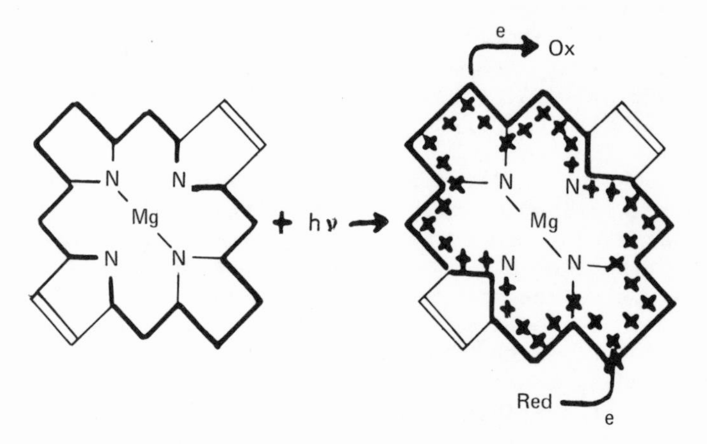

Figure 2. Excited (singlet or triplet) porphyrin molecule reversible interaction with electron donors (Red) and acceptors (Ox). To visualize: electron which is excited leaves an electron vacancy (+) behind.

In the case of more complicated protobionts having lipoproteinaceous membranes, the ability of porphyrins to incorporate the lipophilic tail into pigment molecules in this membrane is required. In the case of chlorophyll this tail is phytol, in the case of bacterioviridin, farnesol.

It should be noted that the lipophilic tail does not influence the photochemical properties of the pigment that are determined mainly by the porphyrin ring of the molecule.

EVOLUTION OF PIGMENTS AND EXCITATION ENERGY MIGRATION

The ancestors of photosynthetic organisms probably possessed small quantities of pigments directly participating in photochemical electron transfer as the effective pigment biosynthetic system was not yet developed. As far as the catalysts-enzymes are concerned they are active in very low concentrations; on the contrary, the quantity of pigment-photosensitizer must be very high to absorb effectively the Sun's radiation so as to win in the competition for survival.

In contemporary organisms, the energy of light quanta absorbed by aggregated pigments migrate to a minor part of the pigment functioning in the so-called "active center."

The development of metabolism in the course of biological evolution gradually led to the appearance of effective biosynthesis of chlorophylls, phycobilins and carotenoids.

When protobionts were formed, different types of pigment could enter into their structure. Various primary prophotobionts possessed different sets of pigments--carotenoids, linear or cyclic tetrapyrroles and various combinations of these pigments.

But only aggregated magnesium-porphyrins possessed the most longwave absorption band prerequisite to playing a role as a final trap of energy migration. Phycobilins have absorption bands in the shorter wavelengths of the spectrum, their high photochemical stability is favorable to their participation as an energy donor in the energy migration chain.

The organisms that survived acquired a combination of all types of pigment (chlorophylls, phycobilins and carotenoids), effectively absorbing most of the energy of the solar spectrum channelling the excitation energy to the longest wave trap--active centers. Such an organism is the blue-green alga.

It is remarkable that in the case of inorganic photosensitizers (TiO_2, ZnO) there exists a mechanism of energy migration from the bulk crystal lattice to the active center where the primary photoprocess is realized.

COUPLING OF PHOTOCHEMICAL AND CATALYTIC REACTIONS

Thus it is possible to suppose that a primary photocatalytic block of chemical and biological evolution comprised the ternary system: electron donor-pigment-electron acceptor. This system works efficiently not only in models but also in biologically organized structures (21).

The photochemical and catalytic electron transport chains were improved by diverse coupling of the primary photocatalytic blocks into the primary protobiont membranes. In the processes of biological evolution diverse combinations of photocatalytic blocks were realized. It is possible to trace several types of primitive photocatalytic electron transfer; we begin with noncyclic electron transfer. The excited pigment molecule reduces the activation energy and the overall reaction is thermodynamically possible ($-\Delta F$); here the photosensitizer acts as a photocatalyst.

$$\text{Red} + \text{Ox} + (\text{P}) \rightarrow \text{Red}_{\text{ox}} + \text{Ox}_{\text{red}} + (\text{P})$$

Another type of reaction which is more important for the use of an inactive hydrogen donor, that is the water molecule, is the overall process accompanied by increase of free energy of the system ($+ \Delta F$). This case can be reproduced in a model system--in solutions of chlorophyll; here pyridine nucleotides and other electron acceptors are photoreduced at the expense of ascorbic acid as an electron donor. This case was studied in our laboratory (22, etc.).

$$\text{Ascorbate} + \text{NAD} + (\text{CHL}) \xrightarrow{\text{red light}} \text{Ascorbate}_{\text{ox}} + \text{NAD}\cdot\text{H} + (\text{CHL})$$

The modelling of Hill's reaction photosensitized by inorganic semiconductors also proceeds with a gain of free energy (see above).

It is possible to presume a primitive cyclic electron transfer coupled with phosphorylation. In the cyclic processes, Mg or Fe porphyrin may participate; chlorophyll and porphyrins photosensitize effectively oxido-reductive transformations of cytochromes. But the coupling of such cyclic electron transfer with ATP formation has not yet been achieved in any model systems.

On the other hand, the formation of high energy phosphate esters in model reactions was described by Ponnamperuma, Wang, *et al*. In the experiments of Lohrmann and Orgel (23) the urea and inorganic phosphate mixture was an efficient phosphorylating agent.

It was found by Arnon (24) and Kamen (25) that ancient photosynthetic organisms (photosynthetic bacteria) possess both types of electron transfer, noncyclic and cyclic coupled with photophosphorylation. So, we may conclude that the primitive "bricks" of electron transfer may be used in biological evolution.

If the metabolism of contemporary photosynthetic organisms possesses rudiments of primary metabolic pathways, we must take into account their ability to heterotrophic existence and to use molecular hydrogen as an electron donor.

The use of water as an ultimate electron donor in more perfect photosynthetic organisms required coupling of two photoreactions for electron transport from water to pyridine nucleotides via ferredoxins (24), in other words, the coupling of two photocatalytic "bricks" which are denoted, according to contemporary ideas, as photosystems I and II.

The chemical and biological evolution from simple photocatalytic reactions to the perfectly organized systems of photosynthetic electron transfer required probably more than two billion years. The model experiments represent a plausible explanation of the various stages of chemical evolution; however, only future experimental data derived from geology and exobiology will probably reveal the real nature of the process.

SUMMARY

The primary metabolism of protobionts was probably based on the electron transfer reactions regulated by catalysts or photosensitizing pigments. The action of photoreceptive pigments was inevitable in the case of electron transfer leading to light energy storage in the reaction products. The primitive tetrapyrrolic pigments formed abiogenically (porphin, chlorin) as well as their more complicated biogenic analogs (chlorophylls) are capable of photosensitizing electron transfer in systems, having various degrees of molecular complexity. The inorganic photosensitizers (titanium dioxide, zinc oxide, etc.), being excited in the near ultraviolet, are able to perform the same reactions as porphyrins--electron transfer from donor to acceptor molecule (including photoreduction of viologens) or water molecule photooxidation (oxygen liberation), coupled with reduction of ferric compounds and quinones. The inorganic photosensitizers are not used in biological evolution; actually the inorganic ions entered into tetrapyrrolic cycle, forming effective photocatalysts. Inclusion of pigments into primary membranes led to elaborated coupling between pigments and enzymatic systems. The involvement of the excited pigments into the biocatalytic electron transfer chain was prerequisite to effective function of photosynthetic organisms.

REFERENCES

1. Oparin, A. I., "Origin of Life on Earth," Izd. AN SSSR, Moscow, 1957.

2. Gaffron, H., in: "Horizons in Biochemistry" (Kasha, M., and Pullmann, B., eds.), p. 59, Academic Press, New York, 1962.

3. Calvin, M., "Chemical Evolution," Clarendon Press, Oxford, 1969.

4. Pasynsky, A. G., and Pavlovskaya, T. E., Uspekhii khimii 33, 1198 (1964).

5. Ponnamperuma, C., Quart. Rev. Biophys. 4, 77 (1971).

6. Krasnovsky, A. A., in: "Origin of Life on Earth" (Oparin, A. I., *et al*., eds.), p. 606, Pergamon Press, New York, 1959.

7. Krasnovsky, A. A., and Brin, G. P., review in "Problems of Evolution and Technical Biochemistry" (Kretovich, V. L., ed.), p. 221, Izd. Nauka, Moscow, 1964.

8. Krasnovsky, A. A., and Brin, G. P., Dokl. AN SSSR 163, 761 (1965).

9. Krasnovsky, A. A., and Mikhailova, E. S., Dokl. AN SSSR 212, 237 (1973).

10. Kritzky, M. S., in this volume.

11. Shuvalov, V. A., and Krasnovsky, A. A., Molekularnaja biologia 5, 698 (1971).

12. Krasnovsky, A. A., and Brin, G. P., Dokl. AN SSSR 147, 654 (1962).

13. Krasnovsky, A. A., and Brin, G. P., review in "Molecular photonics," p. 161, Izd. Nauka, 1970.

14. Fomin, G. V., Brin, G. P., Genkin, M. V., Lubimova, A. K., Blumenfeld, L. A., and Krasnovsky, A. A., Dokl. AN SSSR 212, 424 (1973).

15. Krasnovsky, A. A., and Brin, G. P., Dokl. AN SSSR 213, 207 (1973).

16. Rothemund, P., J. Am. Chem. Soc. 58, 625 (1936).

17. Krasnovsky, A. A., and Umrikhina, A. V., in: "Molecular Evolution: Prebiological and Biological" (Rohlfing, D. L., and Oparin, A. I., eds.), p. 141, Plenum Press, New York, 1972.

18. Krasnovsky, A. A., Biophys. J. 12, 749 (1972).

19. Krasnovsky, A. A., Bokuchava, E. M., and Drozdova, N. N., Dokl. AN SSSR 211, 981 (1973).

20. Dose, K., and Zaki, L., in: "Chemical Evolution" (Buvet, R., and Ponnamperuma, C., eds.), Vol. 1, p. 263, North-Holland, Amsterdam, 1971.

21. Krasnovsky, A. A., in: "Chemical Evolution" (Buvet, R., and Ponnamperuma, C., ed.), Vol. 1, p. 279, North-Holland, Amsterdam, 1971.

22. Krasnovsky, A. A., and Brin, G. P., Dokl. AN SSSR 67, 325 (1949).

23. Lohrman, R., and Orgel, L. E., Science 171, 494 (1971).

24. Arnon, D. I., Proc. Nat. Acad. Sci. U.S. 68, 2883 (1971).

25. Kamen, M. D., "Primary Processes in Photosynthesis," Academic Press, New York, 1963.

COMPARATIVE STUDY OF PLANT ORGANISM GLUTAMINE SYNTHETASE

W. L. Kretovich and Z. G. Evstigneeva

A. N. Bakh Institute of Biochemistry

USSR Academy of Sciences, Moscow, USSR

In his work, "The Origin of Life and the Origin of Enzymes," A. I. Oparin (1) writes: "In the evolution process as a result of the natural selection numerous variants of polypeptides possessing a low catalytic activity, were discarded. Modern organisms, ranging from bacteria to animals, possess strong enzymes, the synthesis and activity of which are rigidly controlled. Modern biochemistry, however, can detect in these catalysts certain typical evolution features."

Glutamine synthetase (GS), one of the basic enzymes in nitrogen metabolism of animals, plants, and microorganisms, catalyzes the synthesis of glutamine.

The aim of this paper is to consider the results of the work carried out in our laboratory in comparative study of glutamine synthetase of higher plants, chlorella, and yeast.

The process of glutamine synthesis in plants and certain microorganisms is the most important assimilation route of ammonia and its transformation into a labile nitrogen compound, used easily in metabolism (2,3). Glutamine amide nitrogen is used for the synthesis of the most important cell metabolites: tryptophan, histidine, asparagine, glucosamine-6-phosphate, carbamoyl phosphate, CTP, AMP, CMP, NAD, aminobenzoate, α-ketoglutaramate, alanine, glycine, and a number of other amino acids (4).

Glutamine synthesis is produced with the participation of the glutamine synthetase [E.C.6.3.I.2] according to the following equation:

$$\text{L-glutamate} + NH_4^+ + \text{ATP} \xrightarrow{Me^{2+}} \text{L-glutamine} + \text{ATP} + P_i \quad (I)$$

Glutamine synthetase (GS), being the key enzyme in the synthesis of a number of nitrogenous metabolites, is rigidly controlled in the living cell.

The enzymes secreted from E. coli and some other bacteria, as well as the animal GS, have been at present studied better than anything else (4).

The properties of plant organism GS have been studied inadequately, though Elliott (5) isolated GS from pea seeds almost 20 years ago.

In our work, GS activity was determined by the amount of γ-glutamylhydroxamate (γ-GH), which is formed in the assay mixture instead of glutamine when NH_4^+ was replaced by NH_2OH, and also by the amount of P_i, produced in the synthetic reaction of both glutamine and γ-GH in quantities, equimolar to the amount of glutamine (6-8). The first method was used only during operations with unpurified enzyme. In order to make a standard curve, a preparation of the γ-glutamylhydroxamate was synthesized. The second method was utilized in experiments with purified GS preparations of peas, chlorella, and yeast.

Specific activity was expressed as μmoles P_i or γ-GH per milligram protein per min.

The purification of chlorella glutamine synthetase (Chlorella pyrenoidose, thermophyllic strain; 9) was done in the following ways: precipitation of nucleic acids with streptomycin sulphate, salting-out with ammonium sulphate, and chromatography on DE-cellulose 22. The pea GS (Pisum sativum) was purified, without the utilization of ammonium sulphate, by the reprecipitation of the enzyme with changes of pH and subsequent chromatography on DE-cellulose 22 (10). Several methods (11,12) were developed for the purification of yeast (Candida tropicalis) GS. The highest degree of purification was achieved by the adsorption on the $Al(OH)_3$ Cγ gel, precipitation by $(NH_4)_2SO_4$, and reprecipitation by changes of pH.

The GS preparations from mature pea seeds (10), chlorella (9), and yeast (11,12) were homogeneous. This is shown by polyacrylamide gel electrophoresis and sedimentation in a Spinco model E ultracentrifuge at 52,600 rev/min (Fig. 1). The sedimentation constant of pea GS=16.95; sedimentation coefficient of chlorella GS=15.0; molecular weight, calculated by the Halsall equation, constitutes 450,000.

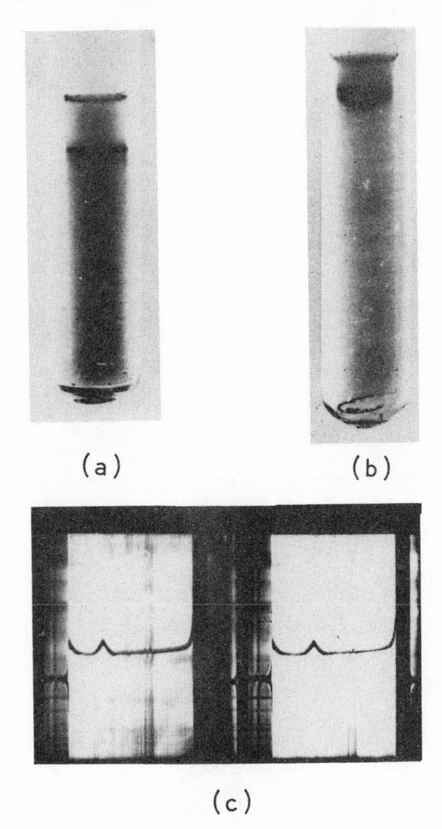

Figure 1. Electrophoretogram (a), zymogram (b), and sedimentogram (c) of glutamine synthetase for pea seeds.

It follows that, according to the molecular weight, pea and chlorella GS are closer to the GS or animals S_{20w}=15.0, rather than to that of bacteria S_{20w}=20.3 (13). Experiments with pea GS showed that during the dilution of enzyme solution and during freezing-defreezing, it is broken down to nonactive subunits.

SPECIFICITY AND KINETIC FEATURES OF PLANT ORGANISM GS

As shown by our experiments, certain properties of GS available in all of the objects studied--chlorella, plants, pea seeds, and yeast--make it similar to glutamine synthetase of animals and bacteria. GS of plant organisms catalyzes the synthetic reactions of glutamine (1) and γ-glutamylhydroxamate: L-glutamate + NH_2OH + ATP $\xrightarrow{Me^{2+}}$ γ-GH + ADP + P_i (II).

Glutamine synthetases studied by us are capable of catalyzing the synthetic reaction of D-glutamine, though only in the presence of Mg^{2+} (see Table I; 9,36). ATP, an essential component in the

Table I

Specificity of Plant Organism GS

(Specific activity of GS in the assay mixture: L-glutamate, NH_4, ATP, Mg^{2+} taken as 100%)

Enzyme source	D-Glutamate	NH_2OH	CTP	GTP	Mn^{2+}	Co^{2+}	Fe^{2+}
Peas	10.0	100	0	0	30.0*	12.1	2.1
Chlorella	6.0	74.2	0	0	15.0*	40.0	10.0
Yeast		97.5	85	80	42.5	35.0	4.7

*Under optimum conditions for Mn^{2+} activation.

reaction of glutamine synthesis, cannot be replaced by any other nucleoside triphosphate. GS of plant organisms is most active with Mg^{2+} as a cofactor, but shows certain activity with other divalent cations.

Obtained are similar data concerning the GS specificity for certain bacteria and animals (13). However, Bacillus licheniformis GS is active only with Mn^{2+}, while Bacillus subtilis GS shows equal activity both with Mg^{2+} and Mn^{2+}, and its activity with hydroxylamine (instead of ammonium) constitutes only 30% of that with NH_4^+ (15).

The pH optimum of plant organism GS as well as that of GS of bacteria and animals (16,17) depends on the cofactor; with Mg^{2+} the optimum activity is at pH close to 7, with Mn^{2+} it lies in the region of pH 5-6 (Fig. 2) (9-12,18).

Nonhyperbolic curves showing the dependence of the reaction rate on the substrate concentration and cofactors (19-22) indicate in some cases the GS specificity of the investigated organisms. Besides this, the pattern of the curves showing the dependence of the reaction rate on the concentration of glutamate and ammonium varies with the presence of Mg^{2+} and Mn^{2+}. Kinetic curves for GS of certain plant organisms are shown in Fig. 3. It should be pointed out that the GS of peas, chlorella, and yeast is characterized, as a rule, by a higher affinity of enzyme for ammonium, rather than for glutamate (Table II, Fig. 3). Lack of purified enzyme inhibition by ammonium either with Mg^{2+} or Mn^{2+} is a peculiarity of pea and yeast GS, while the nonpurified GS of these organisms is inhibited by NH_4. The chlorella GS is inhibited by ammonium (with Mg^{2+}), when its concentration exceeds $3 \cdot 10^{-3}$M. At the same time, the chlorella GS is

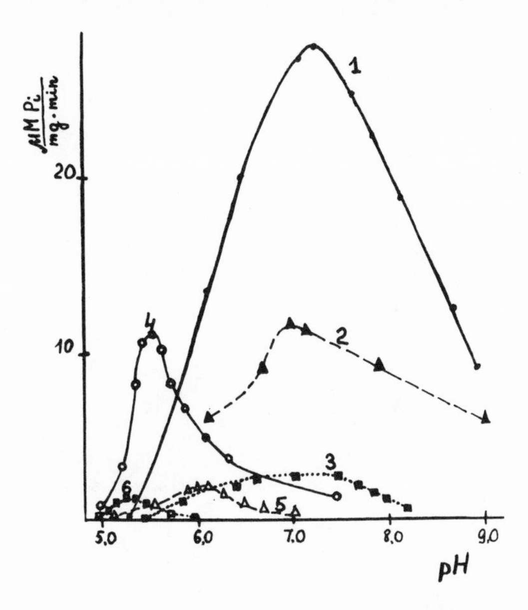

Figure 2. Activity of pea, chlorella, and yeast glutamine synthetases at various pH values. Pea GS: 1 with Mg^{2+}, 4 with Mn^{2+}; chlorella GS: 2 with Mg^{2+}, 5 with Mn^{2+}; yeast GS: 3 with Mg^{2+}, 6 with Mn^{2+}.

extremely active when the ammonium concentration varies in the range of $1 \cdot 10^{-4}$ to $1 \cdot 10^{-3}$M.

The assimilation of ammonium in chlorella under the conditions of its low content in the nutrient medium is effected by the synthesis of glutamine. It is known that the GS activity of chlorella,

Table II

$S_{0,S}$ Constants for the GS of Peas, Chlorella, and Yeast ($S_{0,S}$ are calculated graphically according to the Lineweaver-Burk method)

Enzyme source	For L-glutamate		For ammonium	
	with Mg^{2+}	with Mn^{2+}	with Mg^{2+}	with Mn^{2+}
Peas	$1.4 \cdot 10^{-2}$	$7.3 \cdot 10^{-3}$	$3.9 \cdot 10^{-5}$	$1.1 \cdot 10^{-3}$
Chlorella	$7.5 \cdot 10^{-3}$	$3.4 \cdot 10^{-3}$	$1.5 \cdot 10^{-3}$*	$7.5 \cdot 10^{-4}$
Yeast	$1.9 \cdot 10^{-2}$	$3.7 \cdot 10^{-2}$	$2.4 \cdot 10^{-4}$	$2.1 \cdot 10^{-4}$*

*Substrate inhibition

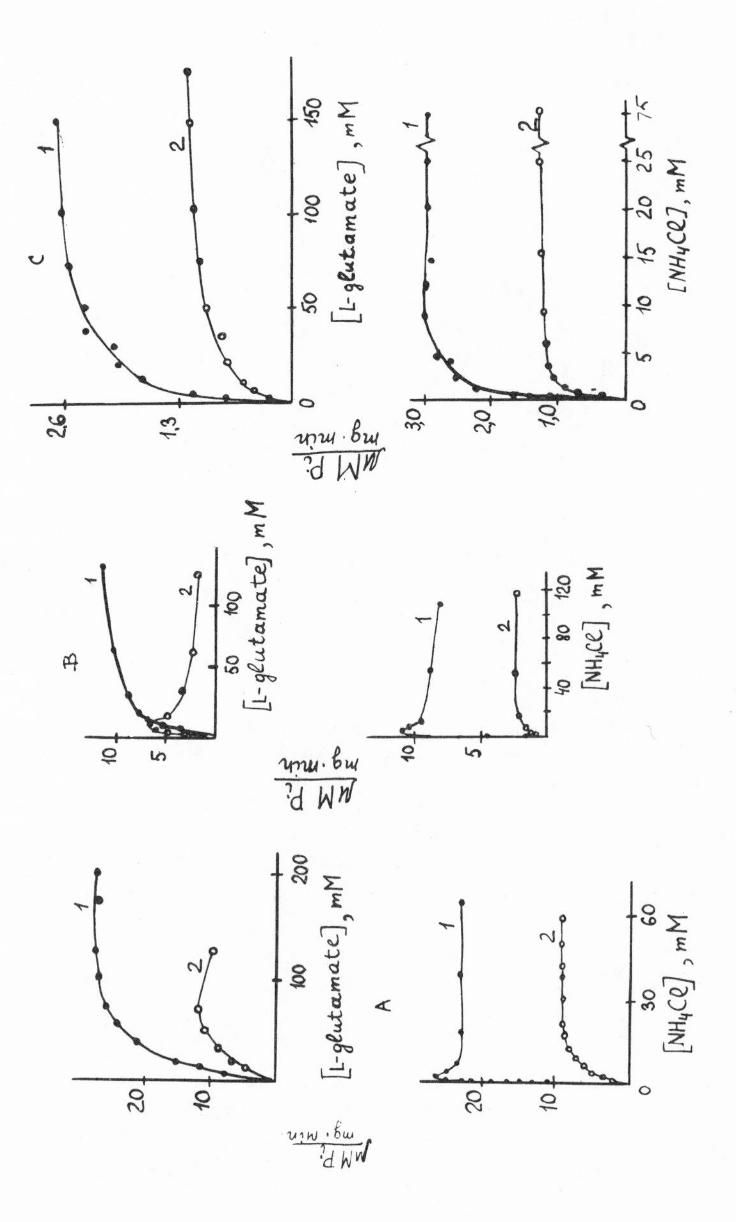

Figure 3. Kinetic curves for GS of peas, chlorella, and yeast. A, peas; B, chlorella; C, yeast. 1, Mg^{2+} system; 2, Mn^{2+} system.

grown on ammonium as the only nitrogen source, is low (23). However, under these conditions the synthesis of NADP-specific glutamate dehydrogenase (24,25) is induced. The assimilation of ammonium is effected by intensive synthesis of glutamic acid. The GS activity of plant organisms depends considerably not only on the availability and concentration of divalent cations of metals and ATP, but also on the concentrations of these compounds (Table III; Fig. 4; 18,19,22,26).

In any case, there should be an excess of Mg compared to ATP, while the Mn concentration should be equal to that of ATP. Changes in the concentration ratios cause a decrease of activity in all the cases excepting the ratio of Mg^{2+}:ATP for yeast (Fig. 4). This decrease is seen most vividly when the Mn^{2+}:ATP concentration ratio is violated. Similar data have been obtained for GS from Bacillus subtilis (15). The experiments with pea and yeast GS have shown

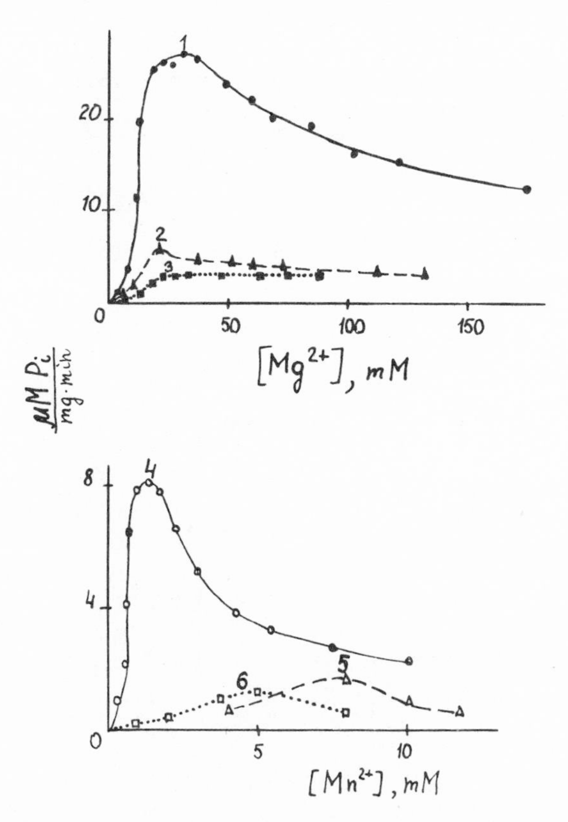

Figure 4. Mn^{2+}:ATP concentration ratio and its influence on the activity of pea, chlorella, and yeast GS: 1.4 pea GS; 2.5, chlorella GS; 3.6, yeast GS. ATP: 1, 6.25 mM; 2, 12 mM; 3, 7.5 mM; 4, 1.6 mM; 5, 7.5 mM; 6, 5 mM.

Table III

Optimum Concentration Ratios of Mg^{2+}:ATP and Mn^{2+}:ATP for the Activation of Pea, Chlorella, and Yeast GS

Enzyme Source	Mg^{2+}:ATP	Mn^{2+}:ATP
Peas	5:1	1:1
Chlorella	5:1	1:1
Yeast	3:1*	1:1

*With further increase in the Mg^{2+} concentration, inhibition is not observed.

that the addition of Mn^{2+} into the assay mixture containing Mg^{2+} leads to an abrupt inhibition of these enzymes (18,19,22; Fig. 5).

The evaluation of Mg^{2+} and Mn^{2+} content in peas done by the emission spectral analysis showed that the Mg^{2+} content in the seeds of this plant is about 150 times as high as that of Mn^{2+}. It is known that the content of Mg^{2+} is 100-1000 times higher than Mn^{2+} in plants. The optimum content of Mg^{2+} sufficient for pea GS to show

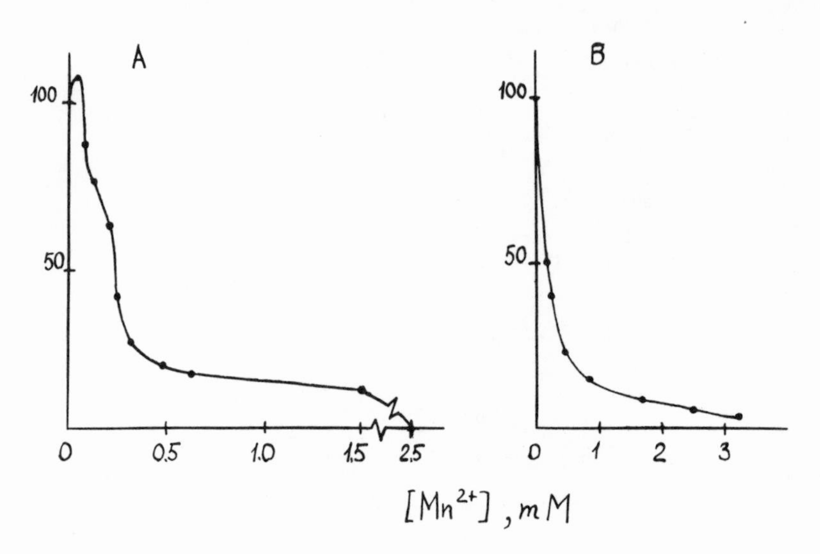

Figure 5. Inhibition of Mg^{2+}-bound activity of pea GS (A) and yeast GS (B) with the addition of Mn^{2+} into the assay mixture.

activity in the assay mixture is 100-200 times higher than that of Mn^{2+} in the assay mixture optimum for the display of GS activity with this cofactor.

The experiments to investigate the influence of both Mg^{2+} and Mn^{2+} on the activity of pea and yeast GS have shown (Fig. 6) that GS may display an optimum activity only with a certain qualitative ratio of these cations and a narrow pH range. However, this activity is higher than that of GS with only one of the cations, and the optimum values of pH are shifted towards a more physiological region (18).

The data concerning the influence of Mg^{2+} and Mn^{2+} on the activity of plant organism GS, particularly on the kinetic curve character, S-shaped character of the curves showing the saturation of GS with these cations (Fig. 6), provide the opportunity to assume that the role of Mg^{2+} and Mn^{2+} in the activation of plant organism glutamine synthetase is not confined to the formation of a complex with ATP at the enzyme active center. Moreover, as well as in the case of *E. coli* GS (4), Mg^{2+} and Mn^{2+} participate in the formation of a certain enzyme conformation.

CONTROL OF GS ACTIVITY

Like similar enzymes of bacteria and animals, GS of plants is regulated through feedback inhibition by certain products of glutamine metabolism. AMP is a strong inhibitor of pea, chlorella, and yeast GS. Amino acids, including glutamine, and a number of keto-acids inhibit this enzyme as well. Inhibitory effect of these metabolites on pea (Fig. 7) and chlorella GS is observed, mainly in the presence of Mn^{2+}, while yeast GS is inhibited more strongly in the presence of Mg^{2+}. It is known (13) that GS of bacteria and animals is inhibited by feedback inhibition, mainly in the presence of Mn^{2+}. Thus, the control of yeast GS possesses a characteristic property.

The control of plant organism GS displays other peculiar properties. For example the inhibiting effect on pea GS of the feedback inhibitors studied by us (27,28) is of a cooperative character, displaying a most powerful synergism under the joint action of aspartic acid and any other amino acid (Table IV). Along with AMP, the metabolites of asparagine metabolism, oxaloacetic acid, aspartate, and asparagine, are the strongest inhibitors of pea GS (Fig. 7), while α-ketoglutarate activates this enzyme like the GS of animals. Unlike the pea GS, regulation of chlorella (26) and yeast (29,30) GS by feedback inhibitors is of a cumulative character (Tables V,VI), as had already been pointed out with respect to *E. coli* GS (4).

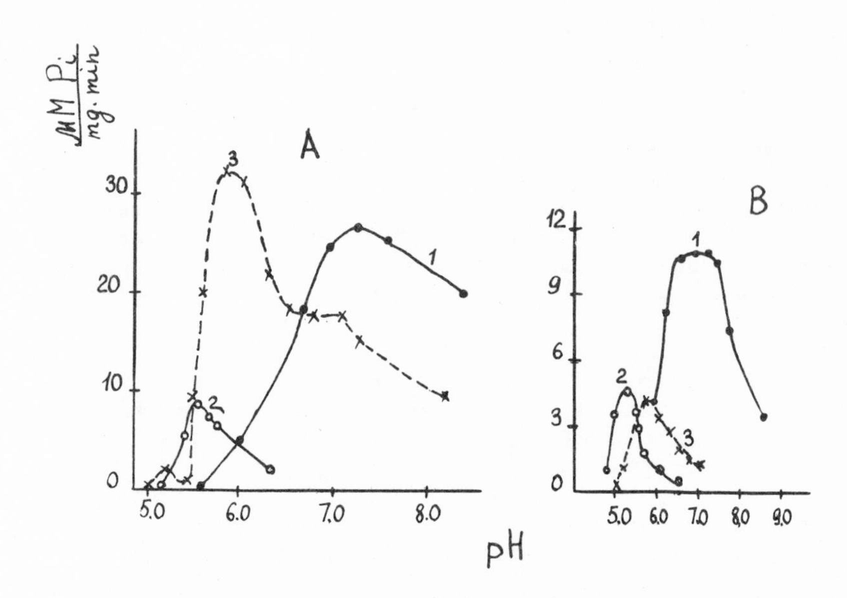

Figure 6. Activity of pea GS (A) and yeast GS (B) in the presence of Mg^{2+} and Mn^{2+}:

A. 1 - Mg^{2+} = 30 mM; Mn^{2+} = 0
2 - Mg^{2+} = 0 ; Mn^{2+} = 1.6 mM
3 - Mg^{2+} = 30 mM; Mn^{2+} = 0.12 mM

1 & 2 - ATP = 6.25 mM; 3 - ATP = 1.6 mM

B. 1 - Mg^{2+} = 50 mM; Mn^{2+} = 0
2 - Mg^{2+} = 0 ; Mn^{2+} = 7.5 mM
3 - Mg^{2+} = 50 mM; Mn^{2+} = 1 mM

ATP = 7.5 mM

The comparison of regulatory properties inherent in the GS of peas, chlorella, and yeast is indicative of considerable discrepancies in the regulation of GS of peas, on the one hand, and that of chlorella and yeast, on the other. The cumulative character of the influence exerted by feedback inhibitors on the GS of chlorella and peas brings these enzymes close to the GS of bacteria, while the GS of higher plants is regulated like that of animals (13). It should be pointed out that peas, yeast and, apparently, chlorella do not show activity regulation of the adenylylation-deadenylylation type, characteristic for E. coli and some other representatives of Enterobacteriaceae as well as Pseudomonas (31).

Table IV

The Influence of Amino Acids, Taken in Pairs, on the Activity of Pea GS (Mn^{2+}-Activated System)

Amino Acids	Glycine	L-Histidine	L-Glutamine	L-Aspartic acid	L-Valine	L-Tryptophan	Concentration of amino acids, M^{-3}
Glycine	55	67 $\frac{51}{46}$	68.5 $\frac{53}{51}$	10 $\frac{42}{31.5}$	80.5 $\frac{65}{73.5}$	61 $\frac{51}{48}$	3.1
L-Histidine	67 $\frac{51}{46}$	91	79 $\frac{87}{87}$	9 $\frac{70}{67.5}$	93 $\frac{108}{109.5}$	74 $\frac{85}{84}$	6.25
L-Glutamine	68.5 $\frac{53}{51}$	79 $\frac{87}{87}$	96	11.5 $\frac{73}{72.5}$	99 $\frac{114}{114.5}$	89 $\frac{89}{89}$	6.25
L-Aspartic acid	10 $\frac{42}{31.5}$	9 $\frac{70}{67.5}$	11.5 $\frac{73}{72.5}$	76.5	14 $\frac{91}{95}$	10.5 $\frac{71}{69.5}$	3.1
L-Valine	80.5 $\frac{65}{73.5}$	93 $\frac{108}{109.5}$	99 $\frac{114}{114.5}$	14 $\frac{91}{95}$	118.5	100 $\frac{110}{111.5}$	3.1
L-Tryptophan	61 $\frac{51}{48}$	74 $\frac{85}{84}$	89 $\frac{89}{89}$	10.5 $\frac{71}{69.5}$	100 $\frac{110}{111.5}$	93	3.1

Table V

Regulation of Chlorella GS by Feedback Inhibitors

(Mg^{2+}-Activated System)

Inhibitor	Percent inhibition	Inhibitor pairs	Percent inhibition	
			Found	Calculated*
AMP	42	AMP + alanine	53	48
GMP	10	AMP + glycine	56	49
Glucosamine	0	AMP + tryptophan	58	18
		CMP + alanine	18	19
L-Histidine	25	CMP + glycine	21	21
		CMP + tryptophan	16	20
Glycine	12	glycine + alanine	20	21
L-α-Alanine	10	glycine + tryptophan	13	22
L-Tryptophan	11	alanine + tryptophan	9	22
		Total of 6 inhibitors	62	68

Concentration: histidine and other amino acids: 0.01 M, GMP and glucosamine: 0.015 M; AMP: 0.00 M. In case of histidine, inhibition is observed only in the presence of Mn^{2+}.

*Calculated assuming the inhibitors have a cumulative effect.

Table VI

Regulation of Yeast GS by Feedback Inhibition

(Mn^{2+}-Activated System)

Inhibitor	Percent inhibition	Inhibitor pairs	Percent inhibition	
			Found	Calculated*
L-α-Alanine	25	L-alanine + glycine	40	40
Glycine	20			
L-Histidine	61	L-alanine+L-histidine	70	71
L-Tryptophan	4			
AMP	29	L-alanine+L-tryptophan	23	28
		L-alanine + AMP	40	47
		glycine+L-histidine	67	69
		glycine+L-tryptophan	24	23
		glycine + AMP	41	43
		L-histidine + AMP	69	72
		L-histidine+L-tryptophan	63	63

*As in Table V

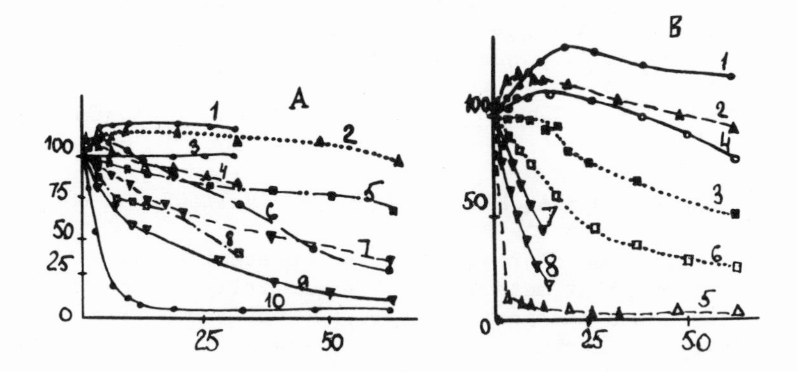

Figure 7. Influence of certain feedback inhibitors on pea GS (activity without the addition of feedback inhibitor set at 100%). A. 1) valine, 2) leucine, 3) threonine, 4) alanine, 5) glutamine, 6) asparagine, 7) histidine, 8) tryptophan, 9) aspartic acid, 10) glycine, Mn^{2+} system. B. 1,4: α-ketoglutarate; 3,6: oxaloacetate; 7,8: AMP. 1,2,3,7: with Mg^{2+}; 4,5,6,8: with Mn^{2+}.

INDUCTION AND REPRESSION OF GS SYNTHESIS

Induction and repression of GS synthesis in higher plants and animals is practically not investigated.

Unlike leaves and seeds, pea roots possess a very low GS activity (23,32). The GS activity, however, increases during the infiltration of ammonium, glutamate, or ammonium + glutamate (Table VII). As this phenomenon is not observed in the presence of cycloheximide, an assumption may be made that ammonium and glutamate induce the GS synthesis *de novo*.

Table VII

Influence of Ammonium and Glutamate Infiltration on GS of Germinated Pea Roots

Infiltrated Substances	Specific GS Activity
H_2O	0.75
$(NH_4)_2$ HPO_4 0.025 M	6.04
L-Glutamate 0.05 M	2.98
L-Glutamate 0.05 M + $(NH_4)_2$ HPO_4, 0.025 M	9.35
L-Glutamate + cycloheximide 30 μg/ml	0
Intact roots	0.10

On the contrary, with the addition of ammonium into a nitrogen-free nutrient medium for growing chlorella and yeast, GS activity decreases (Table VIII). This is, apparently, related not only to enzyme inhibition, but also to the repression of its synthesis, observed previously in *Candida utilis* (33) and *E. coli* (34).

DIURNAL AND SEASONAL VARIATION ACTIVITY OF PEA GS

Clearly expressed diurnal variations present a characteristic feature of higher plant metabolism. The activity of pea GS during the twentyfour hours changes both in roots and leaves (Fig. 8). Maximal GS activity in leaves is observed in the daytime, when photosynthesis is not intense. On the contrary, the activity in the roots is almost negligible in the daytime, but it appears in the evening when photosynthesis declines.

Experimental data (Fig. 9) show that while the activity in roots is not great during the entire vegetation period, the GS is very active in leaves from the moment of their appearance till they start turning yellow. GS activity appears in the young seeds and continues to increase till the moment of ripening. During storage, the ripe pea seeds practically do not lose activity, but GS activity disappears during germination, when the reserve proteins of the seeds disintegrate. GS activity appears subsequently in leaves and germinated roots, apparently due to the resynthesis of enzyme protein.

* * * *

The comparison of control and properties of GS in the organisms investigated by us shows similar features: very close pH optimums (with Mg^{2+} and Mn^{2+}), identical specificity in respect to substrates, cofactors, nucleoside triphosphates, and equal sedimentation

Table VIII

Influence of NH_4^+ on Chlorella GS

(specific activity)

		0	5	15	240
1	Growth in light on nitrogen-free medium (starvation), 16 hrs	18.0	-	-	-
2	After starvation 2 hrs on NO_3^-	-	20.2	-	23.0
3	After starvation 2 hrs on NH_4^+	-	11.5	4.9	3.7

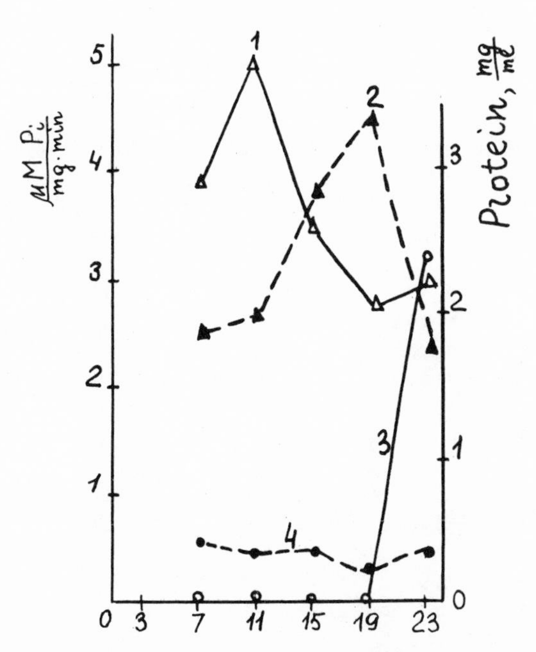

Figure 8. Diurnal activity variations of GS in pea plants. 1, 2, leaves; 3, 4, roots. 1, 3, GS activity; 2, 4, protein.

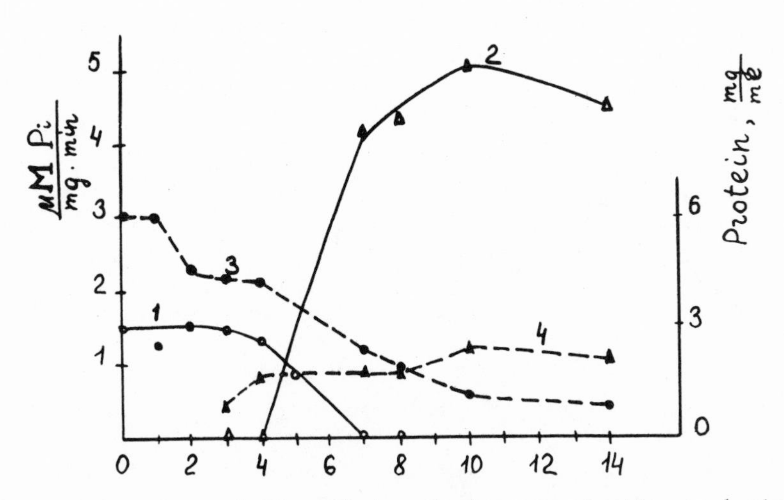

Figure 9. Changes of GS activity during germination and ripening of pea seeds and in the leaf. 1, 2, GS activity; 3, 4, protein; 1, 3, seeds; 2, 4, leaves.

coefficients. However, there are certain differences, which appeared in the process of evolution and are exhibited most distinctly in control mechanisms of this enzyme.

First of all, it should be pointed out that increased concentrations of NH_4^+ in the nutrient medium for growing chlorella and yeast diminishes enzyme activity, while in the case of peas the infiltration of ammonium into germinated roots results in a higher GS activity. This growth of activity is, apparently, related to GS synthesis *de novo* (23). Similar data were obtained for GS of rice plant roots (35).

All these data testify that the activity of GS studied depends first of all on the cofactor and upon the cofactor:ATP ratio. This form of regulation is common for plant organism glutamine synthetases. The pattern of the curves showing the saturation of GS with Mg^{2+} and Mn^{2+} is indicative of conformational changes in the GS molecule during the addition process of these cations.

As far as the action of feedback inhibition is concerned, the properties of the GS studied vary. For example, control of chlorella and yeast GS activity by feedback inhibitors is cumulative, unlike that of peas. The characteristic feature of pea GS control consists in the inhibition of its activity by asparagine, oxaloacetic and aspartic acids; the inhibitory action of the latter is increased in the presence of certain amino acids. Very clear changes of pea GS activity during the day may be connected with the content of these metabolites.

The inhibitory action of high concentrations of NH_4^+ in the nutrient medium and cumulative action of feedback inhibitors on chlorella and yeast GS indicates that the GS control mechanisms of these organisms and a number of bacteria may be similar.

The control of pea GS synthesis and activity shows certain similar features of this enzyme with glutamine synthetases of rice plant roots (35) and animals (13).

In conclusion, it should be pointed out that GS alteration in the evolutionary process occurs at the level of amino acid sequences of the polypeptide chains. These chains may arrange themselves to more-or-less active conformational structures of the enzyme protein. Therefore, future work should be aimed at the comparative study of the structure of plant organism GS.

REFERENCES

1. Oparin, A. I., *Advances Enzymol*. *27*, 347 (1965).

2. Meister, A., "Biochemistry of the Amino Acids," Vols. 1, 2, Academic Press, New York, 1965.

3. Kretovich, W. L., "The Metabolism of Nitrogen in Plants," "Nauka," Moscow, 1972.

4. "The Enzymes of Glutamine Metabolism" (Prusiner, S., and Stadtman, E. R., eds.), Academic Press, New York, 1973.

5. Elliott, W. H., J. Biol. Chem. 201, 661 (1953).

6. Evstigneeva, Z. G., Gromyko, E. A., and Aseeva, K. B., Appl. Biochem. Microbiol. (Russ.) 7, 479 (1971).

7. Pushkin, A. W., and Evstigneeva, Z. G., Appl. Biochem. Microbiol. (Russ.) 8, 86 (1972).

8. Auerman, T. L., Serebrenikov, W. M., and Akimova, N. I., in press.

9. Gromyko, E. A., Evstigneeva, Z. G., and Kretovich, W. L., Biokhimia 38, 6, 1261 (1973).

10. Pushkin, A. W., Evstigneeva, Z. G., and Kretovich, W. L., Biokhimia, in press.

11. Generalova, T. G., Auerman, T. L., and Karpilenko, G. P., Appl. Biochem. Microbiol. (Russ.) 9, 26 (1973).

12. Kretovich, W. L., Serebrenikov, W. M., and Auerman, T. L., Proc. Acad. Sci. USSR, in press.

13. Tate, S. S., and Meister, A., Proc. Nat. Acad. Sci. U.S. 68, 781 (1971).

14. Hubbard, J. S., and Stadtman, E. R., J. Bact. 94, 1007 (1967).

15. Deuel, T. F., and Stadtman, E. R., J. Biol. Chem. 245, 5206 (1970).

16. Woolfolk, C. A., Shapiro, B., and Stadtman, E. R., Arch. Biochem. Biophys. 116, 177 (1966).

17. Faserkas, S., and Denes, G., Acta Bioch. Biophys. Acad. Sci. Hung. 1, 45 (1966).

18. Evstigneeva, Z. G., Pushkin, A. W., and Kretovich, W. L., Proc. Acad. Sci. USSR 212, 502 (1973).

19. Evstigneeva, Z. G., Pushkin, A. W., and Kretovich, W. L., Plant Physiol. (Russ.) 19, 729 (1972).

20. Kretovich, W. L., Gromyko, E. A., and Evstigneeva, Z. G., Proc. Acad. Sci. USSR 207, 1479 (1972).

21. Kretovich, W. L., Generalova, T. G., and Auerman, T. L., Mikrobiologia 39, 767 (1970).

22. Serebrenikov, W. M., Auerman, T. L., and Kretovich, W. L., Mikrobiologia 42, 418 (1973).

23. Evstigneeva, Z. G., Aseeva, K. B., Mochalkina, N. A., and Kretovich, W. L., Biokhimia 36, 388 (1971).

24. Kretovich, W. L., Evstigneeva, Z. G., and Tomova, N. G., Can. J. Bot. 48, 1179 (1970).

25. Shatilov, W. R., Kaloshina, G. S., and Kretovich, W. L., Proc. Acad. Sci. USSR 194, 964 (1970).

26. Gromyko, E. A., Evstigneeva, Z. G., and Kretovich, W. L., Biokhimia 39, 1 (1974).

27. Pushkin, A. W., Evstigneeva, Z. G., and Kretovich, W. L., Biokhimia, in press.

28. Evstigneeva, Z. G., Pushkin, A. W., and Kretovich, W. L., Plant Physiol. (Russ.), in press.

29. Kretovich, W. L., Generalova, R. G., and Auerman, T. L., Proc. Acad. Sci. USSR 200, 237 (1971).

30. Kretovich, W. L., Serebrenikov, W. M., and Auerman, T. L., Mikrobiologia, in press.

31. Holzer, H., and Duntze, W., Ann. Rev. Biochem. 40, 345 (1971).

32. Radukina, N. A., Pushkin, A. W., Evstigneeva, Z. G., and Kretovich, W. L., Plant Physiol. (Russ.) 20, 376 (1973).

33. Sims, A. P., Folkes, B. F., and Bussey, A. H., "Mechanisms Involved in the Regulation of Nitrogen Assimilation in Microorganisms and Plants," p. 91, Academic Press, New York, 1968.

34. Wu, C., and Yuan, L. H., J. Gen. Microbiol. 51, 57 (1968).

35. Kanamori, T., and Matsumoto, H., Arch. Biochem. Biophys. 125, 401 (1972).

36. Kretovich, W. L., Pushkin, A. W., and Evstigneeva, Z. G., Proc. Acad. Sci. USSR 208, 988 (1973).

THE LIGHT-DEPENDENT PROCESSES IN FUNGI: A POSSIBLE APPROACH TO SOME PROBLEMS OF PHOTOBIOLOGICAL EVOLUTION

M. S. Kritsky

A. N. Bakh Institute of Biochemistry
USSR Academy of Sciences
Moscow, USSR

According to the basic principles of the theory of the origin of life formulated in the early twenties by A. I. Oparin (1), the origin and development of systems capable of photochemical transformations was extremely important for the general problem of biopoesis. In fact, the whole process of both chemical and, especially, biological evolution had to depend upon the production of organic matter through utilization of solar energy. During chemical evolution, ultraviolet light served as a principal source of energy (2), whereas biological evolution was mainly related to utilization of rather long-wave visible light through the photosynthetic processes.

The fundamental concepts in the process of evolution of photobiological phenomena have been developed in the early sixties, mainly by the works of A. A. Krasnovsky (3) and H. Gaffron (4). According to these concepts, a transition occurred from utilization of ultraviolet radiation by prebiotic, and possibly early biological, systems to photosynthetic processes consuming visible radiation energy *via* compounds that are photochemically active in long-wave ultraviolet and short-wave visible regions, *e.g.*, pyridine nucleotides, flavins, etc. Indeed, some experiments have demonstrated possible photochemical activity of such compounds in model prebiotic systems (*e.g.*, cf. 5).

At present, the attention of those studying the origin and evolution of photobiological processes is mainly focused on two groups of phenomena: first, photochemical reactions connected with the utilization of ultraviolet radiation, and, second, detailed studies of photosynthetic reactions in various systematic groups of photoautotrophic organisms. However, a rather broad group of photochemical reactions in living systems is almost ignored; these are the ones related to photoregulatory processes in plants and microorganisms.* At the same time, these processes cover a broad spectral zone and are widely represented both among photoautotrophic and heterotrophic organisms.

The study of photoregulatory processes appears to be significant for the problems of photobiological phenomena, especially since some of them are associated with the absorption of comparatively short-wave light by flavin photoreceptors. In this connection, an attempt to search among them for possible analogs, or rudiments, of the ancient photobiological phenomena seems very attractive.

It is important to note that the photoregulatory processes are peculiar not only to photoautotrophic, but as well to heterotrophic, organisms. We may point out, for example, the broad group of photoregulatory processes which are known to be present among various fungi (cf. the review of 6). The light-dependent processes in fungi, in most cases, are connected with the regulation of growth and development, e.g., phototropism, triggering of morphogenetic events, as well as with the induction of pigment formation.

The study of action spectra has shown that among various representatives of fungi there are present photobiological reactions dependent on different zones of the spectrum, from ultraviolet to red and far-red. Unfortunately, we have not succeeded in establishing any relation between the active spectral zone and the systematic position of the organisms. Based upon the studies of action spectra, we can draw some conclusions on the nature of photoreceptor mechanisms. As to the processes induced by ultraviolet light, the attempts to identify their action spectra with the absorption spectra of chemical compounds have not yet brought any success (7,8).

The processes induced by the blue-violet region are the most well-studied today. Their action spectra are rather similar. In the visible zone, they have two principal maxima at 360-385 nm and 470-485 nm, as well as an additional maximum, or a shoulder, at 440-465 nm. Based upon action spectra, it has been suggested that flavins or carotenoids serve as photoreceptor molecules (9-12).

---

*We do not touch in this paper on the problem of vision. Certainly, photochemical and physiological studies of vision have provided much interesting information concerning behavior of photobiochemistry of visual pigments, role of membranes, etc. However, this problem is, in our opinion, a special subject of photobiology that is related to very special functions of animal organisms - the activity of the nervous system.

The inhibitor studies indicate that participation of flavins is possible (14,15). On the other hand, the role of carotenoid pigments in photoreception seems not yet to be completely excluded (15,16). In my opinion, the photoreceptor may include both flavins and carotenoids associated with a membrane complex. It is of interest that the processes having similar action spectra are distributed as the regulatory ones among the green plants (17,18), as well as among procaryotic microorganisms (19).

The photobiological phenomena in fungi are not limited, however, by rather short-wave zones of visible spectra. There is some evidence that the light-dependent suppression of respiratory adaptation in yeast cells is connected with the absorption of light up to 650 nm, which is according to the opinion of authors of this work, the result of photochemical transformations of the cytochrome photoreceptor (20). In addition, the developmental processes of some fungi have been shown, in some works, to be controlled by the longest wave region of the visible spectrum - red and far-red (23,24). In this connection, the reversion of the morphogenetic effect of red light, 650-700 nm, by long-wave radiation, 690-750 nm, as demonstrated in these works, seems very interesting. This view allows the suggestion of a possible participation of phytochrome, not only in the photochemical processes of green plants, but in some fungi also.

Thus, the photobiological phenomena in fungi cover, generally speaking, the same spectral range as the photobiological phenomena in green plants. What is more, it seems very likely that the photobiological processes in fungi are connected with the photochemical activity of the same principal classes of compound that are related to the photobiological activity of green plants e.g. flavins, possibly, carotenoids, metalloporphyrins, and tetrapyrroles similar to the chromophore group of phytochrome.

Such similarity poses for the investigator a very interesting question - whether it is a result of the same evolutionary roots of the photobiological processes in both photosynthetic and nonphotosynthetic organisms, or it has appeared due to convergence and is based rather on the unique capability of the mentioned groups of compounds to serve in photochemical reactions in living systems.*

*This question must be inevitably considered in connection with the general problem of the evolutionary origin and relationship of fungi. Already for some decades, two alternative hypotheses are discussed in literature. According to one of them, fungi developed from photosynthetic ancestors, likely algae, after the loss of plastids. According to the other, they originated from nonphotosynthetic organisms, some protists. In recent literature, these two views have been discussed by R. Klein and A. Cronquist (27) (cont. on next p.)

The answer to this problem would, no doubt, help us in understanding not only the pathways of photobiological evolution, but the course of evolution of the organic world as a whole. However, the attempt to find today a sure answer to this question seems to be premature. This is connected, first of all, with the fact that, contrary to photobiological phenomena in photosynthetic organisms, the molecular mechanisms of the light-dependent processes of nonphotosynthetic nature are yet studied rather poorly. Only during recent years, there have been some attempts to clarify the molecular mechanism of the nonphotosynthetic phenomena, in particular, in heterotrophic organisms.

At the same time, just the heterotrophic organisms may serve as interesting objects for the study of the mechanisms of photoregulatory phenomena, because it may be expected that, contrary to photoautotrophic organisms, the experimental study of such processes is not influenced by the photosynthetic reactions. In this respect, there seems especially promising the investigations of those photobiological processes, which depend upon ultra-violet and blue-violet spectral zones, since the study of just those processes might help us in understanding the mechanism of the preceding stages of photobiological evolution.

The studies carried out in our laboratory have demonstrated that in fungal cells, in the course of light-dependent processes which are triggered by blue-violet light (9,10); photoinduction of the colored carotenoid formation in *Neurospora crassa* and light-induction of morphogenesis of the fruiting body of the basidiomycete *Lentinus tigrinus*, there occurs an increase of reduction level of nicotinamide coenzymes, first of all, NAD. It proceeds at the early stages of light-induction, prior to the changes on the transcriptional level participating in the final manifestation of the effect (25,26).

We may speculate that the reduction of nicotinamide, in the course of light-dependent process in fungi, might be considered as a phenomenon somewhat similar to photosynthetic reactions, where generation of reduced nicotinamide coenzyme is one of the principal results. Under modern conditions, such a process is not certainly related to the accumulation of solar energy and leads only to the light-dependent redistribution of chemical energy among the different compounds, and to its accumulation in a form accessible for various biochemical processes - in the reduced nicotinamide coenzymes.

-- and by L. Margulis-Sagan (28,29). Since the latter concept seems to be more valid from the viewpoint of biochemical evolution, I use it in my considerations.

Nevertheless, it seems possible that in the early stages of the development of life, when the portion of blue-violet light in solar radiation on the surface of the Earth was higher, the role of these processes was more significant too. Just in that period, the primitive systems capable of absorbing the short-wave visible light might have acquired a notable place among other photobiological reactions. However, in the course of further evolution they were incapable of competing with the much more energetically effective photosynthetic apparatus based upon the photochemical activity of chlorophyll. The processes dependent upon blue-violet light were thus pushed aside from the fulfillment of an energetic role in living systems, but were retained by the modern organisms both photosynthetic and nonphotosynthetic as regulatory reactions representing, certainly in a very modified way, the rudiments of the ancient photobiological processes.

In my opinion, we should not exclude the possibility that the primitive photobiological reactions might in some way have been associated with the system of photoactivation of metal-porphyrin molecules that led to the dramatic increase of efficiency of such processes and the development of the contemporary photosynthetic apparatus. The possibility of such association was noticed, in particular, by A.A. Krasnovsky (5). We may hope that the study of the molecular mechanism of the nonphotosynthetic photobiological phenomena will lead us to better understanding of the origin of various photobiological processes, and will throw more light on the transition from chemical evolution to biological evolution.

REFERENCES

1. Oparin, A.I., "The Origin of Life on the Earth", Oliver and Boyd, Edinburgh-London, 1957.

2. Pavloskaya, T.E., in "Evolutionary Biochemistry", Moscow, Znanie, p.21, 1973.

3. Krasnovsky, A.A., in "Origin of Life on the Earth", Proc. First, Intern. Symposium, Moscow (1957), p.606, Pergamon Press, 1959.

4. Gaffron, H., in "Horizons in Biochemistry", (Kasha, M., and Pullman, B., eds.), p.59, Academic Press, New York, 1962.

5. Krasnovsky, A.A.,and Brin,G.P.,in "The Problems of Evolutionary and Applied Biochemistry", p.221, Nauka, Moscow, 1964.

6. Carlile, M.J., in "Photobiology of Microorganisms", (Halldall, P., ed.), p.309, Wiley-Interscience, New York, 1970.

7. Leach, C.M., and Trione, E.J., Plant Physiol. 40, 808 (1965).

8. Leach, C.M., and Trione, E.J., Photochem. Photobiol. 5, 621 (1965).

9. Zalokar, M., Arch. Biochem. Biophys. 56, 318 (1955).

10. Rau, W., Planta 72, 14 (1967).

11. Perkins, J.H., and Gordon, S.A., Plant Physiol. 44, 1712 (1970).

12. Delbruck, M., and Shropshire, W., Plant Physiol. 35, 194 (1960).

13. Page, R.M., Mycologia 48, 206 (1956).

14. Lukens, R.J., Am. J. Botany 50, 720 (1963).

15. Curry, G.M., and Gwen, H.E., Proc. Nat. Acad. Sci. U.S. 45, 797 (1959).

16. Thimann, K.V., and Curry, G.M., in "Comparative Biochemistry", (Florkin M., and Mason, H.S., eds.), vol. I., p. 2 & 3, Academic Press, New York, 1960.

17. Shropshire, W., and Withrow, R.B., Plant Physiol. 33, 360 (1958).

18. Zurzycki, J., Acta Soc. Botan. Polon. 36, 134 (1967).

19. Rilling, H.C., Biochim. Biophys. Acta 79, 464 (1964).

20. Guerin, B., and Jacques, R., Biochim. Biophys. Acta 153, 138 (1968).

21. Nair, P., and Zabka, G.S., Mycopath. Mycol. appl. 26, 123 (1965).

22. Lukens, R.J., Phytopathology 50, 720 (1963).

23. Rakoczy, L., Acta Soc. Bot. Polon. 36, 153 (1965).

24. Fraikin, G.Y., Verkhoturov, V.N., and Rubin, L.B., Vestnik MGU (Proc. Moscow University)Biology, Soil Scien. Seria, 4 51 (1972).

25. Kritsky, M.S., Belozerskaya, T.A., Soboleva, I.S., and Mazo, A.M., in Proc. First All-Union Conference on the Regulation of Biochemical Processes in Microorganisms, Pustchino, 1972.

26. Belozerskaya, T.A., and Kritsky, M.S., Biokhymia 38, 5 (1973).

27. Klein, R., and Cronquist, A., Quart. Rev. Biol. 42, 105 (1967).

28. Margulis, L., Science 161, 1020 (1968).

29. Margulis, L., in "Molecular Evolution", Vol. 1, Chemical Evolution and the Origin of Life", (Buvet, R., and Ponnamperuma, C., eds.), p. 480, North-Holland, Amsterdam, 1971.

THE ROLE OF INORGANIC POLYPHOSPHATES IN CHEMICAL AND BIOLOGICAL EVOLUTION

I. S. Kulaev

Institute of Biochemistry and Physiology of Micro-organisms, USSR Academy of Sciences, Pushchino, USSR

The problem of the possible participation of inorganic polyphosphates in abiogenesis and evolution of phosphorus metabolism has become a subject of wide discussion.

Inorganic high-molecular weight polyphosphates (Poly P) (Fig. 1) are the simplest macroergic compounds consisting of a certain number of phosphorus residues and interconnected with phosphoanhydride bonds. The number of phosphorus residues may vary from 2 to 500.

G. Schramm and a number of other researchers assume that high-polymeric polyphosphate deposits could be formed in the conditions existing on the primeval Earth (1-6).

On the other hand, pyrophosphate, the lowest homologue of this series of compounds, could also be formed on the primeval Earth.

The model experiments of Miller and Parris (7), M. Calvin (8), M. Halmann (9), and others (10-12) demonstrated how pyrophosphate could be formed from orthophosphate either at the expense of inorganic redox reactions or the expense of prior activation of phosphate by cyanogen. According to F. Lipmann, pyrophosphate, regardless of the mode of its formation, could be "the simplest compound involved in accumulation and transfer of energyrich bonds on the primeval Earth" (13).

It should also be mentioned that J. Lowenstein (14-16) and later R. Buvet (17) carried out model experiments in which they succeeded in finding conditions for nonenzymatic synthesis of pyro-

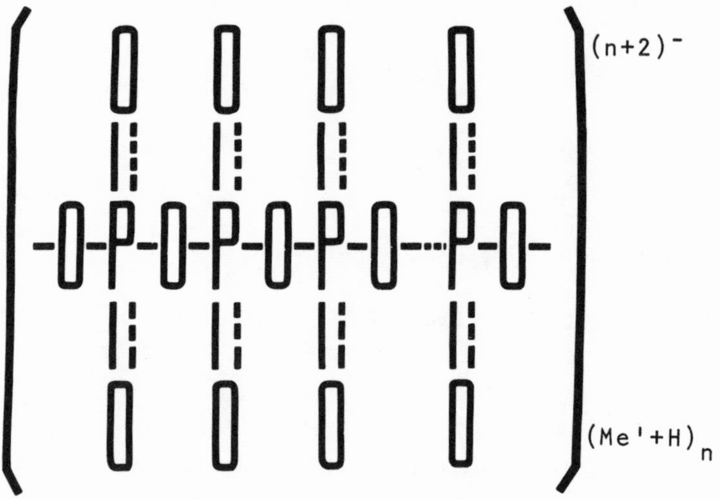

Figure 1. The structure of inorganic polyphosphates. Me' = univalent cations; n = the number of condensed molecules of orthophosphate.

phosphate at the expense of phosphorylation of orthophosphate in the presence of certain cations with the help of ATP and inorganic tripolyphosphate, respectively. Experiments conducted by us together with K. G. Skryabin (18) showed that this reaction may run far more intensively at the expense of high polymeric polyphosphates (Fig. 2). Thus we have found that nonenzymatic formation of a considerable quantity of radioactive pyrophosphate occurs at the expense of phosphorylation of labeled P^{32}-orthophosphate with high molecular weight

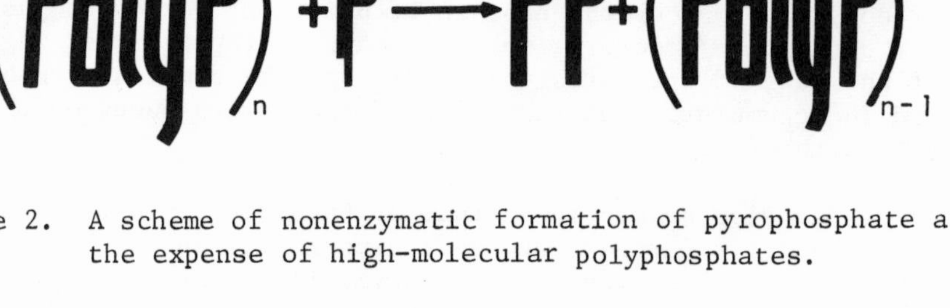

$$(PolyP)_n + P^{*} \xrightarrow{Me^{+2}} P^{*}P + (PolyP)_{n-1}$$

Figure 2. A scheme of nonenzymatic formation of pyrophosphate at the expense of high-molecular polyphosphates.

polyphosphate ($\bar{n}$ = 40) in an aqueous solution at 37°C during 15 hours of incubation at pH = 9 and in the presence of certain bivalent cations.

As one can see in Table I, from 15 to 30% of the initial high polymeric polyphosphate was utilized as phosphoryl group donor. Out of the cations tested: Mg^{++}, Mn^{++}, Ca^{++}, Cd^{++}, Ba^{++}, the highest percentage of P^{32} incorporation was observed to take place in the presence of Ba^{++} (33%) and the lowest in the presence of Mg^{++} (15%). It is interesting to note that in the experiments of Lowenstein and of Buvet, in which organic and inorganic tripolyphosphates were used, only 0.2-0.6% of the initial donor phosphate was utilized.

Thus, our experimental data allow us to assume that upon the cooling of the Earth and the formation of the hydrosphere, various processes of transphosphorylation, particularly formation of pyrophosphate at the expense of phosphorylation of orthophosphate by high-polymeric polyphosphates, could have occurred in the primordial ocean.

Numerous model experiments that have been conducted, particularly those of Schramm (1-4), Ponnamperuma (19-21), Fox (22-24), Schwartz (25,26), and others (27) have shown that high-polymeric polyphosphates, unlike pyrophosphate, could be involved as condensing agents in reactions forming nucleosides, nucleotides, including

Table I

Activation of Transphosphorylation by Metal Ions

Cations	$\frac{\text{Pyrophosphate}}{\text{Polyphosphate}} \cdot 100$
Mg^{++}	16
Ca^{++}	22
Mn^{++}	23
Cd^{++}	26
Ba^{++}	33
--	0

adenosine triphosphate (ATP), the simplest polynucleotides, polypeptides, and even primitive proteinlike compounds on the primeval Earth. It was also discovered that condensation reactions with high-polymeric polyphosphates as activating agents can take place both at high temperatures in a waterless medium and at room temperatures in an aqueous solution (18,28-30). Thus, one may assume that such polyphosphates could have participated in the syntheses of various macromolecules that later became part of living cells both prior to and after the hydrosphere on the Earth was formed.

However, the reaction of transphosphorylation at the expense of high-polymeric polyphosphates with subsequent formation of pyrophosphate, according to our data, could occur only in an aqueous solution at 37°C (18,29). Thus, one may assume that the formation of a substantial quantity of pyrophosphate at the expense of transphosphorylation could take place only after a hydrosphere containing previously synthesized macromolecules had been formed on the primeval Earth. It could well be that the initial macromolecules could then be spontaneously organized into primitive types of structures, such as coacervates of A. I. Oparin (31-33) or microspheres of Fox (24,34). This circumstance might have been of primary importance for the initiation and further evolution of pyrophosphate metabolism.

However, the model experiments have not made it possible to obtain any reliable information about the functioning of high-polymeric polyphosphates and pyrophosphate in the earliest living beings. Nevertheless, certain conclusions about the role of these primitive macroergic compounds in the metabolism of protobionts can be drawn on the basis of the data of comparative biochemistry.

By studying metabolism in the earliest living organisms and that in comparatively lower forms of the contemporary organisms, one can detect, according to Lipmann (13,35), some features of "antedeluvian metabolism" and can observe "fossil" biochemical reactions that have remained unchanged since the most ancient times.

Research in this field of biochemistry, which may be termed "biochemical paleontology," can lead and has led to detection of archaic features of metabolism that have originated in primitive forms of life. For example, Baltscheffsky *et al*. (36-38) and Keister (39-41) have shown that much more pyrophosphate than ATP is formed as macroergic phosphate through photosynthetic phosphorylation in *Rhodospirillum rubrum*, phylogenetically the most ancient and primitive photosynthesizing bacteria. Besides, pyrophosphate synthesis may occur in chromatophores of this bacterium even when ATP formation is completely inhibited.

The data of Baltscheffsky were confirmed in my laboratory by A. Shadi and S. E. Mansurova (42). It was ascertained that in

Rhodospirillum rubrum, pyrophosphate accumulates only in light. But it has not been detected in either whole cells or in chromatophores in the culture grown in dark.

It was established in Baltscheffsky's laboratory that the energy accumulated in a pyrophosphate molecule can be used both for the reverse electron transport and for the active ion transport through the membrane of chromatophores of these bacteria.

Quite recently, Keister has reported ATP synthesis to occur in chromatophores of the same organism at the expense of pyrophosphate energy resulting from hydrolysis of this compound by the respective pyrophosphatase.

The data obtained recently in the laboratories of E. A. Liberman (43), and of V. P. Skulachev (44) are sufficient for the assumption to be made that the utilization of pyrophosphate energy in chromatophores of Rhodospirillum rubrum proceeds via formation of a membrane potential ($\Delta\psi$).

The total sum of reactions discovered in chromatophores of this photosynthesizing bacterium is shown in Fig. 3. It is interesting to point out that chromatophores of Rhodospirillum rubrum are not exceptional with respect to their ability to utilize the pyrophosphate energy resulting from its cleavage by pyrophosphatase for ATP biosynthesis. The existence of a similar process in animal mitochondria has been quite recently shown in our laboratory by S. E.

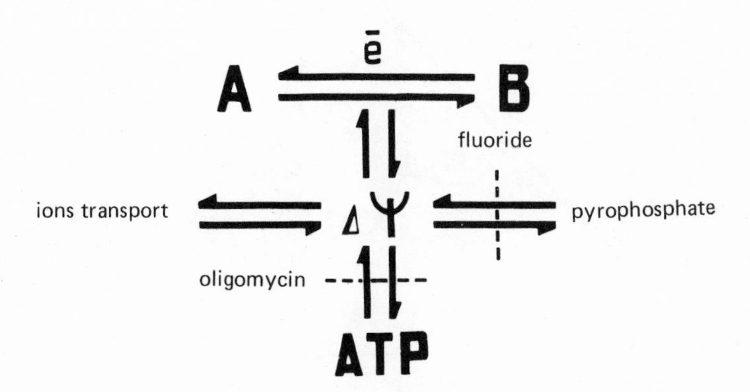

Figure 3. Pyrophosphate metabolism in R. rubrum.

Mansurova and T. N. Belyakova (45). Again, we have recently established, together with S. E. Mansurova and U. Shakhov (46), that not only utilization of pyrophosphate but also its synthesis can take place in animal mitochondria. As is shown in Fig. 4, pyrophosphate is synthesized together with ATP in rat liver mitochondria. Our calculations have shown, however, that there was synthesized ten times less pyrophosphate than ATP and ADP in mitochondria isolated by us (the incubation medium contained AMP).

It was established in the experiments on the effect of respiratory chain inhibitors -- rotenone (2 mg/mg of protein), antimycin (1 mg/mg of protein), and cyanide (10^{-3} M)--as well as of the uncoupler 2,4-DNP (4×10^{4} M), that pyrophosphate biosynthesis occurs in mitochondria not at the expense of transphosphorylation reactions. The data shown in Fig. 5 demonstrate that both the uncoupler and the tested inhibitors of electron transport at the level of all of the three coupling sites of the respiratory chain interrupt pyrophosphate biosynthesis in rat liver mitochondria. One could assume that pyrophosphate formation in mitochondria was secondary, as a part of ATP splits to AMP and pyrophosphate. However, experiments on the effect of oligomycin (2 mg/mg of protein), which inhibits ATP formation in mitochondria (Fig. 6), showed that pyrophosphate synthesis does not decrease in such conditions; on the contrary, it increases considerably. These data testify to the fact that pyrophosphate is synthesized in animal mitochondria at the expense of the respiratory chain

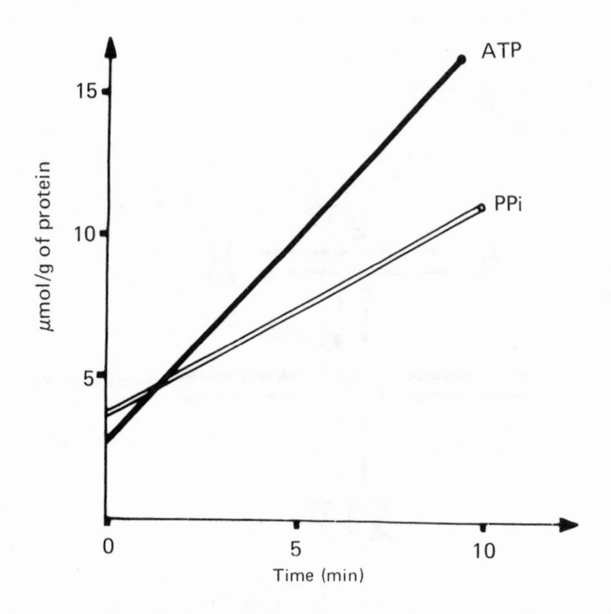

Figure 4. Synthesis of ATP and PP_i in rat liver mitochondria.

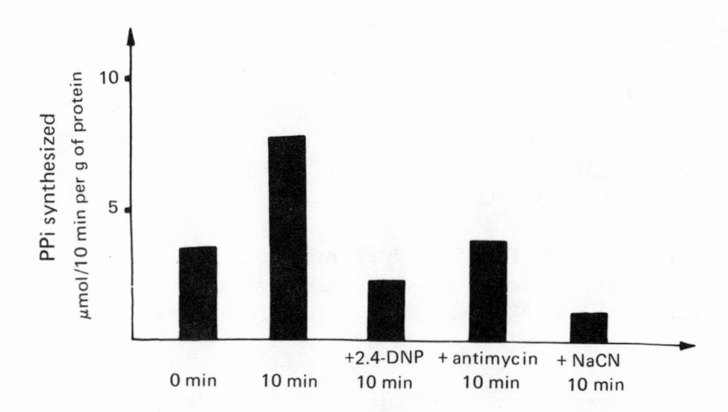

Figure 5. Influence of 2,4-dinitrophenol ($4\text{-}10^{-4}$M), antimycin (1 μg/mg of protein), and NaCN (10^{-3}M) on the synthesis of PP_i.

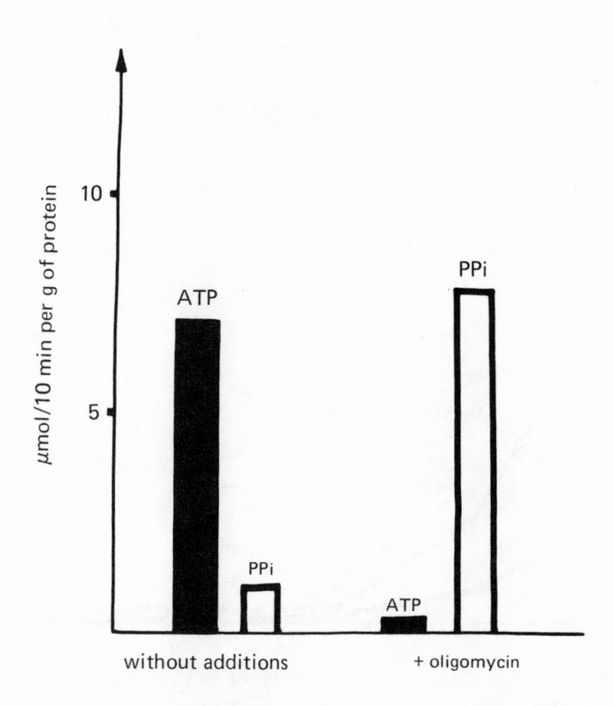

Figure 6. Influence of oligomycin on the synthesis of ATP and PP_i.

functioning independently of ATP and, to a certain extent, counterbalancing it.

The fact that pyrophosphate synthesis in animal mitochondria is independent of ATP was even more convincingly demonstrated in the experiments with ADP and ATP-deprived mitochondria after their preincubation with glucose (4 x 10^{-2} M), hexokinase (0.1 mg/ml), and oligomycin (1 mg/g of protein).

As one can see in Fig. 7, pyrophosphate synthesis in rat liver mitochondria occurs in the complete absence of ADP and ATP. Pyrophosphate increases with addition of orthophosphate to the incubation medium.

It is also important to point out that pyrophosphatase seems to be involved in the process of biosynthesis and utilization of pyrophosphate in animal mitochondria as well as in Rhodospirillum rubrum. In any case, pyrophosphate biosynthesis is inhibited and ATP synthesis increases in the presence of sodium fluoride (10^{-2} M), as one can see in Fig. 8. It seems that the above-given scheme for Rhodospirillum rubrum chromatophores (Fig. 3) can also be used to explain the interconnection of pyrophosphate and ATP synthesis in animal mitochondria. The quantity of pyrophosphate, however, synthesized at the expense of the functioning of the electron transport chain in R. rubrum chromatophores is much greater than that of ATP, whereas the quantity of synthesized ATP in animal mitochondria far exceeds that of pyrophosphate.

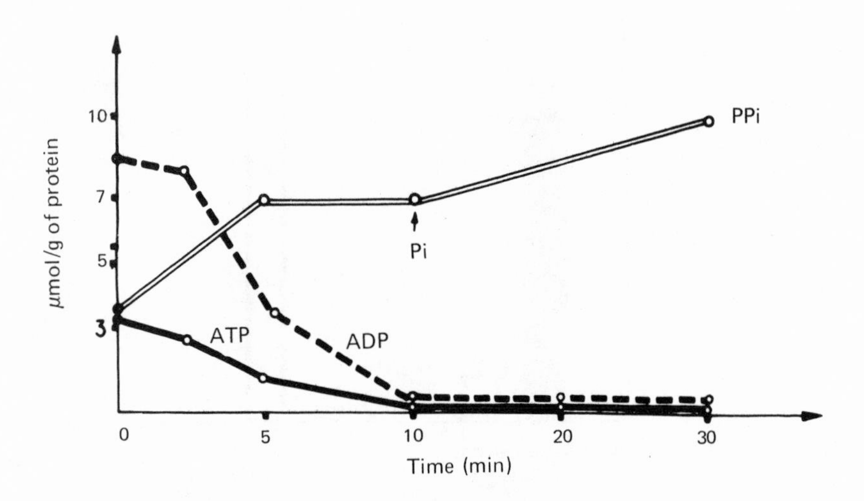

Figure 7. Synthesis of PP_i in mitochondria in the absence of endogenic ATP and ADP.

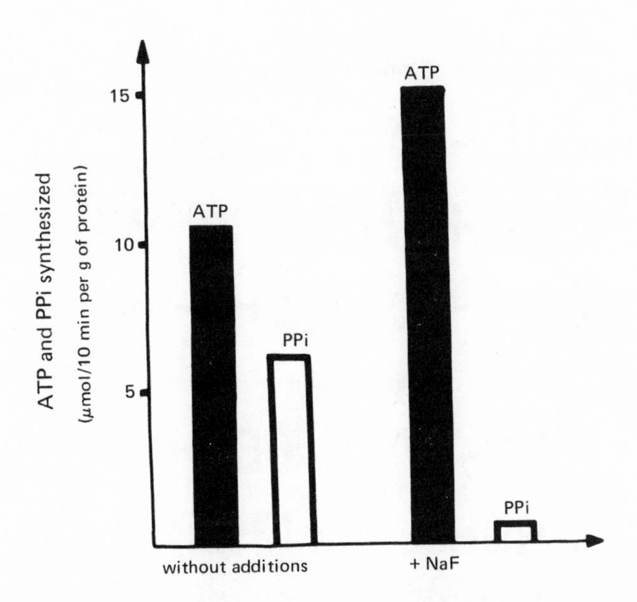

Figure 8. Influence of NaF (10^{-2} M) on the synthesis of ATP and PP_i.

Thus, according to the data of comparative biochemistry, it seems that from time immemorial and probably from the very incipience of the first living organisms, pyrophosphate has been involved in energy processes taking place in the membranes and primarily in the reactions of photosynthetic phosphorylation and phosphorylation at the respiratory chain level. However, according to A. I. Oparin (31-33), the most ancient process supplying life-supporting energy to the first organisms living on the Earth long before oxygen appeared on it was anaerobic fermentation of hexose to lactic acid and ethyl alcohol. At the same time the data obtained by us, together with S. O. Uryson, O. V. Szimona, and M. A. Bobyk, allow the assumption to be made with a high degree of probability that energy-producing processes connected with glycolysis occurred in the primary organisms, not with the help of ATP and pyrophosphate, but due to participation of high molecular weight polyphosphates (28,30).

Thus, we have established that certain contemporary organisms, e.g. some bacteria and fungi (47-50), have an enzyme catalyzing the reaction of transport of phosphate activated at the expense of glycolytic phosphorylation from 1,3-diphosphoglyceric acid, not to ADP with subsequent ATP formation, as is generally believed in accordance with the Meyerhoff-Embden-Parnas scheme (Fig. 9), but straight to high molecular-weight polyphosphate, its chain being by one phosphate residue longer. It is also interesting that another enzyme was found in the cells of some microorganisms and that it is

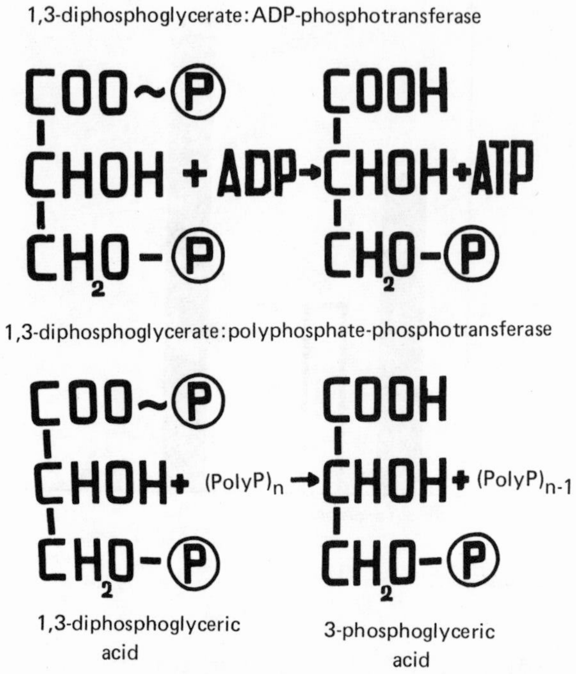

1,3-diphosphoglycerate:ADP-phosphotransferase

$$\begin{array}{lcl} COO{\sim}\text{Ⓟ} & & COOH \\ | & & | \\ CHOH & + ADP \rightarrow & CHOH + ATP \\ | & & | \\ CH_2O-\text{Ⓟ} & & CH_2O-\text{Ⓟ} \end{array}$$

1,3-diphosphoglycerate:polyphosphate-phosphotransferase

$$\begin{array}{lcl} COO{\sim}\text{Ⓟ} & & COOH \\ | & & | \\ CHOH & + (PolyP)_n \rightarrow & CHOH + (PolyP)_{n-1} \\ | & & | \\ CH_2O-\text{Ⓟ} & & CH_2O-\text{Ⓟ} \end{array}$$

1,3-diphosphoglyceric acid   3-phosphoglyceric acid

Figure 9. A scheme of reactions catalyzed by 1,3-diphosphoglycerate polyphosphate-phosphotransferase and 1,3-diphosphoglycerate ADP-phosphotransferase.

polyphosphate-glucokinase which catalyzes the reaction of transport of phosphate of high polymeric polyphosphate to glucose, thus forming glucose-6-phosphate (Fig. 10). This enzyme discovered by M. Szymona (51,52), a Polish biochemist, substitutes the action of the well-known hexokinase, which catalyzes glucose-6-phosphate synthesis at the expense of ATP.

It seems to be a matter of principal importance from the standpoint of comparative evolution to determine the activity of the enzymes which on the one hand use ATP and on the other high molecular weight polyphosphates to form the same glucose-6-phosphate in various organisms. To this end, we tested the activity of the common ATP-hexokinase and polyphosphate-glucokinase in more than 60 species of various microorganisms (53-55). Other researchers carried out similar analyses with 11 more microorganisms (56,57). Thus the presence of ATP-hexokinase was established in all the species analyzed. Polyphosphate-hexokinase, however, was found only in very phylogenetically ancient and closely related microorganisms constituting, according to N. A. Krasilnikov's system (58,59), a separate class of

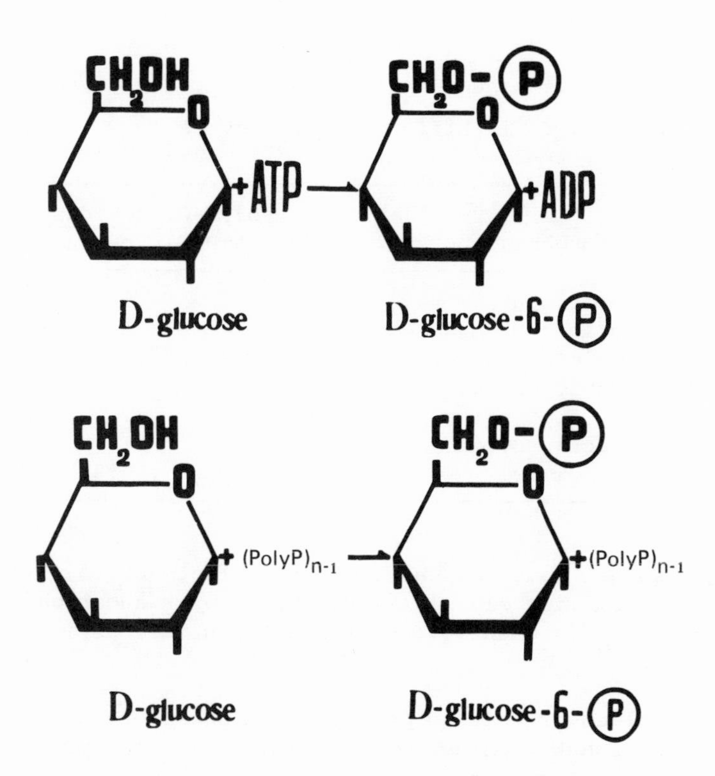

Figure 10. A scheme of reactions catalyzed by hexokinase and polyphosphate hexokinase.

actinomycetes. As one can see in Table II, the activity of polyphosphate-hexokinase is many times higher than that of ATP-hexokinase in such ancient (according to Kluyver and Van Niel; 60) representatives of this group of microorganisms as micrococci, tetracocci, and propionic bacteria. The activity of ATP-hexokinase is much higher than that of polyphosphate-hexokinase in the phylogenetically youngest representatives of this class of microorganisms--the true actinomycetes.

Thus these data indicate that utilization of high molecular weight polyphosphates as activated phosphate donors for glucose phosphorylation might have originated earlier than utilization of ATP for the same purpose.

These experimental data confirm the opinion of A. N. Belozersky (61) who in 1957 expressed the idea that high molecular polyphosphates in the primary living organisms could carry out the functions that are mainly related to ATP in the contemporary organisms. Our data substantiate that idea and testify to the fact that high

Table II

Polyphosphate-Glucokinase Activity:ATP-Glucokinase Activity Ratio in Different Actinomyces

Organism	PolyP-Glucokinase / ATP-Glucokinase
Micrococcus	5
Tetracoccus	4
Propionibacterium	10
Mycobacterium	2
Corynebacterium	2
Proactinomyces	1
Actinomyces	0.3

molecular weight polyphosphates of protobionts could primarily participate in coupling glycolysis reactions with sugar phosphorylation reactions, e.g. in such system of reactions (62) as is shown in Fig. 11.

High molecular polyphosphates could phosphorylate glucose to glucose-6-phosphate which, being subjected to the process of glycolysis, could be converted into 1,3-diphosphoglyceric acid. The latter could lead to polyphosphate synthesis.

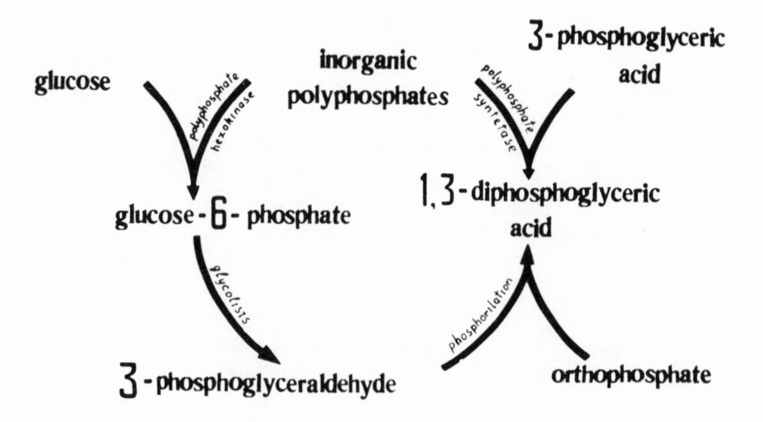

Figure 11. The scheme of polyphosphates participating in coupling of energy-yielding and energy-consuming processes in protobionts.

So, at the early stages of evolution of energy systems in living organisms the function of coupling exergonic and endergonic reactions, which is chiefly associated with ATP in contemporary organisms, was carried out, to a certain extent, by more primitive macroergic compounds, i.e. inorganic polyphosphates.

The hypothetical scheme presented in Fig. 12 sums up all the data obtained by us and the assumptions we made on this basis (63).

First, when the temperatures on the Earth were very high, a considerable part of phosphates existed in the form of high molecular weight polyphosphates. As the temperatures became lower and the hydrosphere was formed, high molecular weight polyphosphates were involved in reactions of abiogenic transphosphorylation, which was followed by polyphosphate formation. Somewhere at this stage, according to Ponnamperuma _et al._, ATP could be formed in the primordial ocean.

In the process of development of the primary organisms the functions which originally were chiefly carried out by pyrophosphate, as in redox processes in the membranes, and by high molecular weight polyphosphates in the reactions in aqueous solutions, were gradually taken over by ATP.

Apparently the far greater polyfunctionality and identifiable adenosine triphosphate structure, as compared with the monotonously composed polyphosphates, accounts for the fact that in the

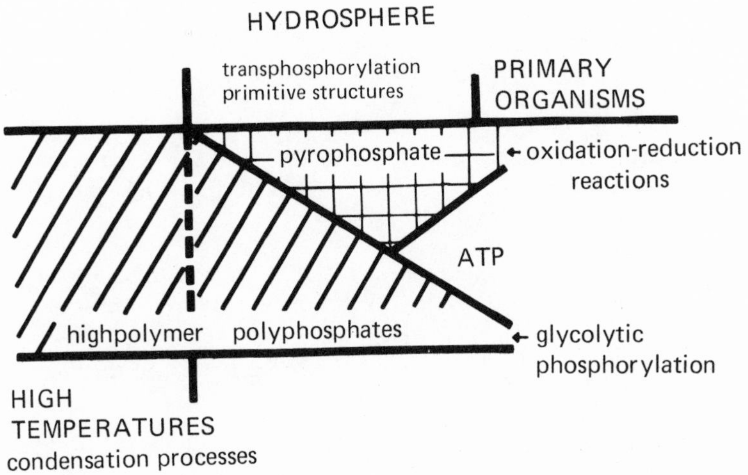

Figure 12. A hypothetical scheme of participation of high-molecular polyphosphates, pyrophosphate, and ATP in phosphorylation reactions at the different steps of origin and evolution of life on the Earth.

process of evolution it was ATP that took over many of the functions that had been carried out by high molecular weight polyphosphates and pyrophosphates in protobionts. One may assume that it was ATP's highly specific and easily identifiable structure that was the decisive factor which led to such wide utilization of this compound in various biochemical processes in contemporary organisms.

REFERENCES

1. Schramm, G., in: "The Origin of Life on Earth" (Oparin, A. I. et al., eds.), p. 312, Pergamon Press, New York, 1959.

2. Schramm, G., in: "The Origins of Prebiological Systems" (Fox, S. W., ed.), Academic Press, New York, 1965.

3. Schramm, G., and Pollmann, W., Angew. Chem. Intern. Ed. 2, 53 (1962).

4. Schramm, G., Grotsch, H., and Pollmann, W., in: Acides Ribonueleigues et Polyphosphates Colloq. Intern. CNRS, N 106, Ed. CNRS, p. 25, Paris, 1962.

5. Schramm, G., Grotsch, H., and Pollman, W., Angew. Chem. Intern. Ed. 1, 1 (1962).

6. Osterberg, R., and Orgel, L. E., J. Mol. Evol. 1, 241 (1972).

7. Miller, S. L., and Parris, M., Nature 204, 1248 (1964).

8. Steinmann, G. D., Lemmon, R. M., and Calvin, M., Proc. Nat. Acad. Sci. U.S. 52, 27 (1964).

9. Degani, Ch., and Halmann, M., in: "Molecular Evolution" (Buvet, R., and Ponnamperuma, C., eds.), Vol. 1, p. 224, North-Holland, Amsterdam, 1971.

10. Gabel, N. W., and Ponnamperuma, C., in: "Exobiology" (Ponnamperuma, C., ed.), p. 95, North-Holland, Amsterdam, 1972.

11. Beck, A., and Orgel, L. E., Proc. Nat. Acad. Sci. U.S. 54, 664 (1965).

12. Lohrmann, R., and Orgel, L. E., Science 161, 64 (1968).

13. Lipmann, F., in: "Proiskozdeniya predbiologicheskich sistem," p. 261, Mir. M, 1966.

14. Lowenstein, J. M., Biochem. J. 70, 222 (1958).

15. Lowenstein, J. M., Nature 187, 570 (1960).

16. Tetas, M., and Lowenstein, J. M., Biochemistry 2, 350 (1963).

17. LePort, L., Etaix, E., Godin, F., Leduc, P., and Buvet, R., in "Molecular Evolution, Vol. 1, Chemical Evolution and the Origin of Life" (Buvet, R., and Ponnamperuma, C., eds.), p. 197, North-Holland, Amsterdam, 1971.

18. Kulaev, I. S., and Skryabin, K. G., Abstracts of Symposium, "Origin of Life and Evolutionary Biochemistry," Varna, p. 22, 1971.

19. Ponnamperuma, C., Sagan, C., and Mariner, R., Nature 199, 222 (1963).

20. Rabinowitz, J., Chang, S., and Ponnamperuma, C., Nature 218, 442 (1968).

21. Gabel, N. W., and Ponnamperuma, C., in "Exobiology" (Ponnamperuma, C., ed.), p. 95, North-Holland, Amsterdam, 1972.

22. Fox, S. W., and Harada, K., Science 128, 1214 (1958).

23. Fox, S. W., and Harada, K., J. Am. Chem. Soc. 82, 3745 (1960).

24. Fox, S. W., Yuki, A., Waehneldt, T. V., and Lacey, J., in "Molecular Evolution," Vol. 1, "Chemical Evolution and the Origin of Life" (Buvet, R., and Ponnamperuma, C., eds.), p. 252, North-Holland, Amsterdam, 1971.

25. Schwartz, A. W., and Fox, S. W., Biochim. Biophys. Acta 134, 9 (1967).

26. Schwartz, A. W., and Ponnamperuma, C., Nature 218, 443 (1968).

27. Zuleski, F. R., and McGrinnes, E. T., Anal. Biochem. 47, 315 (1972).

28. Kulaev, I. S., in "Molecular Evolution" (Buvet, R., and Ponnamperuma, C., eds.), p. 458, North-Holland, Amsterdam, 1971.

29. Kulaev, I. S., and Skryabin, K. G., Zhur. Evol. Biochem. Physiol. 10, 900 (1974).

30. Kulaev, I. S., in "Evoliutsionnaya biokchimiya v serii Novoe v Zhizni, nayke i technike," Izdatelstvo Snanie Moscow, p. 46, 1973.

31. Oparin, A. I., in: "Origin of Life on Earth" (Oparin, A. I., et al., eds.), Pergamon Press, New York, 1959.

32. Oparin, A. I., in: "Proiskozdeniya predbiologicheskich sistem," p. 335, Mir M., 1966.

33. Oparin, A. I., Advan. Enzymol. 27, 347 (1965).

34. Harada, K., and Fox, S. W., "Proiskozdeniya predbiologiches kich sistem," p. 292, Mir M., 1966.

35. Lipmann, F., in: "Molecular Evolution, Vol. 1," "Chemical Evolution and the Origin of Life" (Buvet, R., and Ponnamperuma, C., eds.), p. 384, North-Holland, Amsterdam, 1971.

36. Baltscheffsky, M., Nature 216, 241 (1967).

37. Baltscheffsky, M., VII Intern. Congr. Biochem. Tokyo, Abstracts v. H-67, 897 (1967).

38. Baltscheffsky, M., Biochem. Biophys. Res. Comm. 28, 270 (1967).

39. Keister, D. L., and Minton, N. I., Biochem. Biophys. Res. Commun. 42, N 5 (1971).

40. Keister, D. L., and Minton, N. I., Arch. Biochem. Biophys. 147, 330 (1971).

41. Keister, D. L., and Yike, N. J., Biochemistry 6, 3847 (1967).

42. Kulaev, I. S., Mansurova, S. E., and Schadi, A. I., Biokhimiya 39, 197 (1974).

43. Liberman, E. A., and Tsofina, L. M., Biophysica 14, 1017 (1969).

44. Isaev, P. I., Liberman, E. A., Samuilov, V. D., Skulachev, V. P., and Tsofina, L. M., Biochim. Biophys. Acta 216, 22 (1970).

45. Mansurova, S. E., Beljakova, T. N., and Kulaev, I. S., Biokhimiya 38, 223 (1973).

46. Mansurova, S. E., Shakhov, Y. A., and Kulaev, I. S., Dokl. Akad. Nauk SSSR 213, N5 (1973).

47. Kulaev, I. S., and Bobik, M. A., Biokhimiya 36, 426 (1971).

48. Kulaev, I. S., Bobik, M. A., Nikolaev, N. N., Sergeev, N. S., and Uryson, S. O., Biokhimiya 36, 943 (1971).

49. Kulaev, I. S., Vorobyova, E. I., Konovalova, L. V., Bobik, M. A., Konoschenko, G. I., and Uryson, S. O., Biokhimiya 38, 712 (1973).

50. Nesmeyanova, M. A., Dmitriev, A. D., and Kulaev, I. S., Mikrobiologiya 43, 213 (1974).

51. Szymona, M., Bull. Acad. Polon. Sci. II-5, 379 (1957).

52. Szymona, M., and Ostrowski, W., Biochim. Biophys. Acta 85, 283 (1964).

53. Szymona, O., Uryson, S. O., and Kulaev, I. S., Biokhimiya 32, 495 (1967).

54. Uryson, S. O., and Kulaev, I. S., Dokl. Akad. Nauk SSSR 183, 957 (1968).

55. Uryson, S. O., and Kulaev, I. S., Biokhimiya 35, 601 (1970).

56. Szymona, M., and Szymona, O., Acta Microbiol. Polon. 11, 287 (1962).

57. Dirheimer, G., and Ebel, J. P., Compt. rend. Acad. Sci. 254, 2850 (1962).

58. Krasilnikov, N. A., in: "Opredelitel bacteriy i actinomicetov," Izdatelstvo A. N. SSSR, Moscow, 1949.

59. Krasilnikov, N. A., in: "Rukovodstvo po microbiologii, klinike i epidimiologii infekcionnich zabolevaniy," Izdateelstvo Medicina Moscow, 1962.

60. Kluyver, A. I., and van Niel, C. B., Zbl. Bacteriol. II abt. 94, 369 (1936).

61. Belozersky, A. N., in: "The Origin of Life on the Earth" (Oparin, A. I., et al., eds.), p. 322, Pergamon Press, Oxford, 1959.

62. Kulaev, I. S., in: "Abiogenes i nachalniye stadii evoliutsii zhizm" (Oparin, A. I., ed.), p. 97, Izdatelstvo Nauka, Moscow, 1968.

63. Kulaev, I. S., Abstr. Ninth Intern. Congr. Biochem. Stockholm, 1-7 July, Ba-4, 432, 1973.

SUCROSE SYNTHETASE AND ITS PHYSIOLOGICAL ROLE IN THE STORAGE TISSUES

A. L. Kursanov and O. A. Pavlinova

K. A. Timiryazev Institute of Plant Physiology of the USSR Academy of Sciences, Moscow

The biochemical studies related to the process of sugar accumulation in the sugar beet plant, started by Professor A. I. Oparin at the end of the twenties, were engendered by the great practical value of sugar beet and the necessity of increasing sugar content in the beet roots.

However, in the early thirties the studies resulted in significant scientific discoveries. Thus, for the first time the fundamental concept of the necessity of phosphoric acid for sucrose synthesis was brought forward (1). A new trend in studies related to the investigation of enzyme action in living plant cells (biology of enzymes) was developed (2).

The basic studies of A. I. Oparin were continued in the works of his followers on sugar transport and accumulation in the sugar beet plant (3). In those works it was shown that, in the sugar beet, the sugars are transported in the form of sucrose but not in the form of free hexose or phosphorylated sugars. On the basis of these studies, it was concluded that sucrose is not synthesized in the root tissues but is transported to them from the leaves, in which sucrose is mainly synthesized in the sugar beet plant.

That result set forth the question about the mode of transport and transformation of sucrose in the root,--the use of sucrose not only for accumulation in the storage parenchyma cells, but also as an energetic and "building" material, i.e. the utilization of disaccharide in cell respiration and biosynthesis, which proved to be closely connected with the growth processes in the root. The approach to this problem was based on the study of enzymes participating in sucrose metabolism in the root.

METHODS*

The experiments were carried out with the roots of sugar beet (Verkhnyachskaya 031 variety) grown on an experimental plot of the Institute of Plant Physiology of the USSR Academy of Sciences. The age of the roots varied from 4 days to complete maturity. The activity of invertase was determined by the technique described in two publications (4,5). The root tissue sample (2 g) was ground with distilled water at low temperature. The homogenate was dialyzed for 20 hours at 4°C. The suspension was brought to 10 ml and used to determine the activity of invertase. The incubation mixture contained 1 ml of suspension and 2 ml of 1 M acetate buffer (pH 4.7), containing 0.1% of sucrose. Samples of boiled homogenate were used as control. The incubation period is 1 hour for young roots and up to 24 hours for the roots of adult plants. The sucrose content in the filtrate was determined according to Roe. The enzyme activity was judged by the number of μmoles of sucrose split per hour per g of fresh tissue at 30°C.

Isolation of sucrose synthetase (UDPglucose:D-fructose-2-glucosyl transferase, E.C., 2:4:1:13).

The sample of root tissue (15 g) was ground in a mortar with quartz sand for 20 min with 30 ml of 50 mM K-phosphate buffer, pH 7, at 4°C; β-mercaptoethanol (20 mM) was added as a stabilizing agent. The mixture was filtered through dense cloth and the residue extracted for the second time. The extract was centrifuged at 15000 g, with 20 min of cooling. To the supernatant liquid was added 1 N CH_3COOH up to pH 5 for the precipitation of sucrose synthetase protein. After 30 min, the precipitate was centrifuged at 15000 g for 20 min. The precipitate was dissolved in 25 ml of cooled 10 mM K-phosphate buffer, pH 7.

Solid $(NH_4)_2SO_4$ was added to the solution up to 30% saturation. After 30 min in the ice water, the suspension was centrifuged at 15000 g for 15 min and the precipitate discarded. Solid $(NH_4)_2SO_4$ was added to the supernatant solution up to 55% saturation, and after 30 min centrifuged under the same conditions. The precipitate was dissolved in 2 ml of cooled twice-distilled water or 10 mM of K-phosphate buffer, pH 7, and dialyzed at low temperature overnight against 1 liter of distilled water or 50 mM of K-phosphate buffer. The dialyzed solution served as an enzyme preparation. The protein concentration was determined according to Lowry *et al.*

*The following abbreviations are used: UDP, uridine diphosphate; UDPG, uridine diphosphate glucose; UDPS, uridine diphosphate sugars.

The activity of sucrose synthetase in the sucrose synthesizing reaction (UDPG + fructose → sucrose + UDP) was determined by the amount of sucrose formed for 10 min at 37°C from UDPG and fructose. Sucrose was determined according to Roe. The reaction mixture contained in a total volume of 0.2 ml: UDPG, 0.5 μmoles; fructose, 2 μmoles; Tris-HCl buffer, pH 7.3, 10 μmoles; 0.1 ml of the enzyme solution (0.05-0.1 mg of protein). The activity was estimated in μmoles of sucrose formed per hour per mg of protein.

The activity of sucrose synthetase in the sucrose cleavage reaction (sucrose + UDP → UDPG + fructose) was determined by the amount of fructose released for 10 min at 37°C from sucrose in the presence of UDP or by the amount of UDPG synthesized under the same conditions.

The reaction mixture contained in a total volume of 0.2 ml: UDP, 0.5 μmoles; sucrose, 5 μmoles; citrate-phosphate buffer, pH 6.4, 10 μmoles; 0.1 ml of the enzyme solution (0.05-0.1 mg of protein). Fructose was determined colorimetrically according to Somogyi-Nelson, UDPG and UDP by quantitative paper chromatography. The separation of nucleotides was performed on washed paper with ethanol-1 M ammonium acetate, pH 7.5 (5:2) as solvent. Spots of nucleotides were cut out and eluted by 5 ml of 0.1 N HCl; the absorption spectra were estimated in UV light spectrophotometrically. The enzyme activity was estimated in μmoles of fructose or UDPG formed per hour per mg of protein.

RESULTS AND DISCUSSION

The study of enzymes involved in sucrose metabolism in sugar beet roots has shown (6,7) that sucrose can be utilized in two ways: a) with invertase hydrolyzing sucrose into glucose and fructose and b) with sucrose synthetase showing activity in cleavage of sucrose followed by UDPG formation and fructose release:

$$\text{sucrose} + \text{UDP} \rightleftharpoons \text{UDPG} + \text{fructose}$$

By an activity test of the second sucrose-synthesizing enzyme of plant tissues, the sucrose phosphate synthetase, which catalyzes the synthesis of sucrose from UDPG and fructose-6-phosphate:

$$\text{UDPG} + \text{fructose-6 phosphate} \longrightarrow \text{sucrose-6-phosphate} + \text{UDP}$$

it has been shown that its activity in the root is very low and is discovered only in the zones of the conducting root bundles. In zones of the storage parenchyma the sucrose phosphate synthetase is not active.

According to published data (4,5), the acid invertase (pH 4.0-5.0) is present in plant tissues in two forms: as a soluble (intracellular) and insoluble (bound to cell walls) invertase. The activity of the soluble invertase is much higher than the activity of the insoluble enzyme. In this connection, while studying the changes in invertase activity in root ontogeny, we estimated the total enzyme activity at pH 4.7 by using the dialyzed homogenate of root tissues as an enzyme preparation.

The data in Fig. 1 (curve 1) show that invertase is most active in the embryonic root, which appears at 4 day seed germination. The activity of invertase in young roots equaling 50 μmoles per g tissue per hour was comparable with the enzyme activity in sugar beet leaves. The high activity of invertase correlated with the increased content of hexoses in the tissues of 4 day roots and low concentration of sucrose (see Table I), that being indicative of the type of metabolism characteristic of the meristematic tissue. But as the root was growing and its tissues differentiated, the invertase activity decreased rapidly and it was impossible to reveal enzyme activity in 40-45 day plants with root weight of 1.5-3 g. As is clear from the tests, the roots show a significant increase in sucrose content and a decrease in the content of monosaccharides (Table I).

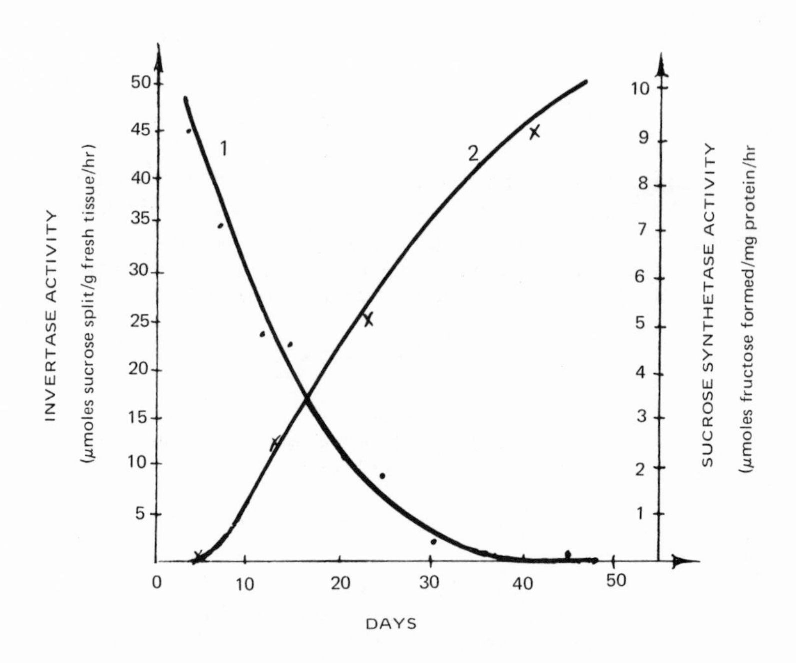

Figure 1. Activity of invertase (1) and sucrose synthetase (2) in early stages of root development.

Table I

Sugar Content in Sugar Beet Roots of Different Ages

(mg/g of fresh weight)

Plant age, days	Root weight g	Sucrose	Glucose	Fructose
4 day shoots	0.01	0.46	9.0	2.5
25	0.10	9.2	2.8	1.15
40	1.5	53.2	0.63	0.58
60	60	87.1	0.47	0.58
100	80	108.0	0.6	-
140	230	163.0	0.7	

In the subsequent periods of root development the invertase did not show any activity either, and thus did not play any important role in sucrose cleavage during the period of the most rapid root growth.

The loss by the root of the ability for sucrose cleavage by invertase is apparently associated with adaptation of its tissue metabolism to accomplish the function of sucrose accumulation.

Fig. 1 (curve 2) and Fig. 2 show age changes in sucrose synthetase activity. As is clear from Fig. 1, the enzyme activity is not revealed in embryonic roots. However, in the roots of 10-15 day shoots (30-50 mg of weight) the activity of sucrose synthetase can be already measured both by sucrose synthesis and UDPG formation. As the root grows, the enzyme activity increases. The highest level of sucrose synthetase activity is observed in the middle of vegetation in the period of rapid growth and intensive sugar accumulation (Fig. 2).

In that period, the difference between the rates of forward and backward reactions reaches its maximum; sucrose cleavage is 2-2.5 times more active than sucrose synthesis. In mature roots, the enzyme activity goes down by the end of the season. Thus, there is a correspondence between the activity of sucrose synthetase and the rate of root growth.

The data obtained in studying the action of invertase and sucrose synthetase in root ontogeny allowed us to conclude that invertase is active in the beet root only until it functions as the root proper rather than the root crop, the organ of sucrose accumulation. When the young root reaches the stage of transformation into a

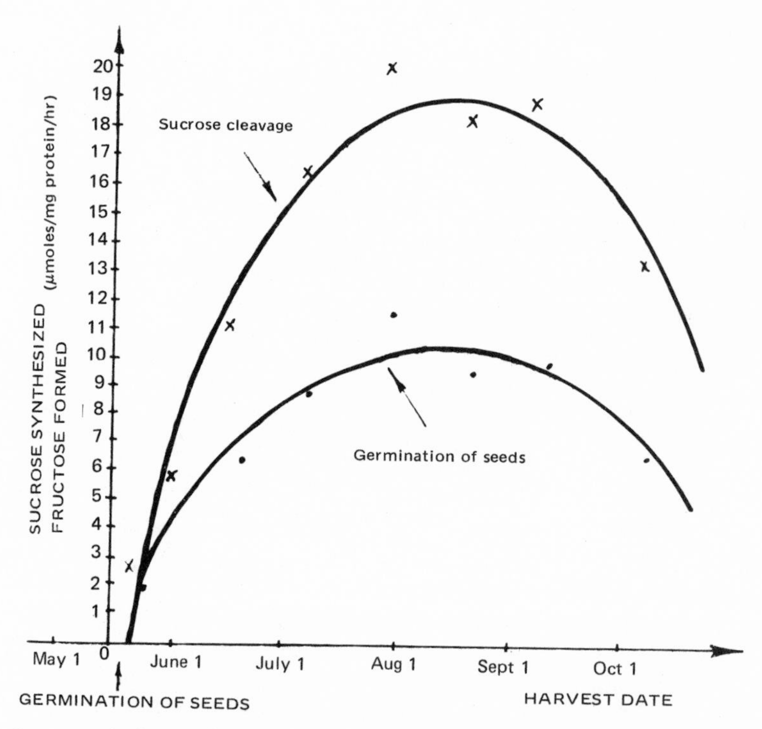

Figure 2. Activity of sucrose synthetase during growth of sugar beet roots.

typical accumulating organ, the synthesis of invertase protein in its tissues stops, or the enzyme is blocked by the invertase inhibitor (8). Invertase retains its activity only in the thin roots performing the absorption function. The transition of the roots to sugar accumulation, i.e. the change of the physiological function of the organ, results in changing the activity of enzymes of carbohydrate metabolism. Beginning with 40 days of age, and in the course of the whole subsequent period of development of the sugar beet plant, the key role in metabolizing the transport sucrose in the root passes to sucrose synthetase. According to recent works (6,7,9-16), this enzyme acts in the nonphotosynthesizing (storage) tissues not in the direction of sucrose synthesis, but in the direction of UDPG or ADPG (in some tissues) formation.

As is clear from analyses, UDPG and fructose released in sucrose cleavage reaction do not accumulate in the root, that being indicative of their intensive consumption in metabolism. We assumed that UDPG is first and foremost used for cellulose synthesis as for that of hemicellulose and pectin, i.e. polysaccharides which are used for cell wall building. It is known that UDPG is a direct precursor in cellulose biosynthesis (17,18). Hemicellulose and pectin are synthesized from other UDPS which are formed from UDPG as a result of enzymatic transformations.

Fructose, the second product of sucrose synthetase reaction, is used mainly for cell respiration. This concept was confirmed by experiments on uptake of ^{14}C-sugars and ^{14}C-UDPG from solutions by root tissue disks. One of these experiments is shown in Table II.

The experiments show that when ^{14}C-sucrose enters the tissues, the radioactive label is found mainly (to 70-75%) in the alcohol-soluble fraction. It indicates that when penetrating into the cells, ^{14}C-sucrose is mainly used for accumulation. It is accumulated in the storage parenchyma cells (vacuole-limited by tonoplast). 25-30% of ^{14}C-sucrose which penetrated into the tissues is used for biosynthesis of polymers. At ^{14}C-UDPG uptake on the contrast, over 70% of the label is found in the alcohol-insoluble fraction (polysaccharides). The experiment with ^{14}C-UDPG was conducted in such a way that root tissues absorbed ^{14}C-UDPG together with non-labelled fructose out of the solution, i.e. we obtained substances which are direct precursors in sucrose synthesis. In spite of that, only about 30% of ^{14}C incorporated into the alcohol-soluble fraction. In the alcohol-soluble fraction, ^{14}C-UDPG was also found. Thus, it can penetrate through cell membranes although the rate of UDP transport is much lower than that of free glucose penetration into the cells.

Hence, it could be concluded that the competition between the enzyme root systems for using UDPG, which can be utilized both for sucrose synthesis (action of sucrose synthetase), and the synthesis of cell wall polysaccharides (action of polysaccharide synthetases) results mainly in UDPG utilization for polysaccharide synthesis. Thus the action of sucrose synthetase *in vivo*, i.e. in the growing root, is mainly directed at UDPG formation.

The study of sucrose synthetase has been advanced, mainly in recent years. Rather recently (19), the necessity of protecting enzymes during tissue homogenization from inactivation by the products of polyphenol oxidase reaction by adding thiols (β-mercaptoethanol, dithiothreitol, cysteine) into the isolation medium was shown. The sucrose synthetase reaction is easily reversible.

Studies of the effect of substrate concentration on the activity of sucrose synthetase of some plant tissues (15,16) have shown essential differences between sucrose synthesis and cleavage reactions catalyzed by this enzyme. In the sucrose-synthesizing reaction the substrate saturation curve (with respect to UDPG and fructose) was hyperbolic in shape. In sucrose cleavage reaction, the substrates (UDP, sucrose) produce saturation curves showing a sigmoidal shape, i.e. the reaction does not follow the ordinary Michaelis-Menten kinetics.

Table II

Utilization of ^{14}C-Glucose, ^{14}C-Sucrose, and ^{14}C-UDPG in Sugar Beet Root Tissue Metabolism at Sugar Uptake from Solutions

(Thousand counts/min per g of tissue for 2.5 hr; concentration of solutions: 12 mM, radioactivity: $1.2 \cdot 10^6$ counts/min)

Sugar uptake	Total amount of ^{14}C absorbed	Incorporation of ^{14}C into fractions		% of ^{14}C in fractions	
		Alcohol soluble	Alcohol insoluble	Alcohol soluble	Alcohol insoluble
^{14}C-Glucose + fructose	47.81	22.65	25.16	47.4	52.6
^{14}C-Sucrose	18.85	14.08	4.77	74.6	25.3
^{14}C-UDPG + fructose	15.78	4.51	11.27	28.6	71.4

Similar features are observed in the sugar beet root enzyme. We have estimated K_m and V_{max} for UDPG in the sucrose synthesizing reaction. K_m = 1.8 mμ, V_{max} = 0.06 μmoles/min. The estimation of the effect of UDP concentration on the rate of UDPG synthesis in the sucrose cleavage reaction has shown that the substrate saturation on curve does not follow Michaelis-Menten kinetics, and K_m was not determined for UDP. Thus the root sucrose synthetase shows allosteric properties.

The root sucrose synthetase is not strictly NDPS- or NDP-specific and shows the highest rate of reaction in the presence of UDPG (sucrose synthesis) and UDP (sucrose cleavage). However, the enzyme acts also in the presence of ADP, showing 30% activity as compared to UDP. Other NDP (GDP, CDP) were not active. Root sucrose synthetase does not cleave raffinose (Table III) and thus differs, for example, from invertase which can hydrolyze oligosaccharides, though at a low rate.

The study of regulatory properties of sucrose synthetase *in vivo*, i.e. in the whole root, growing and accumulating sucrose is of interest. Sucrose synthetase is the key growth enzyme which uses part (approximately 30%) of sucrose transported from leaves to roots for growth processes. We assumed that sucrose synthetase can in this way control the alternative between growth and sugar accumulation, which is of

Table III

Utilization of Different Sugars in the

UDPG-Synthesizing Reaction*

	Control		UDPG Formed	
Substrates	UDP	UDPG	UDP	UDPG
	μmoles		μmoles	
Sucrose + UDP	0.48	0	0.30	0.14
Raffinose + UDP	0.48	0	0.41	0

*The reaction mixture contained: UDP, 0.5 μmoles; sucrose (raffinose), 2 μmoles; citrate buffer, pH 6.5, 10 μmoles; 0.1 mg of protein. Total volume = 0.2 ml, 10 min, 30°C.

great importance for sugar beet productivity. Phytohormones, ions, and some metabolites are the factors which can control the enzyme activity.

We found (6) that the washing of root tissue discs in aerated water and in solutions of indolyl acetic acid (IAA) at 10^{-5}-5.10^{-5} M ("aging") resulted in increase in the activity of sucrose synthetase by 20-30%. Apparently, it was caused not by the formation of sucrose synthetase protein *de novo*, but by activation of the enzyme already present in the tissues (20). Indeed, the adding of IAA into the reaction mixture (to estimate the enzymic activity), caused activation of sucrose synthetase. The result indicated that in the sugar beet root the sucrose synthetase is to a certain extent controlled by IAA, which while stimulating the growth processes and attracting assimilates intensifies through sucrose synthetase the formation of metabolites (UDPG, fructose) necessary for growth. The data show also that sucrose synthetase is not repressed in the root. Apparently, sucrose synthetase belongs to constitutive enzymes of plant cells.

It has been shown that some ions can influence the activity of sucrose synthetase. The study of the action of NH_4^+, Mg^{2+}, Mn^{2+}, B^{3+}, have shown that a clear activating effect on sucrose synthetase *in vitro* is only produced by Mn^{2+}. The effect is produced in the sucrose synthesizing reaction. The sucrose cleavage reaction is inhibited by Mn^{2+}. The nature of this inhibition is being studied in our laboratory. The study of the controlling effect of ions on the functioning of the key enzymes of the sugar beet root carbohydrate metabolism can be of practical importance for the regulation of the relationship between its growth intensity (crop yield) and sugar content.

In the evolution of the sugar beet from the wild forms to the contemporary varieties, the development of a system of transferring enzymes in the roots was, apparently, of great importance. Among them, sucrose synthetase (glycosyl-transferase) replaced invertase in a certain stage of root ontogenesis. That made possible the deposition of sucrose in the sugar beet root cells as a stable reserve substance close to the basic routes of carbohydrate metabolism. On the other hand, since the reactions catalyzed by transferring enzymes proceed with energy conservation of the glycosyl bond, the storage sugar beet tissue has the possibility of using sucrose for its current requirements at much lower losses of energy. That has become another important step in the evolution of wild forms with thin roots into contemporary varities with large sugar-accumulating root crops.

REFERENCES

1. Oparin, A. I., and Kursanov, A. L., Biochem. Z. 239, 1 (1931).

2. Oparin, A. I., Izv. Akad. Nauk. USSR, Ser. biol. No. 6, 1733 (1937).

3. Kursanov, A. L., "Relationship of Physiological Processes in Plants," XX Timiryazev Readings Akad. Nauk USSR (1960).

4. Turkina, M. V., and Sokolova, S. V., Fiziol. Rastenii, USSR 15, 5 (1968).

5. Engel, O. S., and Kholodova, V. P., Fiziol. Rastenii, USSR 16, 973 (1969).

6. Pavlinova, O. A., and Prasolova, M. F., Fiziol. Rastenii, USSR 17, 295 (1970).

7. Pavlinova, O. A., and Prasolova, M. F., Fiziol. Rastenii USSR 19, 920 (1972).

8. Kursanov, A. L., Dubinina, I. M., and Burakhanova, E. A., Fiziol. Rastenii, USSR 18, 568 (1971).

9. Pavlinova, O. A., Fiziol. Rastenii, USSR 18, 722 (1971).

10. de Fekete, M.A.R., Planta 87, 311 (1969).

11. de Fekete, M.A.R., Ber. Dtsch. bot. Ges. 83, 161 (1970).

12. Pressey, R., Plant Physiol. 44, 759 (1969).

13. Delmer, D. P., and Albersheim, P., Plant Physiol. 45, 782 (1970).

14. Hawker, J. S., Phytochemistry 10, 2313 (1971).

15. Murata, T., Agr. Biol. Chem. 35, 1441 (1971).

16. Murata, T., Agr. Biol. Chem. 36, 1815 (1972).

17. Ordin, L., and Hall, M. A., Plant Physiol. 43, 473 (1968).

18. Tsai, C. M., and Hassid, W. Z., Plant Physiol. 51, 998 (1973).

19. Anderson, J. W., Phytochemistry 7, 1973 (1968).

20. Lavintman, N., and Cardini, C. E., Plant Physiol. 43, 434 (1968).

AMINO AND FATTY ACIDS IN

CARBONACEOUS METEORITES

Keith A. Kvenvolden
Chemical Evolution Branch-Planetary Biology Division
Ames Research Center
NASA
Moffett Field, California 94035

The presence of organic substances in carbonaceous meteorites has been known for at least one hundred and forty years. In 1834, for example, Berzelius (1) extracted complex organic substances from the Alais meteorite, and speculated about the significance of his discovery and its relationship to the possibility of extraterrestrial life. Since that time many studies of the organic chemistry of meteorites have been undertaken (2), and during this period controversies have developed concerning whether the organic material was a product of some extraterrestrial life. Within the last few years, however, new discoveries have narrowed the speculation. It is now generally believed that the indigenous organic compounds in meteorites are not derived from an extraterrestrial biota, but rather have been produced extraterrestrially by non-biological, chemical syntheses (3-8). In fact, these recent discoveries have provided persuasive, naturally occurring evidence in strong support of the theory of chemical evolution as first proposed by A. I. Oparin (9) in 1924.

Analyses of two C2 carbonaceous meteorites have provided much of the latest evidence which seems to support Oparin's theory on the origin of life. These meteorites, falling to Earth nineteen years apart, are remarkably similar in chemical composition (10). Murray meteorite fell in Murray County, Kentucky on September 20, 1950; Murchison meteorite fell on September 28, 1969, near Murchison, Australia. Both of these meteorites have a complex mixture of simple organic molecules along with polymeric organic substances which contain most of the more than two percent carbon of these specimens.

Two classes of organic compound of particular interest in contemplations about the origin of life are amino acids and fatty acids, because both of these kinds of molecule are important constituents of living systems. In the laboratories at Ames Research Center we have discovered that these same kinds of molecule are present in Murray and Murchison meteorites. The amino acids were discovered in 1970 (3) and the fatty acids were found in 1973 (11).

Amino acids in Murray and Murchison carbonaceous meteorites are similar in composition and distribution. The structures for eighteen amino acids have been described (4,8). They contain from two to six carbon atoms, and for molecules containing up to four carbon atoms most of the theoretically possible amino acid isomers are present (Table I). Fig. 1 shows a histogram of the approximate distribution of amino acids in the Murchison meteorite; Murray meteorite shows a similar pattern. To the original eighteen amino acids found in Murchison meteorite can now be added leucine and isoleucine (12). Continuing studies have demonstrated that seventeen additional amino acids are present in the Murchison meteorite (13), but these have only been identified by structural class and not yet as specific molecular species. In addition, there are a multitude of minor nitrogen-containing compounds which remain to be identified. Thus, there are in carbonaceous meteorites twenty amino acids whose identities have been well-established. But only eight of these are the same as found in the twenty amino acids of proteins. The other amino acids in meteorites are classified as nonprotein amino acids.

The suite of amino acids found in these carbonaceous meteorites differs greatly from the population of amino acids commonly found in living systems. Also, the enantiomeric distribution of amino acids in meteorites and living systems differ; meteorites contain racemic mixtures (DL) (3), while the proteins of living systems contain only one enantiomer (L). For these reasons, amino acids in meteorites are believed to have been synthesized by nonliving processes. If the abiotic processes which created the amino acids in carbonaceous meteorites resemble the processes which created the original collection of amino acids from which life eventually evolved, then, protoproteinaceous materials may have been characterized by compositions different from those found in today's proteins. Early evolution may have required that organisms synthesize many of the fundamental amino acid building blocks now used in the construction of proteins.

In laboratory experiments testing the concepts of chemical evolution, Miller (14) demonstrated that amino acids could be produced from the interaction of an electric discharge on a mixture of gases which was presumed to simulate the atmosphere of the primitive Earth. In this work nine amino acids, of which four are found in contemporary protein, were identified. In later experiments (15,16) thirty-three amino acids have been identified, including the twenty now known to be indigenous to carbonaceous meteorites.

Table I

Amino Acids in Murray and Murchison Meteorites

(identified as of December 1973)

Carbon number	Name
2	Glycine
3	α-Alanine β-Alanine N-Methylglycine
4	α-Aminoisobutyric acid α-Amino-*n*-butyric acid β-Aminoisobutyric acid β-Amino-*n*-butyric acid γ-Aminobutyric acid N-Methylalanine N-Ethylglycine Aspartic acid
5	Valine Isovaline Norvaline Proline Glutamic acid
6	Pipecolic acid Leucine Isoleucine

Not only were amino acids found in Miller's now classic experiment (14), but also the three monocarboxylic fatty acids, formic, ethanoic, and propanoic acids, were reported. In a later simulation experiment, Allen and Ponnamperuma (17) identified ethanoic, propanoic, butanoic, 3-methylbutanoic, petanoic, and 4-methylpentanoic acids. Now all of these compounds, except formic acid, along with about ten other monocarboxylic fatty acids have been discovered in Murray and Murchison meteorites (11). Concentration of total monocarboxylic fatty acids (about 200 μg/g) recovered from these meteorites exceeds the concentration of amino acids by an order of magnitude.

Identification of monocarboxylic fatty acids in the meteorites was based on coincidence of gas chromatographic retention times and equivalence of mass spectra between the sample compounds and

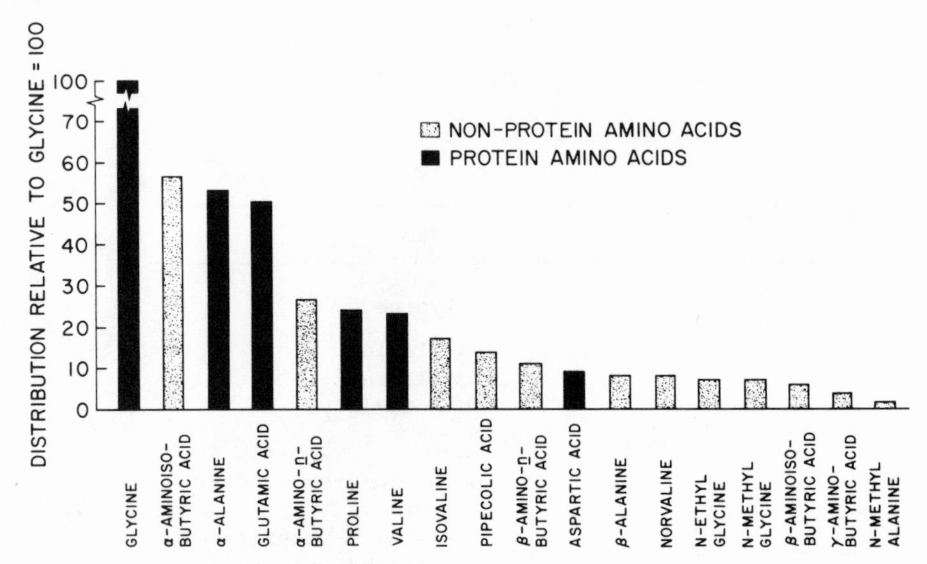

Figure 1. Approximate distribution of eighteen amino acids in Murchison carbonaceous meteorite (abundances relative to glycine = 100).

standards. Fig. 2 shows a gas chromatogram of the fatty acids in the Murray meteorite; chromatograms for the fatty acids in the Murchison meteorite are similar. Table II lists the monocarboxylic fatty acids identified thus far in the study. Formic acid has not yet been identified by mass spectrometry, but gas chromatographic evidence suggests its presence.

Mass spectra of these fatty acids are relatively easy to interpret (18). The molecular weights of the homologous series are two mass units greater than the molecular weights of a homologous series of hydrocarbons having two carbon atoms more than the corresponding fatty acid. Significant peaks at mass 45 due to loss of $COOH^+$ are common. Characteristic rearrangement ions occur at mass 60, which is the base peak for a number of these acids, as well as at masses 74 and 88. Molecular ions are usually present, but are generally weak. Fig. 3 shows a comparison of the mass spectra of standard 3-methylbutanoic acid and the equivalent compound in the Murray meteorite.

Both straight- and branched-chain fatty acids are present in the meteorites. All possible monocarboxylic fatty acid isomers containing two, three, four, or five carbon atoms are accounted for. Of the eight possible fatty acids containing six carbon atoms, at least five have been identified. (The enantiomeric distribution of the asymmetric fatty acids has not yet been determined.) The presence of such a collection of isomers of low molecular weight

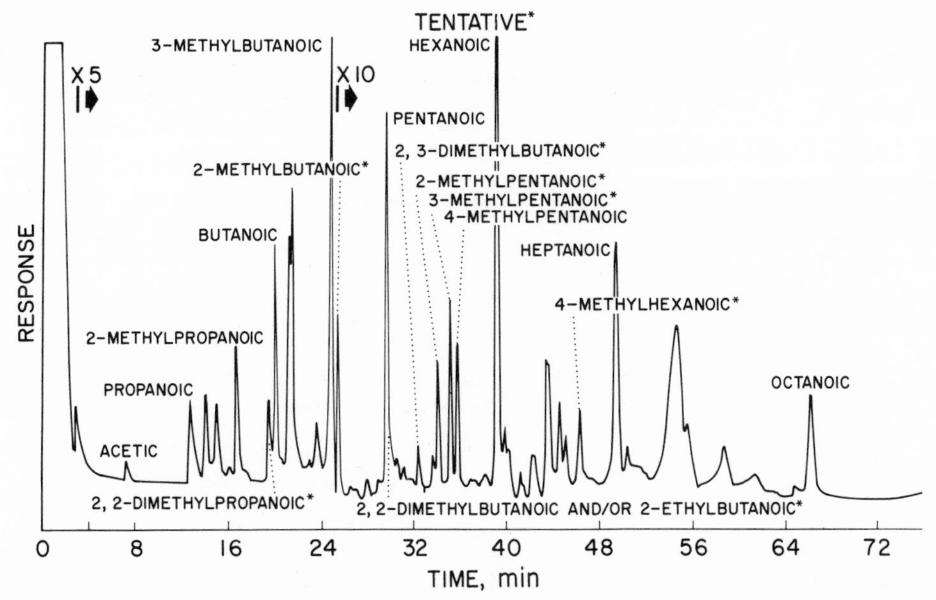

Figure 2. Gas chromatogram of monocarboxylic fatty acids in Murray meteorite. The flame ionization detector was located on the backside of the molecular separator which is between the gas chromatograph (Perkin Elmer Model 990) and the mass spectrometer (Dupont Model 491). Chromatographic column: stainless steel 46m x 0.05cm coated with UCON LB 550-X+10% H_3PO_4; temperature programmed from 70-120°C at 1° per min.

*Indicates tentative identification

fatty acids is comparable to the similar isomeric distribution of amino acids having two to four carbon atoms (4). A random chemical synthesis was probably responsible for producing both populations of molecules. Whether a relationship exists between the two populations of molecules is uncertain, but some of the fatty acids and amino acids possibly could have been derived from a common aldehyde precursor.

Within the Murray and Murchison carbonaceous meteorites are some of the basic building blocks for natural, fat- and wax-containing systems (19) which are present in, and apparently necessary for, most living organisms. The biological synthesis of long chain fatty acids requires acetyl and propionyl moieties associated with coenzyme A, and acetic and propionic acids are both present in the meteorite. Perhaps the very common fatty acids in living systems, i.e., myristic, palmitic, and stearic acids, were initially constructed from simpler fatty acids after life had originated.

Table II

Monocarboxylic Fatty Acids in Murray and Murchison Meteorites

(identified as of December 1973)

Carbon number	Name
2	Ethanoic (acetic)
3	Propanoic (propionic)
4	Butanoic (butyric) 2-Methylpropanoic (isobutyric)
5	Pentanoic (valeric) 3-Methylbutanoic (isovaleric) 2,2-Dimethylpropanoic 2-Methylbutanoic*
6	Hexanoic (caproic) 4-Methylpentanoic (isocaproic) 2-Methylpentanoic* 3-Methylpentanoic* 2,3-Dimethylbutanoic** 2,2-Dimethylbutanoic** or 2-Ethylbutanoic**
7	Heptanoic acid (heptoic acid) 4-Methylhexanoic acid*
8	Octanoic acid (caprylic acid)

*Tentative identification (based on comparison with published mass spectra, but without comparison of chromatographic retention times.

**Identification based on rationalization of mass spectra only.

Other studies at Ames Research Center are showing that both straight- and branched-chain dicarboxylic fatty acids are also present in the Murchison meteorite (20). Table III lists some of the compounds identified. Succinic acid is the most abundant dicarboxylic acid yet found, and this same compound was observed by Miller (14) in his original "prebiotic" experiments. In the meteorite, methylsuccinic acid appears to be present as a racemic mixture, an observation which complements the previous finding with regard to the racemic nature of amino acids in meteorites (3). As is the case with the monocarboxylic acids, the dicarboxylic fatty acids are biologically significant. They have been found in numerous organisms

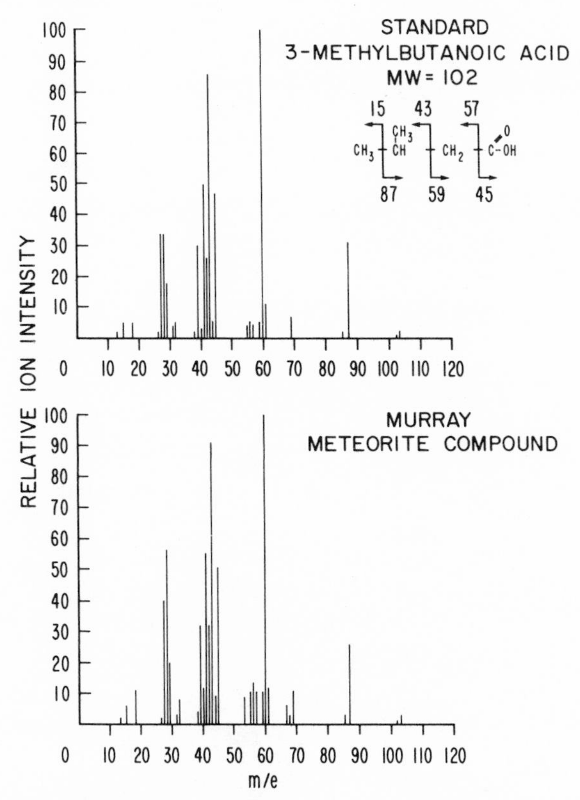

Figure 3. Mass spectra of standard 3-methylbutanoic (isovaleric) acid compared with spectra of meteoritic fatty acid. Base peak m/e 60 results from the following rearrangement:

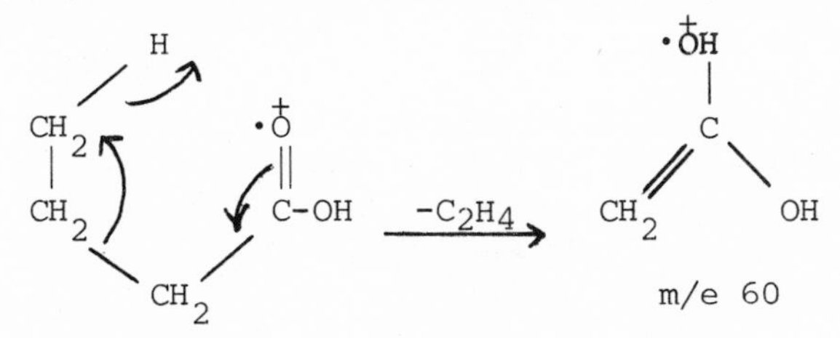

(19); succinic acid, for example, is an important intermediate in the Kreb's cycle.

It is now apparent that carbonaceous meteorites, at least C2 chondrites, contain a rich assemblage of molecules of biological significance which were most likely generated extraterrestrially in abiotic chemical syntheses. Carbonaceous meteorites, harboring some of these important building blocks of life, may have played a significant role in the origin of life. They may have transported these compounds through the reducing atmosphere of the primitive Earth to

Table III

Dicarboxylic Fatty Acids in Murchison Meteorite

(identified as of December 1973)

Carbon number	Name
2	Ethanedioic (oxalic)
3	Propanedioic (malonic)
4	Butanedioic (succinic) Butenedioic (maleic and/or fumaric)
5	Pentanedioic (glutaric) 2-Methylbutanedioic (methylsuccinic)
6	Hexanedioic (adipic) 2,2-Dimethylbutanedioic (α,α-dimethysuccinic) 2,3-Dimethylbutanedioic (α,β-dimethysuccinic) 3-Methylpentanedioic (β-methylglutaric)
7	3-Methylhexanedioic (β-methyladipic) Heptanedoic (pimelic)
8	Octanedioc (suberic)
9	Nonanedioc (azelaic)

the sites where life began, and through evolution the complex of molecules now found in living systems was formed. Even if meteorites did not play such a direct role in the origin of life, their content of organic compounds indicates that there were, and likely are, extraterrestrial chemical processes capable of generating many biologically significant molecules. Indeed, the early stages of Oparin's theory of chemical evolution find strong support in the organic chemistry of meteorites.

REFERENCES

1. Berzelius, J. J., Ann. Phys. Chem. 33, 113 (1834).

2. Hayes, J. M., Geochim. Cosmochim. Acta 31, 1395 (1967).

3. Kvenvolden, K. A., Lawless, J., Pering, K., Peterson, E., Flores, J., Ponnamperuma, C., Kaplan, I. R., and Moore, C., Nature 228, 923 (1970).

4. Kvenvolden, K. A., Lawless, J. G., and Ponnamperuma, C., Proc. Nat. Acad. Sci. U.S. 68, 486 (1971).

5. Cronin, J. R., and Moore, C. B., Science 172, 1327 (1971).

6. Oró, J., Nakaparksin, S., Lichtenstein, H., and Gil-Av, E., Nature 230, 107 (1971).

7. Oró, J., Gibert, J., Lichtenstein, H., Wikstrom, S., and Flory, D. A., Nature 230, 105 (1971).

8. Lawless, J. G., Kvenvolden, K. A., Peterson, E., Ponnamperuma, C., and Moore, C., Science 173, 626 (1971).

9. Oparin, A. I., "Proiskhozhdenie zhizni," Izd. Moskovskii Rabochii, Moscow, 1924.

10. Jarosewich, E., Meteoritics 6, 49 (1971).

11. Yuen, G. U., and Kvenvolden, K. A., Nature 246, 301 (1973).

12. Pereira, W. E., Summons, R. E., Rindfleisch, T. C., Duffield, A. M., Zeitman, B., and Lawless, J. G., Geochim. Cosmochim. Acta, submitted (1973).

13. Lawless, J. G., Geochim. Cosmochim. Acta 37, 2207 (1973).

14. Miller, S. L., Science 130, 245 (1959).

15. Ring, D., Wolman, Y., Friedman, N., and Miller, S. L., Proc. Nat. Acad. Sci. U.S. 69, 765 (1972).

16. Wolman, Y., Haverland, W. J., and Miller, S. L., Proc. Nat. Acad. Sci. U.S. 69, 809 (1972).

17. Allen, W. V., and Ponnamperuma, C., Currents Modern Biol. 1, 24 (1967).

18. Budzikiewicz, H., Djerassi, C., and Williams, D. H., "Mass Spectrometry of Organic Compounds," P. Holden-Day, Inc., San Francisco, 1967.

19. Markley, K. S., ed., "Fatty Acids," Pt. 1, Interscience, New York, 1960.

20. Lawless, J. G., Zeitman, B., Pereira, W. E., and Duffield, A. M., Nature, submitted (1974).

ISOLATED MICROSYSTEMS IN EVOLUTION

James C. Lacey, Jr. and Dail W. Mullins, Jr.

Laboratory of Molecular Biology
University of Alabama in Birmingham
University Station, Birmingham, Alabama 35294

This chapter is concerned with the subject of isolated microsystems and their importance in the evolution of the living state from purely chemical systems. We have chosen to discuss this particular topic for two main reasons. First, although the subject of the role of microsystems in evolution has been previously examined by several authors, a brief review of the available literature has revealed that certain basic concepts relative to this topic have not been, for the most part, adequately developed. Second, Professor A. I. Oparin (1), the man whose work we honor here, and his colleagues have spent a considerable amount of time and effort in exploring the many facets of one particular type of chemical microsystem, namely, coacervate droplets. Their studies, coupled with those of Professor S. W. Fox (2), and his co-workers on proteinoid microspheres, have served to emphasize the great importance of isolated microenvironments with regard to the appearance and evolution of living systems.

Rather than attempt a critical review of the many and varied types of experiments which have been performed on the different kinds of modelled protocells, we have chosen, instead, to consider microsystems within the context of two major conceptual questions:

1) Why, and at what point, were isolated microsystems required in the transition from purely chemical systems toward the living state?

2) Why are contemporary cells and, presumably, the isolated microsystems from which they were evolved, necessarily so small?

Although we have examined certain aspects of question one previously

(3), some additional comments can be added at this time.

We certainly agree, as do most workers in the origin of life field, that the appearance of living systems on this planet was preceded in time by a rather lengthy period of chemical evolution, characterized principally by the accumulation, first, of various low molecular weight compounds and, subsequently, of polymers of certain of these substances. In essence, as stated by Calvin (4), the current period of biological evolution was preceded in time by a period of chemical evolution. Calvin imagines that these two incompletely defined periods overlap in time, resulting in a rather diffuse boundary within which the living state arose. Assuming such a boundary did exist, we would like to explore the questions as to why and when, in this diffuse transition from chemical to biological evolution, isolated microsystems first appeared and what role they might have played, and also, to inquire into the reason for their retention in all contemporary life forms. What factors, in other words, necessitated such small packages for both developing and contemporary life?

It is quite easy to make some definitive statements about the extremes of the above continuum. Early chemical evolution, for example, simply involved the formation of various gases in the primitive atmosphere, and did not produce microsystems. Conversely, contemporary living systems all seem absolutely dependent on a very specialized type of microsystem, the cell. Because of this fact, it is obvious that, in the transition from one extreme to the other, the appearance of isolated microsystems did take place. The major question, then, is why were (are) such small microenvironments essential for the maintenance of the living state?

Let us now try to piece together some possible steps in the transition between chemical evolution and biological evolution in an attempt to determine if there is a point where the continued development of living systems would appear to have necessitated an isolated microsystem.

Living systems, while necessarily based only upon the set of all possible physicochemical reactions and interactions are, nevertheless, restricted to that subset of those reactions which can be generated by using those chemicals available to them in the environment. These early available substances would have included, among others, amino acids, which are capable of thermal polymerization (5) and, subsequently, of spontaneous coalescence of polymers to form isolated microsystems. That small, isolated microenvironments are, indeed, capable of forming spontaneously from many different polymeric substances has been unequivocally demonstrated by Professor Oparin and his co-workers (1), and by Professor Fox and his colleagues using thermal proteinoids (5). From the available experimental evidence, it thus seems certain that microsystems did appear spontaneously, and that the smaller ones, presumably being more stable, were thus more likely to have persisted.

At this early stage, of course, the microsystems were generated from polymeric material furnished by the environment, and could have continued to grow and divide through accretion and spontaneous fission. So long as such structures remained entirely dependent upon the environment for such polymeric material, however, their continued existence would have been tenuous, at best.

Thus, we can imagine a time when the temperature of the Earth dropped and polymer production decreased. At this juncture, only those microsystems which had acquired the capability of synthesizing peptide bonds using energy and monomeric precursors furnished by the environment would survive. It has been adequately demonstrated that peptides can readily be formed from mixtures of aminoacyl adenylates (6). Initially, of course, these intramicrosystemically synthesized polypeptides would have had a composition of amino acids which reflected their individual compositions in the environmental milieu. In those areas of the Earth's surface where the amino acid composition of the environment was such as to allow for the synthesis of peptides suitable for incorporation into the microsystem structure, the latter would persist and continue to evolve higher orders of structure and functionality. Certainly the composition of the peptides was important, but perhaps the sequence of amino acids in the peptides would not have been critical. For example, peptides containing relatively large amounts of hydrophobic amino acids might have given rise to membranous structures.

Achievement of the first stage of somewhat refined independence from the environment, however, would not have greatly diminished the number and severity of external selection pressures. In the complete absence of externally available structural peptides, for example, only those microsystems would persist which were capable of synthesizing such structural elements more rapidly than they were lost by degradation. Similarly, by relying now upon monomeric precursors rather than pre-synthesized polymers from the environment, microsystems would, of necessity, have become highly dependent upon the processes of diffusion. In fact, given a rate of peptide synthesis sufficient to counter spontaneous degradation as explained above, the supply of monomeric polymerization precursors to all parts of the microsystem (as well as the subsequent diffusion of the peptide product to the boundary) became _the_ limiting factor in the maintenance and continued evolutionary expansion of these structures.

Size restrictions, too, act as selection pressures for microsystems which have attained independence from the process of external polymer accretion, since the probability of finding a molecule at any point in a microsystem at one time is a function of, among other things, the microsystemic volume. For example, if an isolated microsystem contains only a very few copies of a particular functional molecule, then the probability of one of these functional entities diffusing to a particular site of action is certainly a function of, among other things, the volume of the microsystem. As an illustration of why molecular size and diffusion represent selection pressures for small size, we have calculated the probability of finding a hemoglobin molecule within a spherical structure (hypothetical microsystem) as a function of the radius of the sphere. These data are shown in Table I. As can clearly be seen, molecular size and diffusional requirements would have imposed an additional selection pressure favoring small size, in addition to the stability requirement mentioned above.

Up to this point in our scenario, isolated microsystems have had to contend only with the synthesis of peptides having roughly similar compositions rather than specific sequences. Once such microsystems had become adapted to a particular environmental niche with respect to, for example, amino acid composition, sudden changes in that composition could prove extremely deleterious in the absence of some enzymatic means of adjusting to the change. One can imagine,

Table I

Probability of Finding a Single Hemoglobin Molecule (P) and Two Adjacent Hemoglobin Molecules (P^2) in Spheres of Various Radii

Sphere radius (μ)	Sphere volume (μ^2)	P	P^2
.0625	1.0×10^{-3}	1.1×10^{-3}	1.2×10^{-6}
.0125	8.2×10^{-3}	1.3×10^{-4}	1.7×10^{-8}
.250	6.5×10^{-2}	1.7×10^{-5}	2.9×10^{-10}
.500	5.2×10^{-1}	2.1×10^{-6}	4.4×10^{-12}
1.0	4.2	2.6×10^{-7}	6.8×10^{-14}
2.0	33.5	3.3×10^{-8}	1.1×10^{-15}
5.0	523	2.1×10^{-9}	4.4×10^{-18}

The volume of the hemoglobin molecule, an oblate spheroid of dimensions 64 x 55 x 50 Å, was calculated to be $1.1 \times 10^{-6}\ \mu^3$. $P = 1.10^{-6}$ /sphere volume, and represents the probability of locating a macromolecule (i.e., hemoglobin) at a particular point in the spherical microsystem. P^2 represents the probability of finding two such molecules at adjacent points in the sphere, and is simply the product of the individual probabilities (P).

for example, the growing necessity of peptide catalysts concerned with carrying out reproducible modifications of certain of the incoming monomeric compounds so that the compositional requirements of the structural polymers could be maintained. Now, however, the structural restrictions imposed upon the boundary peptides, which could be met through the synthesis of polymers of constant composition, are replaced by more functionally dynamic restrictions which demand that protoenzymes be synthesized which have nearly constant sequences. This, in turn, demands that isolated microsystems possess a molecular coding mechanism capable of insuring that peptides be synthesized which exhibit a minimal amount of gross sequence heterogeneity.

The introduction of a rudimentary coding apparatus into an isolated microsystem for the purpose of contending with an environment of changing chemical composition, however, necessitates the concomitant inclusion of at least two additional enzymatic functions. In addition to the coding molecule itself (which must, it should be stressed, contain the information necessary for the synthesis of all functional enzymes), it is necessary to postulate the existence of some means of converting the stored information into functional or structural material. We have chosen to call this decoding entity a "convertase," since it is involved in converting information. In addition, the presence of an informational source necessitates that some means exist whereby the information can be replicated with some fidelity; consequently, there must exist what we have termed a "replicase" for achieving this end.

In summary, we envision the minimal coded evolving system to consist of the following:

1. The structure itself, an isolated microsystem separated from the environment by some type of subunit boundary.

2. An informational molecule which contains information for the synthesis of:

3. A convertase, for informational decoding (synthesis of all enzymes, including itself),

4. A replicase, for informational replication, and

5. A structural molecule (or enzymes for the synthesis of a structural molecule).

Pictured schematically in time, we visualize this minimal system operating as shown in Fig. 1.

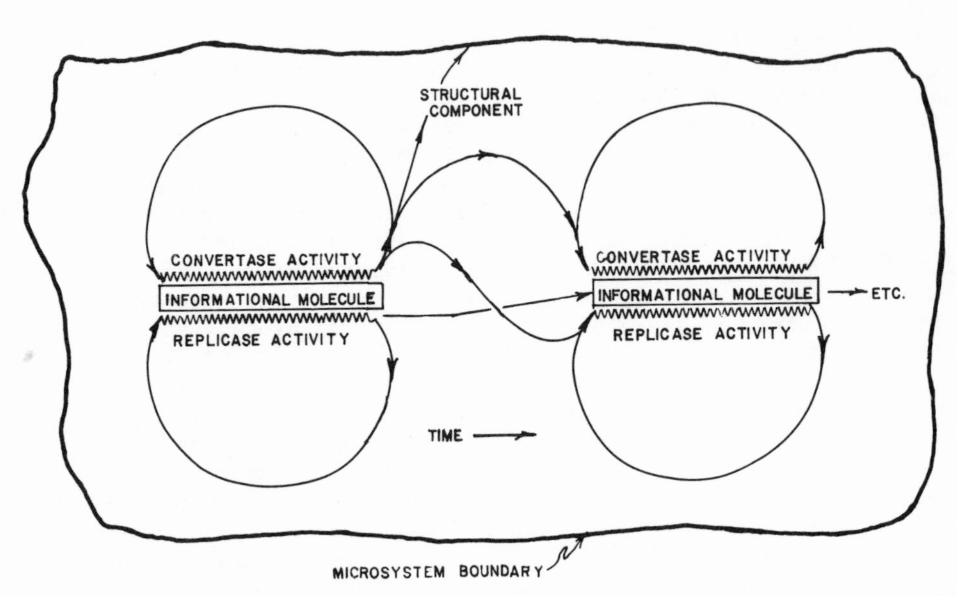

Figure 1. Diagrammatic representation of the minimal coded evolving system as described in the text. The informational molecule on the left is shown being operated upon by a replicase, resulting in the appearance of a new informational source which overlaps the former in space and time. At the same time, the first informational source is shown being translated by a convertase, resulting in the production of the three other obligatory microsystem components--a structural component, a replicase, and another convertase. These components, like the replicated informational source, continue to function in another space and another time.

It should be stressed at this point that we have not attempted, nor is it necessary, to define the precise nature of the proposed informational molecule, nor have we chosen to speculate as to the chemical nature and complexity of the convertase and replicase systems.

Previously (3), we postulated the most primitive coding system to be a protein-copying enzyme in which the enzyme functions as both "convertase" and "replicase." Such a system, however, would have been replaced rather early by one employing a coding molecule of low functionality (for example, a nucleic acid), since a protein template could itself be functional and might compete with the template product.

Regardless of its precise nature, we believe that the appearance of a coding molecule within an isolated microsystem was an event of crucial importance with respect to both the retention and size limitation of these structures. While such a molecule may contain information necessary for the survival of a microsystem, it (and, hence, its information) can only be selected in an evolutionary context if its informational content is expressed--that is, if it is decoded. Such expression requires that a "convertase" must be able to find, and act upon, the coding molecule. To make this event probable, a closed system of small size is required. In Table I, we have attempted also to illustrate the importance of this concept by showing how the probability that two hemoglobin molecules will be found at the same place in a spherical microsystem varies with the radius of the microsystem. As can be seen, this probability is phenomenally low, even for very small microsystems. It is apparent from this that, following the appearance of a coding system, only those informational molecules could persist which were continued within a closed system small enough to insure frequent physical contact between them and a decoding "convertase."

Similarly, only those informational microsystems could persist over long periods of time which contained a replication function for the information source. Again, in order for the "replicase" to locate and act upon the coding molecule, it must be confined within the same microsystem.

In addition to the probability factors mentioned above, small microsystems would also have been selected for on the basis of macromolecular diffusion rates as, for example, the time necessary to furnish membrane components to the microsystem boundary from some internal point of synthesis.

We believe, in conclusion, that there is something basic to the phenomenon of molecular coding which results in the selection of small isolated microsystems. Our basic tenet has been that, while molecular coding allows for the storage of vast amounts of information in an extremely small volume in space, this small volume must be capable of being located, and the information source decoded, before it can be expressed, and consequently, operated upon by natural selection. Each informational molecule must be enclosed within a structural microenvironment sufficiently small to insure its being located, and operated upon, by appropriate functional molecules if it is to survive and offer selective advantage to the isolated microenvironment.

In summary, we have attempted to show in Fig. 2 below how the introduction of molecular coding influenced the evolution of living microsystems.

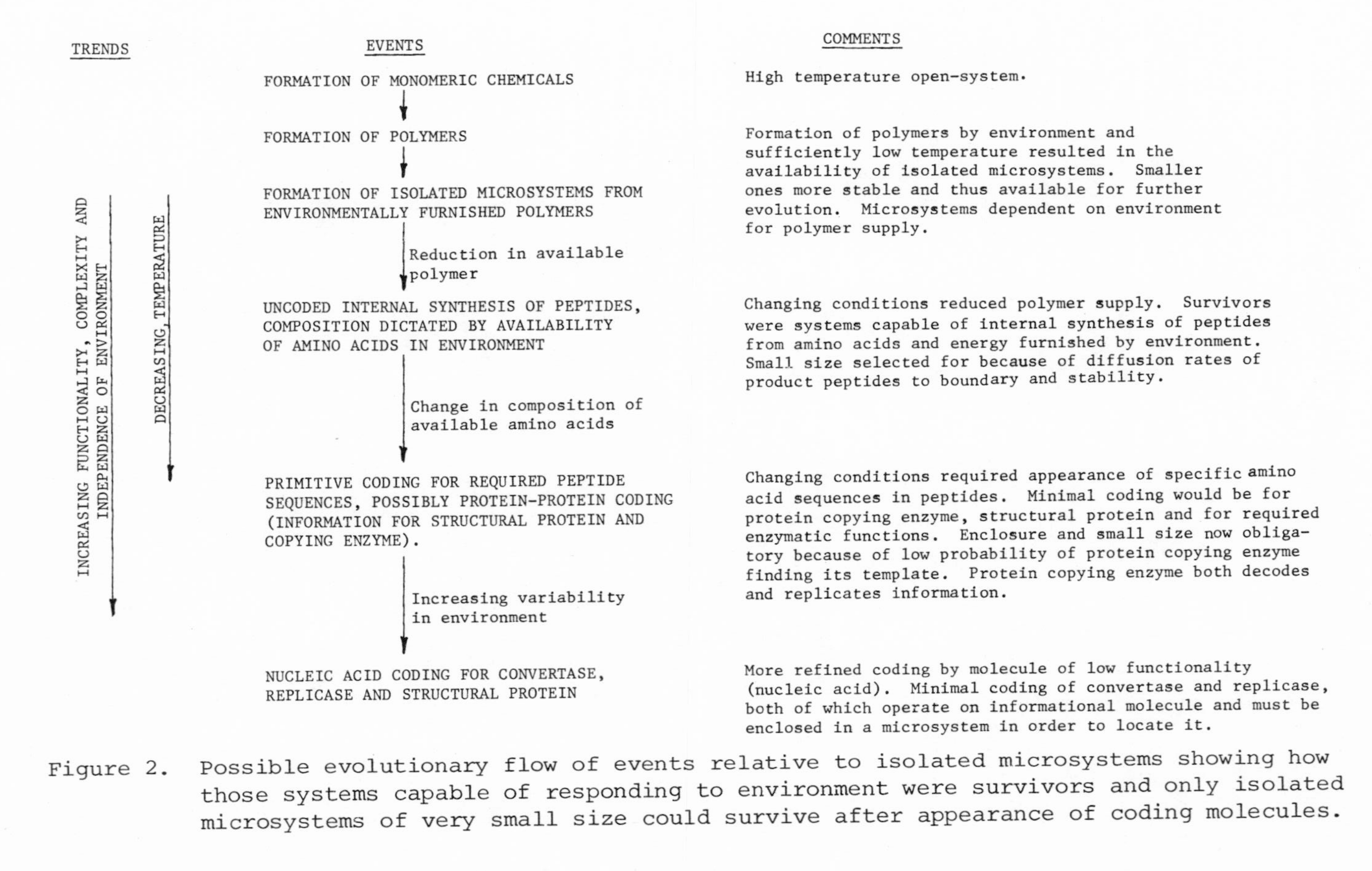

Figure 2. Possible evolutionary flow of events relative to isolated microsystems showing how those systems capable of responding to environment were survivors and only isolated microsystems of very small size could survive after appearance of coding molecules.

REFERENCES

1. Oparin, A. I., "Life: Its Nature, Origin, and Development," Academic Press, New York, 1964.

2. Fox, S. W., Naturwissenschaften 60, 359 (1973).

3. Lacey, J. C., Jr., and Mullins, D. W., Jr., in: "Molecular Evolution: Prebiological and Biological" (Rohlfing, D. L., and Oparin, A. I., eds.), p. 171, Plenum Press, New York, 1972.

4. Calvin, M., "Chemical Evolution," Oxford University Press, New York, 1969.

5. Fox, S. W., Naturwissenschaften 56, 1 (1969).

6. Krampitz, G., and Fox, S. W., Proc. Nat. Acad. Sci. U.S. 62, 399 (1969).

THE SEARCH FOR REMNANTS OF EARLY EVOLUTION IN PRESENT-DAY METABOLISM

Fritz Lipmann

The Rockefeller University

New York, NY 10021

As a newcomer to the field, it makes me quite proud to have been asked to participate in celebrating the 50th anniversary of the appearance of Oparin's "Origin of Life." This date truly marks the birth of a new branch of bioscience, which I have only lately begun to appreciate. In this book, Oparin established a base from which one could begin to explore the prebiotic phase of evolution.

I had been much involved in the analysis of the mechanism of protein, i.e. polypeptide, synthesis in modern organisms. Looking towards the past, I started searching for a less complex mechanism that might promise to present us with a simpler alternative of sequential polymerization of amino acids into polypeptides, hoping to find a possible predecessor of the ribosomal mechanism.

In the ribosomal system, specification of the amino acid position in ribosome-tRNA-linked polypeptide synthesis is a function of amino acid-specific enzymes that catalyze an ATP-linked amino acid activation (1,2). This made me propose (3), before the discovery of tRNA and mRNA, a model for polypeptide synthesis that consists of an assemblage of activating enzymes that fix amino acids to specific sites, from which they are collected vectorially in a specified order into polypeptides (Fig. 1). It is just such a mechanism using an assembly of activating enzymes as a template that we found to operate in the synthesis of several antibiotic polypeptides (4). The most fully analyzed synthesis of this type is that of the decapeptide, tyrocidine (5), schematically pictured in Fig. 2. After ATP-linked activation, the amino acids are fixed in thioester links to the polyenzymes (Fig. 3) and appear to be collected into polypeptides, through alternate transpeptidation and transthiolation, by an SH-pantetheine-peptide-carrier protein (6) built into each of the

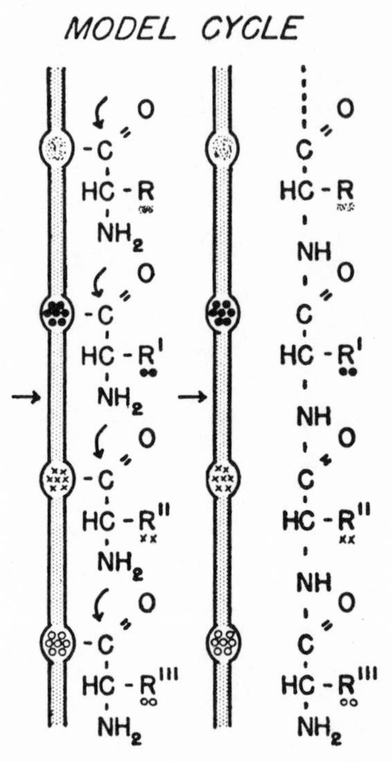

Figure 1. Model cycle. In this model of amino acid polymerization, the amino acids are supposed to be bound to specific activation centers lined up on a polyenzyme structure, from which they are polymerized in prescribed sequence (3).

polyenzymes (Fig. 4). Crude extracts were found to contain a factor that separates the polyenzymes irreversibly into three and six 70,000 Dalton subunits of activating enzymes and one 17,000 Dalton pantetheine-peptide-carrier protein. A similar separation of purified enzymes into subunits was obtained by sodium dodecyl sulfate gel electrophoresis (Fig. 5).

In an attempt to go back to possible mechanisms of transpeptidation from thioester-linked amino acids at an earlier stage of biochemical evolution, we became aware of the peculiar structure of a primitive protein, one of the ferredoxin apoproteins (7-9). The ferredoxin isolated from the anaerobic Clostridium pasteurianum is composed of 55 amino acids, with a repeat sequence of 16 amino acids at its center (Fig. 6). This repeating sequence contains four cysteines. In this ferredoxin, eight cysteines complex with two ferrous sulfide molecules to form an electron carrier with a redox potential near the hydrogen potential. The evenly spaced cysteines in apoferredoxin are surrounded by a variety of three to four amino acids.

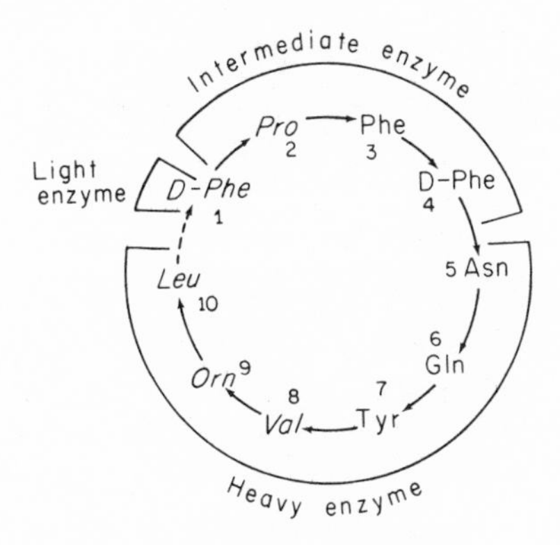

Figure 2. The tyrocidine decapeptide cycle. The amino acids activated by the light, intermediate, and heavy enzymes of 100, 230, 440 x 10^3 Daltons, respectively, are indicated by the brackets. Polymerization is in the direction of the solid arrows from N-terminal D-Phe, 1, to leucine, 10; the latter remains still activated by the carboxyl-thioester link of the decapeptide to the heavy enzyme until it causes the release of cyclic tyrocidine by ring closure through interaction with the free amino group of D-Phe, 1, indicated by the dotted arrow in the figure.

Activation

$$(1)\quad E^{-SH} + aa + ATP \underset{}{\overset{Mg^{++}}{\rightleftharpoons}} E^{-SH}\ldots\ldots aa \sim AMP + PP_i$$

$$(2)\quad E^{-SH}\ldots\ldots aa \sim AMP \rightleftharpoons E^{-S\sim aa} + AMP$$

Figure 3. Amino acid activation in antibiotic synthesis. The ATP-linked amino acid activation produces first--as in ribosomal protein synthesis--a noncovalently linked enzyme-bound aminoacyl∿AMP from which the aminoacyl transfers to an enzyme-bound -SH group. After being thioesterified, the amino acid becomes trichloroacetic acid-precipitable.

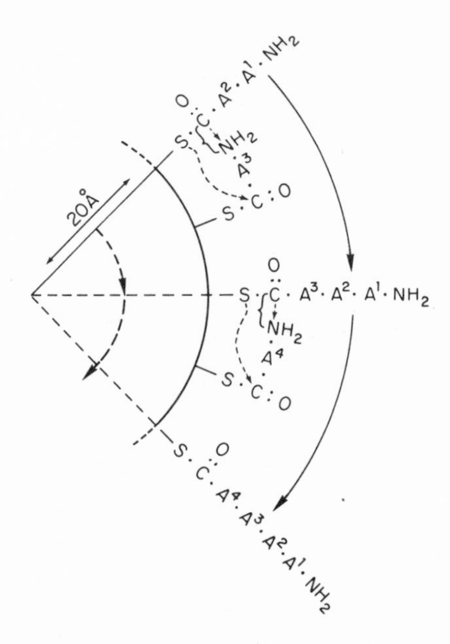

Figure 4. Scheme for the progress of polymerization on a polyenzyme assembly. The enzyme-bound pantetheine, 20 Å long, is shown in the upper part as a solid line ending in a thioester link to a dipeptide (for the numbering of amino acids, see Fig. 2). The first step is transpeptidation of the carboxyl of the dipeptide D-Phe, 1, -Pro, 2, on the intermediate enzyme from pantetheine which donates it to the free amino group of the thioester-linked acceptor amino acid, Phe, 3 (short arrow under bracket). Transpeptidation is followed by transthiolation of the tripeptidyl carboxyl, thus translocating the latter to a donor position on pantethine. The pantetheyl tripeptide then moves in the direction of the dotted inner arrows towards enzyme-bound D-Phe, 4, to repeat the transpeptidation-translocation cycle in order to position the tetrapeptide for continuation (not shown); when enzyme-bound leucine, 10, is reached, it then cyclizes with the free NH_2 of D-Phe, 1, as shown in Fig. 2. The initial dipeptide was found to form by direct transfer of 1 to 2 from light to intermediate enzyme (6,10).

This structure invited us to speculate whether a peptide similar to this, by its particular surrounding of the cysteine-SH, might be able to specify amino acids to thioesterify to this -SH, for example by transthiolating from simple thioesters (10).

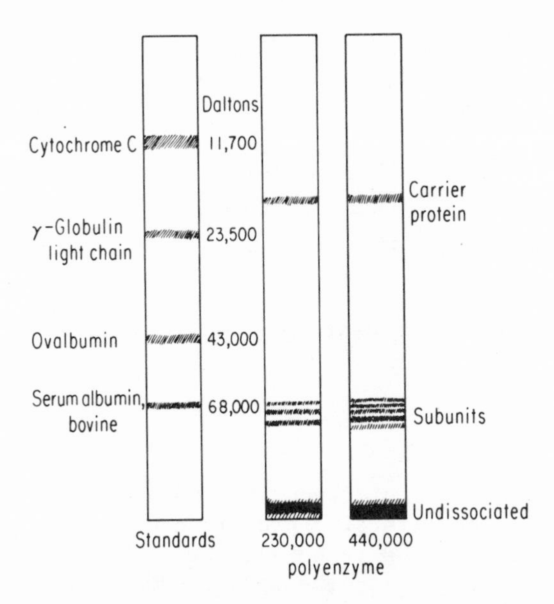

Figure 5. Sodium dodecyl sulfate gel electrophoresis of purified intermediate and heavy enzymes. The polyenzymes partly dissociate to subunits of approximately 70,000 molecular weight, corresponding to the number of amino acids activated (see Fig. 2). The intermediate enzyme yields three evenly stained bands, and the heavy enzyme five bands, one of which is more heavily stained indicating overlap of the two, or a total of six bands.

5 10 15 20

···· Ala· Asp·Ser· *Cys*·Val· Ser· *Cys*·Gly· Ala· *Cys*·Ala·Ser· Glu· *Cys*·Pro·Val·········

35 40 45 50-55

···· Ala· Asp·Thr· *Cys*·Ile· Asp·*Cys*·Gly· Asn·*Cys*·Ala·Asn·Val· *Cys*·Pro·Val·····

Repeat sequence in Ferredoxin, *Clostr. pasteurianum*

Figure 6. The sixteen amino acid repeats in Clost. pasteurianum ferredoxin. The numbers indicate the position of the amino acids in the complete peptide.

Ala	Ser
Ile	Thr
Val	*Cys*
Gly	Asp
Phe	Asn
Pro	Glu
	Gln

Figure 7. Amino acids present in Clost. butyricum ferredoxin.

The Clost. butyricum apoferredoxin, and even more so, the repeating subunit, is a rather primitive protein containing only thirteen relatively simple amino acids (Fig. 7), excluding methionine and tryptophan. In the Clostridia, the apoferredoxins undoubtedly are made by a fully developed ribosomal system. Nevertheless, it seems not quite idle to use this protein, which is the apoenzyme of one of the earliest electron carrier enzymes, as a model for speculation on how a peptide of this type might have been formed and used, possibly for purposes other than oxidation-reduction.

I want to turn now to what I think is a common difficulty in building up to the present from the unknown past or down to the past from the known present: sooner or later, one meets a gap which can only be bridged by imagination. To meet such a gap, however, is not unique for mapping prebiotic events; it is also met with in later stages of the evolutionary process. As an example, I would like to comment briefly on the present state of our understanding of the transition from prokaryote to eukaryote cells.

The recently emerging heterogeneity between organelles, e.g. mitochondria and chloroplasts, and the rest of the eukaryote cells indicates an apparent jump in this transition (11-13), beginning with the discovery in chloroplasts and mitochondria of a prokaryote-like cyclic DNA. Our own interest was aroused by the reactivity of the ribosomal system of the eukaryote mitochondria to inhibitors specific for prokaryote protein synthesis, such as chloramphenicol (14), but with insensitivity to cycloheximide and to diphtheria toxin (15), both of which are specific for eukaryote protein synthesis. At first, this apparent similarity between the mitochondrial and bacterial systems seemed surprising, as it had been shown early (16) that the supernatant factors from mammalian cells were ineffective with E. coli ribosomes, and those from E. coli did not complement mammalian ribosomes. Thus, when reports appeared that yeast (15,17) and mammalian supernatant fractions (18,19) could activate bacterial ribosomes, we became suspicious that this might be due to the presence of mitochondrial, presumably prokaryote-like factors, in the homogenates. This was indeed the case (15), since separating mitochondria from yeast autolysates before homogenization removed from the supernatant fraction the effect on _E. coli_ ribosomes. When this fraction was tested for complementation of the respective ribosomes, those from mitochondria responded only to mitochondrial factors, which were also very active in complementing bacterial ribosomes but were inactive with cytoplasmic ribosomes. The reverse was found with cytoplasmic factors: active only with cytoplasmic but not with bacterial ribosomes.

This seemed strong support for the proposition that the mitochondria of eukaryote cells, and the chloroplasts (20) of green plants are the descendants of prokaryote cells, but particularly in the case of mitochondria are integrated quite extensively with the mother cell. The mitochondrial DNA is rather small, and impresses one as being a remnant after a rather extensive shift to nuclear DNA. We found that the prokaryote-like supernatant factors were still produced in yeast without mitochondrial DNA (21). The origin of the organelles, although still being disputed, at least for mitochondria (22), is, I feel, one of the better understood phases of the transition from prokaryote to eukaryote organization by symbiotic inclusion of prokaryote cells. Rather unresolved remain the origin of the receiver cell, the enclosure of the genetic

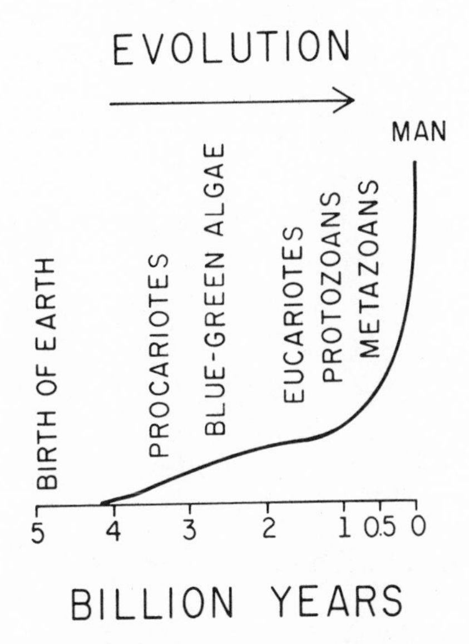

Figure 8. Time chart of evolution, transcribed from Schopf's geologic clock (23). I believe it might be preferable to count time instead from Earth's birth to the present.

transcription system into the nucleus organelle, and the development of the mitotic division mechanism.

Although prokaryotes may have existed for over three billion years, eukaryotes began to appear about 1.3 billion years ago, and metazoae only about 500 million years ago. The time map of evolution in Fig. 8 uses data given in a geologic clock (23). It indicates that the prokaryotes developed an evolutionary plateau creating by amazing metabolic diversification the essentials of intermediary metabolism, most vitamin-coenzymes, and the genetic information transfer system that culminates in the translation of nucleic acid coding into protein reality. However, they did not get out of the unicellular state, and evolution towards more complex organisms clearly only began with the mitotic-meiotic division of the eukaryotic cells about 500 million years ago; this led to the formation of multicellular organisms and then accelerated rapidly towards the present. Thus, true evolution, I propose, began only after the transition to nucleated, mitotic-meiotic dividing and mating organisms that had assimilated prokaryote symbionts. It is this critical transition that is still quite in the dark and shows a gap in our understanding at the truly evolutionary period in the development of life which may be just as difficult to bridge as the one from prebiotic to prokaryotic.

REFERENCES

1. Davie, E. W., Koningsberger, V. V., and Lipmann, F., Arch. Biochem. Biophys. 65, 21 (1956).
   The isolation of a tryptophan-activating enzyme from pancreas.

2. Chapeville, F., Lipmann, F., von Ehrenstein, G., Weisblum, B., Ray, W. J., Jr., and Benzer, S., Proc. Nat. Acad. Sci. U.S. 48, 1086 (1962).
   On the role of soluble ribonucleic acid in coding for amino acids.

3. Lipmann, F., in "The Mechanism of Enzyme Action" (McElroy, W. D., and Glass, B., eds.), p. 599, Johns Hopkins Press, Baltimore, 1954.
   On the mechanism of some ATP-linked reactions and certain aspects of protein synthesis.

4. Lipmann, F., Accts. Chem. Res. 6, 361 (1973).
   Nonribosomal polypeptide synthesis on polyenzyme templates.

5. Lee, S. G., Roskoski, R., Jr., Bauer, K., and Lipmann, F., Biochemistry 12, 398 (1973).
   Purification of the polyenzymes responsible for tyrocidine synthesis and their dissociation into subunits.

6. Lee, S. G., and Lipmann, F., Proc. Nat. Acad. Sci. U.S. 71, 607 (1974).
Isolation of a peptidyl pantetheine protein from tyrocidine-synthesizing polyenzymes.

7. Lipmann, F., in: "Chemical Evolution and the Origin of Life" (Buvet, R., and Ponnamperuma, C., eds.), p. 381, North-Holland, Amsterdam, 1971.
Gramicidin S and tyrocidine biosynthesis; a primitive process of sequential addition of amino acids on polyenzymes.

8. Hall, D. O., Cammack, R., and Rao, K. K., Nature 233, 136 (1971).
Role for ferredoxins in the origin of life and biological evolution.

9. Hall, D. O., Cammack, R., and Rao, K. K., presented at Fourth International Conference on the Origin of Life, Barcelona, June, 1973, to be published.
The iron sulphur proteins: evolution of a ubiquitous protein from model systems to higher organisms.

10. Roskoski, R., Jr., Ryan, G., Kleinkauf, H., Gevers, W., and Lipmann, F., Arch. Biochem. Biophys. 143, 485 (1971).
Polypeptide biosynthesis from thioesters of amino acids.

11. Stanier, R., in: "Organization and Control in Prokaryotic and Eukaryotic Cells," p. 1, Cambridge University Press, Cambridge, 1970.
Some aspects of the biology of cells and their possible evolutionary significance.

12. Cohen, S. S., Amer. Scientist 58, 281 (1970:
Are/were mitochondria and chloroplasts microorganisms?
ibid., 61, 437 (1973).
Mitochondria and chloroplasts revisited.

13. Margulis, L., Scientific American 225, 48 (1971).
Symbiosis and evolution.

14. Rendi, R., Exp. Cell Res. 18, 187 (1959).
The effect of chloramphenicol on the incorporation of labeled amino acids into proteins by isolated subcellular fractions from rat liver.

15. Richter, D., and Lipmann, F., Biochemistry 9, 5065 (1970).
Separation of mitochondrial and cytoplasmic peptide chain elongation factors from yeast.

16. Nathans, D., and Lipmann, F., Proc. Nat. Acad. Sci. U.S. 47, 497 (1961).
Amino acid transfer from aminoacyl ribonucleic acids to protein on ribosomes of Escherichia coli.

17. Richter, D., Hameister, H., Petersen, H. G., and Klink, F., Biochemistry 7, 3753 (1968).
Amino acid transfer factors from yeast. II. Interaction of three partially purified protein fractions with guanosine triphosphate.

18. Canning, L., and Griffin, A. C., Biochim. Biophys. Acta 103, 522 (1965).
Specificity in the transfer of aminoacyl-s-ribonucleic acid to microbial, liver, and tumor ribosomes.

19. Krisko, I., Gordon, J., and Lipmann, F., J. Biol. Chem. 244, 6117 (1969).
Studies on the interchangeability of one of the mammalian and bacterial supernatant factors in protein biosynthesis.

20. Sy, J., Chua, N. H., Ogawa, Y., and Lipmann, F., Biochem. Biophys. Res. Commun. 56, 611 (1974).
Ribosome specificity for the formation of guanosine polyphosphates.

21. Richter, D., Biochemistry 10, 4422 (1971).
Production of mitochondrial peptide-chain elongation factors in yeast deficient in mitochondrial DNA.

22. Raff, R. A., and Mahler, H. R., Science 177, 575 (1972).
The nonsymbiotic order of mitochondria.

23. Schopf, J. W., Biol. Rev. 45, 319 (1970).
Precambrian micro-organisms and evolutionary events prior to the origin of vascular plants.

PRIMITIVE CONTROL OF CELLULAR METABOLISM

M. A. Mitz

National Aeronautics and Space Administration

Washington, D.C.

INTRODUCTION

Life, as we know it, is the natural consequence of the environment and time. Biological laws are only special cases of the general laws of chemistry and physics. One of the first to state these ideas was A. I. Oparin, who also taught that we will understand the origin of life if we can follow the path back to the earliest life forms and in the process unravel the physics and chemistry of the primitive Earth (1). These teachings have influenced me to look at cell regulatory mechanisms in such terms and in this way complement the fundamental work on origin and evolution of the cell.

Regulation and control systems of the cell were sought by early workers. I believe that such mechanisms may have been largely overlooked because there has been a tendency to seek unique compounds, or involved mechanisms which exercise control over the complex chemistry of the modern cell. I have asked myself, "What about the primitive cells? Did early cells have controls? If they did, what was the nature of those primitive controls? Were they indispensable to the function and survival of the early life forms? Are remnants of these controls evident today?" I have come to the conclusion that control substances must have existed from the earliest times in the evolution of life and, furthermore, that these same control mechanisms must exist even today, in various forms. The basis of this conclusion is my thesis suggesting that certain common cellular end-products act as cell regulators. Two examples of such substances are carbon dioxide and urea. In this paper I shall try to develop only the idea that carbon dioxide is a primitive regulator of cell function. Some evidence exists to support the thesis, but a great many of my conclusions are based on projections or extrapolations. Let us look first at the general concept, and then at the bits and pieces of evidence that are available, and finally how they all may fit together.

GENERAL CONCEPT

In 1954, I suggested the general concept that metabolic products can control cell metabolism by the simple mechanism of causing dissociation and thus activating enzymes bound to insoluble structure within the cell (2). This idea was consistent with the basic concepts of A. I. Oparin who claimed that "the degree of association of the enzyme with the cell structure material has a decisive effect not only in determining changes in the rate of the reaction catalyzed by enzymes but also in displacing the dynamic equilibrium of the chemical processes toward a predominance of breakdown or synthesis (3)." The evidence I presented went one step further in that it identified a metabolic material, namely carbon dioxide, as one of the regulators. Subsequently, I published a series of papers demonstrating the ability of carbon dioxide to dissociate (4-6) and/or solubilize enzymes (7) and other substances.

Over the last few decades scientists have been preoccupied with the role of carbon dioxide in regulating hydrogen ion concentration (pH) in biological fluids. More recently, attention has been focused on the enzyme carbonic anhydrase which accelerates the reaction of carbon dioxide and water (8). Carbon dioxide reacts relatively slowly with water, and therefore the presence of carbonic anhydrase is necessary if carbon dioxide is to affect the pH rapidly. Such pH changes may, in turn, stimulate or retard selected metabolic reactions (9). What is often not considered is what happens when there is little, or no, carbonic anhydrase. My own work leads me to believe that carbon dioxide gas has more subtle but also more fundamental effects upon cellular material; these effects may be the basic regulators or controls of cell processes.

Carbon dioxide, which stimulates the respiratory centers in higher organisms, could satisfy the definition of a hormone. Although such centers may be a recent evolutionary phenomenon, the mechanisms of action at these centers may be a key to primitive cell metabolism. The development of carbon dioxide-sensing centers may have arisen because of the inability of mammalian cells to maintain a pressure differential between the inside and outside of the cell. In contrast, the cell wall of plants and bacteria can maintain different pressures between the cell and its environment. In other words, the partial pressure of gases such as CO_2, O_2, and H_2S in the cell can build up to a total pressure greater than ambient. This fact is important in making the transition from my early laboratory experiments with CO_2 partial pressure of up to, and greater than, one atmosphere, to the thesis that CO_2 partial pressure affects cellular metabolism. Another important factor that should be introduced at this point is that carbon dioxide has a solubility in water of approximately 30 times that of oxygen and 200 times that of nitrogen. Even more interesting is the fact that CO_2 has an even higher solubility (3-10 X) in lipids than in water. With high pressures and high solubility of the gas, what effects are possible?

EFFECTS OF CARBON DIOXIDE ON CELLULAR MATERIALS

In reviewing the evidence upon which I wish to make my case, I find that carbon dioxide causes the following effects on cellular materials:

1. Solubilization
2. Dissociation
3. Changes in charge
4. Stabilization
5. Structural changes
6. Increased wetting and increased rate of solution
7. Exclusion of other gases
8. Activation of compounds
9. Changes in plasticity
10. Changes in membrane permeability and selectivity

Most or all of these effects depend solely upon an increased concentration of carbon dioxide in the environment of the material, and are often dramatic and significant. The effects are, for the most part, reversible and respond relatively rapidly to the CO_2 pressure changes. They are not necessarily involved with the pH effects of carbon dioxide reacting with water. All of these effects, including the latter, could act as control mechanisms for the cell as it uses or produces carbon dioxide. Although these effects overlap and are highly interdependent, let us examine each of these effects individually.

<u>Solubilization</u>. Carbon dioxide has differential effects on the solubility of some substances; controlled CO_2 pressure can cause some substances to dissolve while leaving others precipitated (7). Removing the carbon dioxide by reducing the partial pressure of the CO_2 over the solution generally causes the solubilized material to reprecipitate. In most cases solubilization causes an increase in the reactivity of the material with other materials in its environment. For example, if the solute is an enzyme, its rate of reaction is generally accelerated in solution. In some cases the soluble material may be enabled to pass through an interface, or move generally from one part of the cell to another in solution. Thus, substrate and enzyme can be brought together, or a messenger

material may reach its target. All of these effects were suggested by Oparin and co-workers (3).

Dissociation. A similar property of carbon dioxide is its demonstrated ability to cause certain ionically bound materials to dissociate. This, like the solubilization effect, may be independent of changes in hydrogen-ion concentration. The dissociations most carefully studied are the type in which one of the materials is insoluble and the other is normally soluble (4,5). Here again, all the effects of solubilization mentioned above can be expected, except that removing the carbon dioxide causes reprecipitation only if the insolubilizing agent and conditions are present.

Materials in solution may also be dissociated. The activity of one molecule may be masked by another molecule. The interaction between an enzyme and an enzyme inhibitor may represent the association of two proteins. Substances which dissociate these inhibitor complexes could have a pronounced effect on the cellular activity of that enzyme. An interesting bit of speculation is that this kind of mechanism is also active in the release of peptides and other components carried by the blood proteins. Local high concentrations of carbon dioxide in the capillaries could cause a release of these materials which could then be absorbed into the cell.

Changes of charge. Carbon dioxide reacts directly with free amino groups in solution to form a carbamic acid which exists

$$RNH_2 + CO_2 \rightleftharpoons RNHCOOH \underset{H^+}{\overset{M^+}{\rightleftharpoons}} RNHCOO^-M^+$$

primarily as a salt. In contrast with the very slow reaction of CO_2 with H_2O, which necessitates special enzyme catalysts, carbamate forms very rapidly, without help. Thus, a potentially basic material like an amine becomes acidic. The net effect is not only a temporary change in charge, but other properties of the molecule may be changed as well. For example, the barium salts of amino acids are normally insoluble in water but the addition of carbon dioxide causes the formation of an amino acid dicarboxylic acid barium salt which is soluble (10).

Carbamic acid formation in haemoglobin is important in the transportation of CO_2 (11). A major fraction of the CO_2 in the blood is carried by this mechanism. At high CO_2 levels, haemoglobin derivatives have a lower affinity for O_2 and release it. By a similar mechanism, carbamic acid formation may cause important changes in the charged properties of a membrane. For example, amino groups on a membrane may be converted to carbamic acids which may affect the type of ion that migrates across the boundary from acidic to basic. If the carbon dioxide concentration is reduced, the amine is regenerated. A similar effect occurs when the pH shifts towards the acid side because the free carbamic acid is less stable. Heat also destroys the carbamic acid derivatives. Thus CO_2 derivatives are sensitive to a number of different factors in the environment.

Stabilization. Some proteins treated with carbon dioxide have a notably increased stability. This may be temperature stability or surface stability. In any event, the same proteins are less likely to become insoluble (denature) in the presence of carbon dioxide. The mechanism is not understood but I suspect that the CO_2 produces temporary internal cross-links which prevent unraveling of the protein at an interface. This effect may be important for the preservation of subcellular bodies and droplets. Thus the stability and activity of these organelles may be dependent on the presence of moderate levels of carbon dioxide.

Structural changes. At high levels of carbon dioxide certain organelles may be dissociated or dissolved into the protoplasm to change the metabolism of the cell drastically. Under dissociation above, I discussed the possibility of solubilizing certain structures, and parts of structures, within the cell. The same mechanism may alter other structural components to allow the materials within a substructure to leak out into the soluble phase; or it may allow a process equivalent to phagocytosis to take place. The structural changes necessary to allow this to happen are fascinating, but not at all well understood.

Wettability. Another dramatic effect of carbon dioxide is its ability to increase the rate of solution of soluble substances which have been dehydrated. For example, although haemoglobin and globulin are both very soluble proteins, the chances are good that their rate of solution in water or buffer will be increased several hundred times by exposure of the dry solid proteins to carbon dioxide gas just before adding the solvent (7). The carbon dioxide is believed to function by displacing the other, less soluble gases which are trapped in the fine structure of the solid, thus preventing the solvent from reaching the surface. It is believed that carbon dioxide first displaces the other gases from the surface of the solid because it is heavier, and then that the carbon dioxide is absorbed into the solution. This later action has the effect of drawing the solvent into the fine structure of the solid, for the primitive surface dehydration must have been a situation as real as it is today. Rehydration mechanisms thus became important in order to reactivate the cell wall without causing it to rupture when suitable conditions recurred.

Exclusion of other gases. Anaerobic organisms, which are often inhibited by oxygen, have developed mechanisms to remove oxygen and other reactive gases by generation of their own carbon dioxide atmosphere. The generation of carbon dioxide has the effect of excluding light gases. Unless one removes carbon dioxide from the medium in deep fermentation, the growth of anaerobic organisms is favored over that of aerobes. The question is generally not one of a lack of oxygen, but rather of the build-up of carbon dioxide, that effects metabolism.

Activation of compounds. Amino acid carbamic acid salt derivatives on heating in a dry state may form anhydrides. These cyclic amino acid anhydrides are very reactive and form peptides, if provided with an activator containing a free amino or hydroxyl group. Cyclic carbamic acid amino acid anhydrides may be a source of polymeric materials which would be trapped within a dehydrated cell. Ammonium and amine carbonates may also be a source of ureas, cyanates and even carbonyl phosphates which may be looked upon as "high energy" phosphate precursors to ATP (12).

Changes in plasticity. As indicated earlier, carbon dioxide is many times more soluble in the lipid phase than in the water phase. For this reason glycerides and waxes tend to have a lower solidification point in the presence of high concentrations of carbon dioxide, which acts as anti-freeze to the lipid. In natural fats and waxes it is difficult to remove all of the dissolved carbon dioxide without resorting to heat and vacuum. When the crystalline structure of the fat and waxes is decreased, the ability to transport ions and molecules through the lipid is changed. In this way, interaction of a hydrophobic enzyme and its lipid-soluble substrate could be increased because of the higher mobility of the substrates in the lipid phase due in turn to greater fluidity caused by dissolved carbon dioxide.

Changes in membrane permeability. Membrane porosity and selectivity may be changed in the presence of carbon dioxide. The effect will depend on the chemical nature of the membrane. All the factors discussed above have an effect. For example, solubilization and dissociation may open or close pores; charge alteration on the membrane may change its selectivity; increased wettability and the exclusion of other gases may change the rate of permeability of gases; and differences in plasticity may make a membrane more soluble to certain ions and make it possible to absorb particles.

What passes through a membrane is influenced by what happens at the surfaces as well as within the membrane. Environmental changes may be reflected in concomitant changes of the membrane or its surfaces. For example, if the membrane is largely layers of lipid and protein, carbon dioxide has a tendency to make the membrane thinner. The gas reacts with free amino groups of protein and introduces identical charges which repel each other. Any lipid phase under high carbon dioxide concentration might become more elastic and spread out under pressure and consequently become thinner and more permeable at any given point.

SUMMARY AND CONCLUSION

What does this all mean? The above facts and extrapolations can be woven into a model for a primitive, hormone-like, control mechanism, based on certain metabolic products which operate directly on the biochemical and physical properties of the cell. This simple mechanism could have started with the first organized life-like system and may continue to function in cells today. It suggests that certain metabolic products can control (a) the rate of metabolism, (b) the growth and possibly (c) the reproduction of the cell. The metabolic or control effect is easiest to visualize when carbon dioxide participates. Certain enzymes, not necessarily related to carbon dioxide metabolism, may be modulated by the solubilizing and dissociating effects of the rise and fall of CO_2 partial pressure in the cell. Growth could occur through the ability of the CO_2 partial pressure to influence the physical properties of the cellular structure, such as the porosity of the cell membrane and the plasticity of the lipids. The effect of carbon dioxide on reproduction is the phenomenon most difficult to visualize in this respect because so many things are happening at once. But it is conceivable that reproduction is simply a consequence of the build-up of cellular carbon dioxide to a critical value at which the cell changes its surface-to-volume ratio to contain the pressure. Contributing to this latter process may be such effects as changes in charge to enable division of internal cell structures.

Although this paper was directed at the possible hormone-like role of CO_2 in the cell, similar cases might be made for O_2, CO, H_2S, and urea as end-products. Oxygen, CO, or H_2S tension can influence the oxidation-reduction potential, which in turn influences the structure and activity of certain enzymes, and also influences cross-linking in vital structures. If one stretches one's imagination, urea might be thought of as a "hormone" in its ability to help catalyze the phosphorylation of sugars, which would influence the management of carbohydrate, protein, and nucleic acid synthesis (13).

All-in-all, I have tried to develop the general concept that certain metabolic end-products can regulate the cell through a series of reversible physical and chemical changes from isolated studies on individual cell components. It still remains to be seen how well this thesis conforms to more experimentation. If there is any validity to this thesis, a large measure of credit should be given to Professor Oparin, who helped characterize the nature of the primitive cell.
the

REFERENCES

1. Oparin, A. I., "The Origin of Life," original in Russian (1936), English Ed. Macmillan Co., 1938. Reissued Dover Co., New York, 1953.

2. Mitz, M. A., Science 123, 1076 (1956).

3. Oparin, A. I., "The Origin of Life on Earth," p. 379, Academic Press, New York, 1957.

4. Mitz, M. A., and Yanari, S. S., J. Am. Chem. Soc. 78, 2649 (1956).

5. Volini, M., and Mitz, M. A., J. Am. Chem. Soc. 82, 4572 (1960).

6. Yanari, S., Volini, M., and Mitz, M. A., Biochim. Biophys. Acta 45, 595 (1960).

7. Mitz, M. A., Biochim. Biophys. Acta 25, 426 (1957).

8. Carter, M. J., Biol. Revs. 47, 465 (1972).

9. Mitz, M. A., in: "Chemical Evolution and the Origin of Life" (Buvet, R., and Ponnamperuma, C., eds.), p. 355, North-Holland, Amsterdam, 1971.

10. Faurhalt, C., J. Chim. Phys. 22, 1 (1925).

11. Stadie, W. C., and O'Brien, H., J. Biol. Chem. 242, 1579 (1937).

12. Mitz, M. A., "Proceedings of the Fourth International Conference on the Origin of Life," Barcelona, Spain, Abstracts, 1973.

13. Lohrmann, R., and Orgel, L. E., Science 171, 490 (1971).

ON THE EVOLUTION OF

MACROMOLECULES

Hiroshi Mizutani, Hiroyuki Okihana, Masami Hasegawa,
Taka-aki Yano, and Haruhiko Noda
Department of Biophysics and Biochemistry
Faculty of Science
University of Tokyo, Hongo, Tokyo, Japan

INTRODUCTION

After the formation of our planet, the evolution of matter led to contemporary living systems. This process may be divided roughly into three stages, that is: (a) abiogenic syntheses of macromolecules, (b) spontaneous phase-separation of the macromolecules followed by self-organization, and (c) biological evolution after the first production of life. The former two stages are called prebiotic evolution. Each stage has been investigated by many authors from the standpoint of molecular evolution.

For a living system made of organic compounds, a constant turnover of the constituent parts seems to be necessary. A living system requires metabolism to process the energy it requires.

In order for many chemical reactions to occur in a controlled fashion, various chemical pathways have to be turned on and off, or adjusted, by valves.

For the control to be an established design, the living system must hold the information for its own control. Nucleic acid seems to be the unique material for that function. For the readout of the information, macromolecules, mostly proteins, are necessary.

For the first living system to have appeared, a few kinds of macromolecule should have appeared, although we believe that such molecules would have had great difficulty in finding themselves assembled into an organized living system.

In our laboratory we are studying properties of macromolecules from various perspectives, such as spontaneous formation, spontaneous phase-separation, and evolutionary properties of macromolecules in the biological history after the first production of life. These correspond to the above-mentioned three stages of the evolution of matter, although gaps among the stages are still left untouched. The results of our investigation are represented in this article, although discontinuities among the stages may be conspicuous. The gaps should be filled by future investigations, from which the unified understanding of life will be obtained. For the present, it will be of value to study each important stage, although separately.

THE FORMATION AND CHARACTERISTICS OF HCN POLYMERS PRODUCED BY UV IRRADIATION

Many investigators emphasize the importance of HCN in primary chemical evolution (19, 2 and references therein). HCN polymerizes easily when it is an aqueous solution or liquid with a base. Many have studied the structure of HCN polymers and have tried to clarify the mechanism of HCN polymerization (30,18,2). In addition to these studies, Labadie _et al_. (16) showed that the HCN polymers were not attacked by some proteolytic enzymes. Catalytic activities of HCN polymers have been studied by the same authors and slightly positive activities were found in some reactions.

It is known that HCN alone can polymerize in the gas phase when it is irradiated by α-rays (17) or UV rays (28). Recently, the gas phase products, obtained by HCN gas phase polymerization reaction activated by 184.9 nm rays, were identified, and the following reaction mechanism was proposed (20):

$$HCN \xrightarrow{h\nu} \cdot H + \cdot CN \qquad (1)$$
$$\cdot H + H \xrightarrow{M} H_2 \qquad (2)$$
$$\cdot CN + CN \xrightarrow{M} (CN)_2 \qquad (3)$$
$$\cdot H + HCN \longrightarrow H_2CN\text{(radical)} \qquad (4)$$
$$\cdot H + H_2CN \longrightarrow (H_2C{=}NH) \qquad (5)$$
$$\cdot H + (H_2C{=}NH) \longrightarrow H_4CN\text{(radical)} \qquad (6)$$
$$\cdot H + H_4CN \longrightarrow \cdot CH_3 + \cdot NH_2 \qquad (7)$$
$$\cdot H + CH_3 \xrightarrow{M} CH_4 \qquad (8)$$
$$\cdot H + NH_2 \xrightarrow{M+} NH_3 \qquad (9)$$
$$\cdot CH_3 + CH_3 \xrightarrow{M} C_2H_6 \qquad (10)$$
$$\cdot NH_2 + NH_2 \xrightarrow{M} N_2H_4 \qquad (11)$$
$$\cdot CH_3 + NH_2 \xrightarrow{M} CH_3NH_2 \qquad (12)$$

M denotes the third body in a ternary collision and it absorbs excess energy generated by a recombination between two radicals.

The solid phase products of the reaction are highly reactive. Although HCN polymers partly dissolve in methanol and the majority of the rest dissolves in dimethyl sulfoxide (DMSO), methanol solution spontaneously generates insoluble matter adherent on the surface of a glass flask in a few weeks. Polymers of HCN dissolve also in $CHCl_3$ or $C_2H_4Cl_2$, but the solutes in these solvents react spontaneously and show change in R_f for paper chromatography in a few days.

We hydrolyzed methanol-soluble and DMSO-soluble fractions after evaporating the solvents. Table I shows the amino acid analyses of hydrolyzates of these fractions as well as those of HCN tetramer (diaminomaleonitrile, decomposes at 182∿184°C) and a pentamer (adenine). The identification of glycine was confirmed by paper chromatography with two different solvents. The yields of glycine and NH_3 decrease in the order of diaminomaleonitrile>adenine>methanol-soluble>DMSO-soluble. The number of the minor products detected as only traces increases in the same order. This order may reflect the increasing complexity of structures of these HCN polymers.

Figs. 1 and 2 show the mass spectra of diaminomaleonitrile and the methanol-soluble fraction, respectively. The prominent peaks in

Table I

Amino Acid Analyses of the Hydrolyzates of HCN Polymers

Product	DMSO-soluble	Methanol-soluble	Adenine	Diamino-maleonitrile
NH_3	0.85	3.2	7.2	18
Gly	0.33	0.62	3.2	6.1
His	*	*	*	
Asp	*	*		
Glu	*	*		
Ala		*		
Lys		*		
Arg	*			
Ser	*			
Thr	*			
Val	*			
Ile	*			
Leu	*			

The unit of figures is μmole/mg polymer and * is less than 0.005 μmole/mg polymer. The conditions of hydrolysis are: 6 N HCl, 24 hrs, 109°C for DMSO- and methanol-soluble fractions and adenine; 6 N HCl, 24 hrs, 102°C for diaminomaleonitrile.

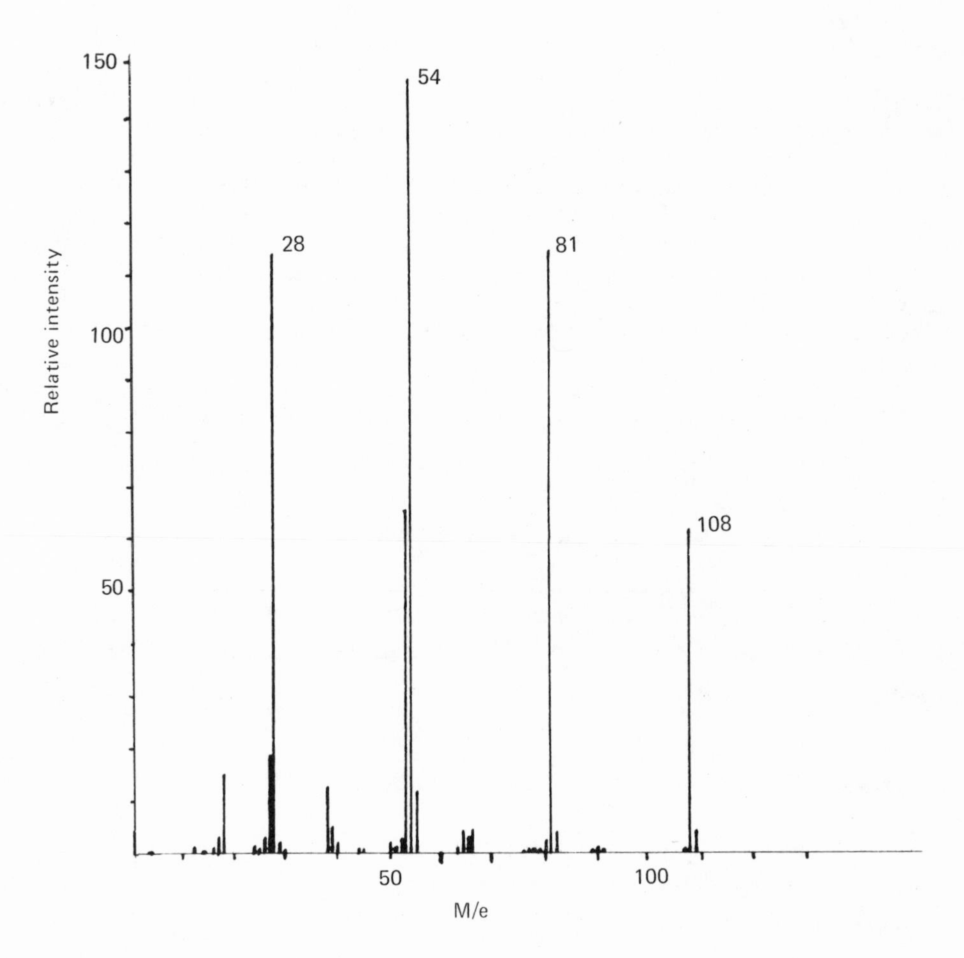

Figure 1. Mass spectrum of diaminomaleonitrile.

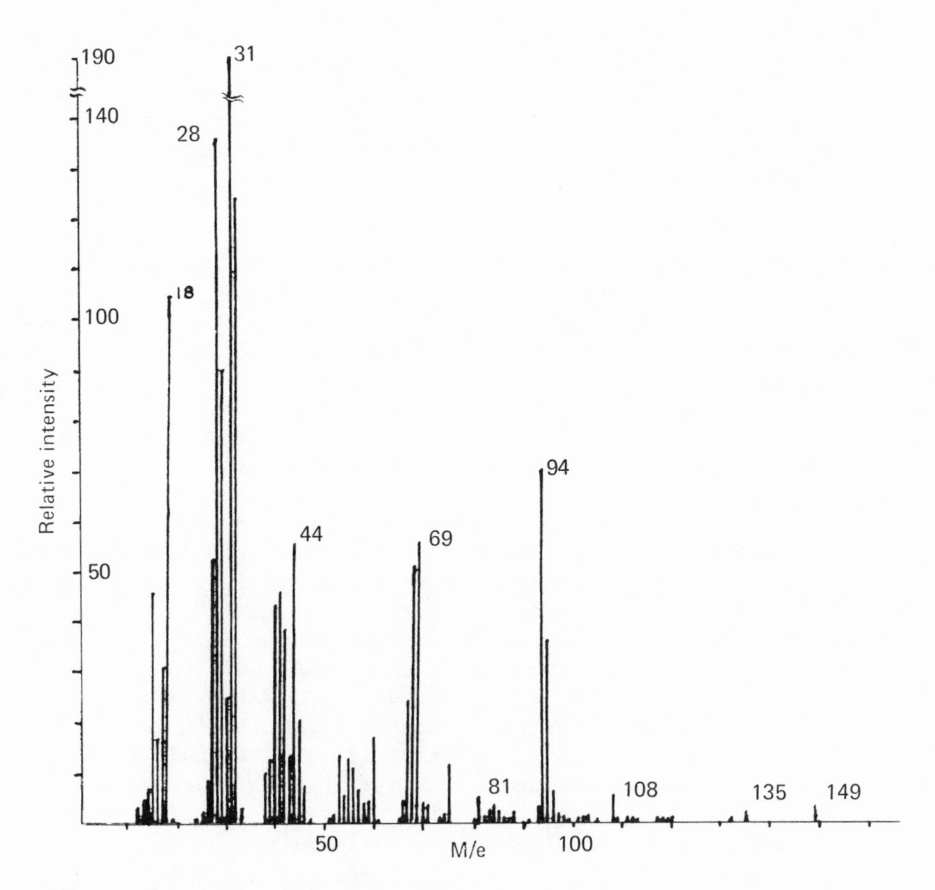

Figure 2. Mass spectrum of the methanol-soluble fraction.

Fig. 1 are at the mass numbers of H_2CN, HCN-dimer, HCN-trimer, HCN-tetramer (Me = 28, 54, 81, 108 respectively). Only small peaks are also observed between these peaks. The peaks for HCN oligomers are also found in Fig. 2. The main difference between Fig. 1 and Fig. 2 is the existence of the large peaks whose mass numbers are at the middle value between multiples of HCN molecular weight. The largest mass number in Fig. 2 (M/e=149) suggests a molecular formula of $C_6H_7N_5$, which could be of a purine-like structure. According to the temperature-dependence of fragmentation pattern of the methanol-soluble fraction, the fraction begins to decompose above 150°C.

From Table I and the comparison between Figs. 1 and 2, we may be able to presume that the methanol-soluble fraction of HCN polymers is a mixture of compounds, whose molecular weights are a little larger than that of HCN pentamer and whose structures are more complex than those of HCN tetramer or pentamer; some fragments are easily released when it is heated.

SPONTANEOUS PHASE-SEPARATION OF MACROMOLECULES

Many kinds of small molecule and macromolecule accumulated on the primordial Earth since the Earth had formed (4). The next essential step on the way to life was the appearance of an enclosed space. Substances were concentrated in this enclosed space, their interactions were more efficient, and they might then reproduce themselves more rapidly. Moreover, those spaces were separated from their environment which was their mother liquid, and protected from the unfavorable changes in the environment, and the mean life time of various compounds might be longer than they had been in the solution. This may be one of the important steps of chemical evolution. Some proposed models for the enclosed spaces are coacervates (22) and microspheres (3), etc. What kind of molecule was able to make an enclosed space? Qualification for such a model molecule has been studied by Voorn (31), Veis (27), Okihana (21) with gelatin as an example. The outline of the work will be presented in this chapter. Although gelatin is an evolved protein, it is a suitable macromolecule for study of coacervation, because it has the random-coil configuration, its net charge density is relatively low, and there are similar structures and chemical constitutions, except for net charge.

Gelatins with different isoelectric points are used for model molecules. An acid-precursor gelatin has a basic isoelectric point (pI=8.8), and is denoted by A. An alkali-precursor gelatin has an acidic isoelectric point (pI=5.0), and is denoted by B. These two gelatins were fractionated by the ethanol-salt procedure (26). The amount of each fraction and the result of SDS-disc-electrophoresis are given in Table II and Fig. 3. The average molecular weight is larger for fractions which separate at lower alcohol concentrations. The fractionation was not sharp.

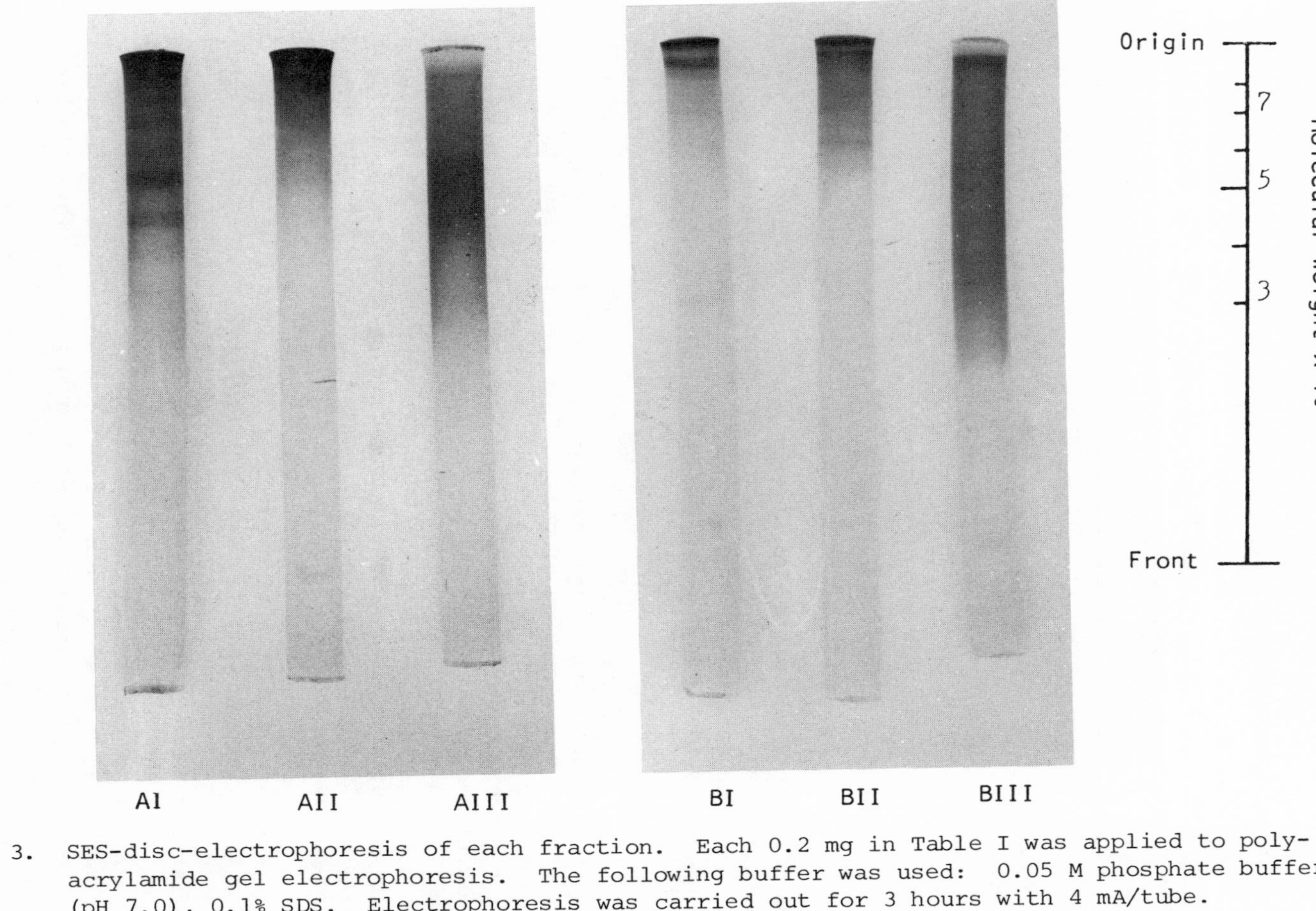

Figure 3. SES-disc-electrophoresis of each fraction. Each 0.2 mg in Table I was applied to polyacrylamide gel electrophoresis. The following buffer was used: 0.05 M phosphate buffer (pH 7.0), 0.1% SDS. Electrophoresis was carried out for 3 hours with 4 mA/tube.

Table II

Fractionation of Gelatins

Fraction	Ethanol/water ratio	Temperature	Weight fraction
AI	2/1	40°C	28%
AII	3/1	40°C	29
AIII	5/1	room temperature	13
BI	2/1	40°C	32
BII	3/1	40°C	29
BIII	5/1	room temperature	5

18 g of gelatin was dissolved in 900 ml of 0.8 M NaCl solution and ethanol was added at 40°C. Each fraction was collected at the indicated ethanol-water ratios. The amount of each fraction is shown by weight percentage; the values did not add up to 100%.

This study has been restricted to symmetrical mixtures, namely mixtures of equal amounts of solutions of A and B of the same concentrations. Each fraction of gelatin had been deionized through the mixed-bed ion exchange resin column. The mixture became opaque and the coacervates could be seen under the microscope (Fig. 4). The mixture was centrifuged for 10 minutes at 1,500 G, and then the volume of coacervates was measured. The results are given in Fig. 5. The coacervates were formed only when fractions AI and BI, fractions of the largest molecular weight, were mixed at a concentration less than 10 mg/ml. At a higher gelatin concentration, self-suppression seemed to have taken place.

Since the difference of fraction I from fractions II and III is mainly the molecular weight, i.e. degree of polymerization, it is supposed that there is a critical degree of polymerization for the occurrence of coacervation. The critical degree of polymerization lies between the degree of polymerization for the fractions I and II (21). As the degree of polymerization increases, interactions between molecules of A and B seem to increase and molecules make their complexes (AB) and separate into two liquid phases. This is coacervation. The interactions which make complexes are mainly based on their charges, because salts suppress coacervation most effectively, and because coacervation occurs when the pH of mixture is in the range between two isoelectric points (1).

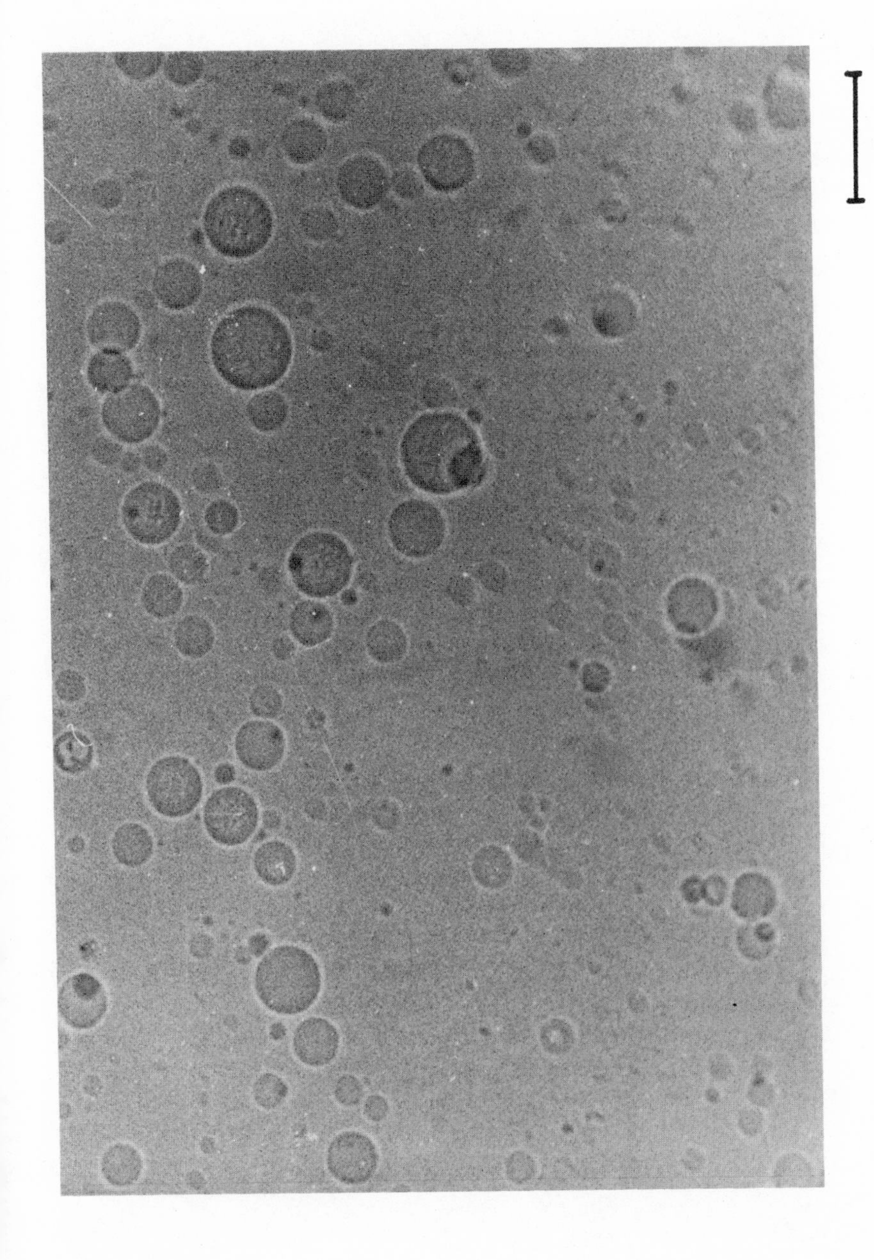

Figure 4. Coacervates.

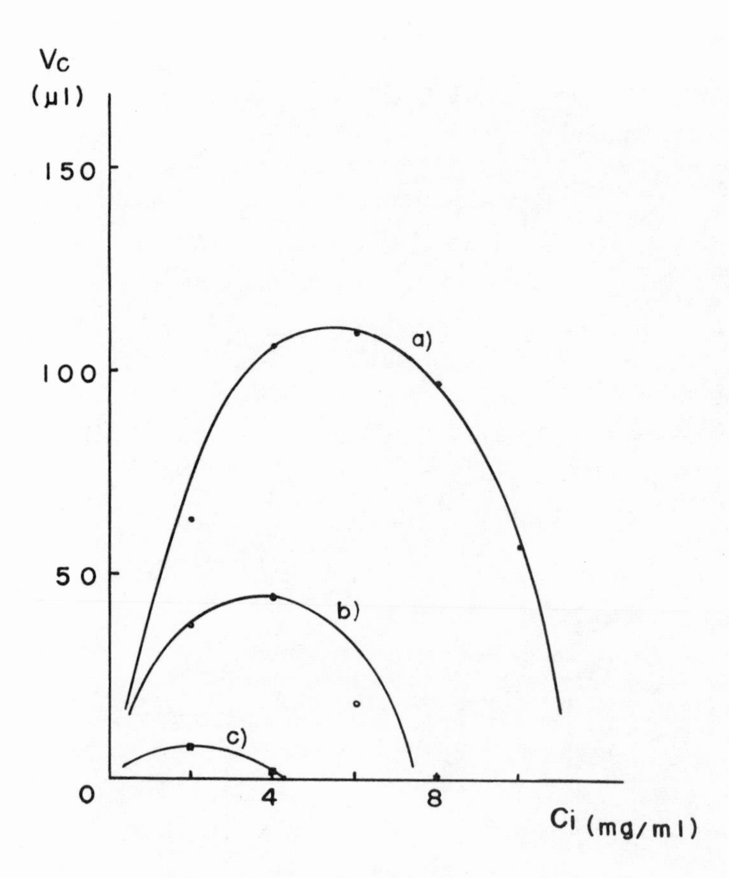

Figure 5. The volume of coacervates, and the effect of smaller peptides. Each 5 ml of deionized fraction AI and BI was mixed at the initial concentration (C_i) at 40°C. a) No salts, b) 2 mg/ml of fractions AII and BII, c) 4 mg/ml of fractions AII and BII, coexisted with the fractions AI and BI. After centrifugation, the volume of coacervates (V_C) was measured. When AII and BII, and AIII and BIII were mixed, no coacervation was observed.

When the degree of polymerization is not high, the ionic repulsion and shielding of A and B due to their own amino acid residues overcome their attractive interactions. In this case, coacervation does not occur. From this viewpoint, it is expected that fractions II and III behave as salts when they are coexistent with fraction I. The results for such mixtures are given in Fig. 5.

On the primordial Earth, small molecules, such as amino acids or mononucleotides, and macromolecules might have had the chance to polymerize to a larger degree beyond the critical one. They were then able to form an enclosed space.

THE GENETIC CODE, AND ENTROPY OF AMINO ACID SEQUENCE IN PROTEIN

The structure of the genetic code (Fig. 6) is now well known, and is considered as universal on this planet. The code is degenerate, and the degeneracy is not evenly distributed. Leucine, serine, and arginine are six-times degenerate, and on the other hand, methionine and tryptophan are not degenerate. The four kinds of nucleotide base are not evenly used in the code, and there is a correlation between GC content of a codon and the degree of its degeneracy. Sets of codons [the group of triplets ABd(d=U,C,A,G), ABe(e=A,G), or ABb(b=U,C], which code for the same amino acid, with maximum GC content in the first two positions, are all four-times degenerate, and those lacking GC are generally twice degenerate. Goldberg and Wittes (8) have said that the greater binding energy of GC pairs, as compared to AU pairs, may obviate reading of the third nucleotide in those triplets containing only GC in the first two positions, and this may be the reason why the code has a highly biassed nature.

In this way, the present genetic code is highly biassed, and on the other hand, the non-Darwinian theory of molecular evolution (13) predicts that the amino acid composition should be strongly influenced by the genetic code; that is, amino acids with more codons should appear more frequently than those with fewer codons. King and Jukes (14) have actually indicated a very strong influence of the genetic code on the amino acid composition.

Although King and Jukes' indication may be true, there seems to be another aspect to this problem; that is, in the course of evolution there seems to be some tendency to deviate from the influence of the genetic code. Zuckerkandl _et al_. (33,29) have suggested that in regard to the amino acid substitution during the evolution of proteins, the more frequent amino acids tend, on the average, to become less frequent, and rarer ones tend to increase their frequencies. In other words, if the entropy of the amino acid sequence in protein is defined by

2nd →

1st ↓	U	C	A	G	3rd ↓
U	Phe / Leu	Ser	Tyr / term	Cys / term / Trp	U C A G
C	Leu	Pro	His / Gln	Arg	U C A G
A	Ile / Met	Thr	Asn / Lys	Ser / Arg	U C A G
G	Val	Ala	Asp / Glu	Gly	U C A G

Figure 6. The genetic code.

$$H = -\sum_{i=1}^{20} p_i \log_2 p_i \quad \text{(bits)}, \qquad \sum_{i=1}^{20} p_i = 1,$$

where p_i is the frequency of the i-th amino acid, this entropy function increases. They interpreted this trend as the tendency towards randomness. Although Yano and Hasegawa (32) have found the same tendency also in fibrinopeptide, immunoglobulin, and lysozyme, their interpretation is different from that of Zuckerkandl et al. Insofar as the Central Dogma is correct, the randomization process will occur only in the DNA sequence. Then, because of the biassed nature of the genetic code, the entropy of the amino acid sequence will not necessarily increase, but rather decrease. Thus, the entropy increase of the amino acid sequence cannot be regarded as a result of random processes which occur on DNA. We regard it rather as a trend against randomization. Although the non-Darwinian theory has explained many aspects of molecular evolution, it cannot explain this tendency.

From the amino acid frequency data of various contemporary organisms compiled by Goel et al. (7), we have calculated entropies H, and the results are shown in Fig. 7 as a function of the GC content. The datum of the vertebrates is an averaged one, all vertebrates having about the same GC content of DNA, roughly $42 \pm 2\%$ (24). There is an obvious correlation between H and the GC content. The larger the H value, the more evenly amino acids are distributed.

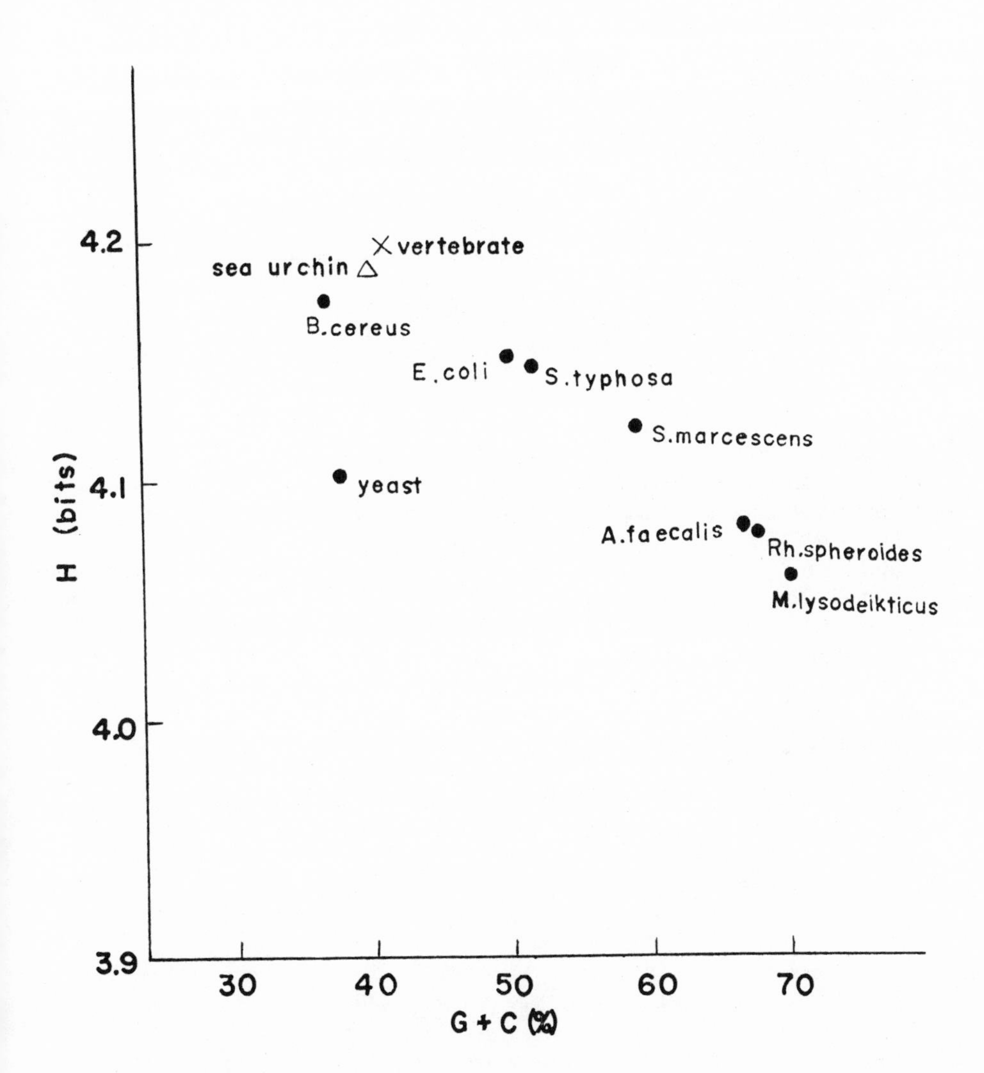

Figure 7. Entropy H, which is calculated from the experimental data of the frequencies of amino acids, as a function of G+C content.

In higher organisms, especially in vertebrates, GC content of DNA seems to converge around 42% as mentioned by Sueoka, and the base doublet of C_pG is highly scarce (12,25). These can be explained if we consider that the evolutional process is accompanied by diversification of proteins so as to carry out various kinds of functions, or more precisely, by increase of the capacity of DNA to code various proteins. The entropy function H is one of the measures of the potential variety of proteins. The larger the H value of the organism, the more diverse proteins it can hold.

Since amino acids with many times degeneracy have GC-rich codons, the GC content of mRNA or DNA should be limited so as to distribute amino acids evenly. Several authors (23,5,6,15,9-11) have shown by detailed calculation that about 42% GC content gives maximum entropy of amino acid sequence, and the DNA of the higher organisms converges around this state.

Furthermore, the rarity of the base doublet C_pG in vertebrate DNA can be interpreted as follows. Under the condition of maximum H, the GC content is less than AT, as is seen above. Arginine has six degenerate codons, CGd (d=U,C,A,G) and AGe (e=A,G), and alanine (GCd), glycine (GGd), and proline (CCd), all with G or C in the first and the second positions of the codons, have only four degenerate codons. Therefore, so that these four kinds of amino acid appear evenly, arginine should prefer to use codons AGe rather than CGd, and the insufficient nucleotide such as G and C should preferentially be used to code alanine, glycine, and proline. This will be the reason why the C_pG is scarce in vertebrates. Hasegawa and Yano (9-11) have explained this by detailed calculation of the entropy function.

The entropy function H of amino acid sequence is not a measure of randomness, but a measure of potential variety of protein.

REFERENCES

1. Bungenberg de Jong, H., in: "Colloid Science," Vol. II (Kruyt, H., ed.), p. 338, Elsevier, New York, 1949.

2. Ferris, J. P., Donner, D. B., and Lobo, A. P., J. Mol. Biol. 74, 499 (1973).

3. Fox, S. W., Harada, K., and Kendrick, J., Science 129, 1221 (1959).

4. Gabel, N. W., and Ponnamperuma, C., in: "Exobiology" (Ponnamperuma, C., ed.), p. 95, North-Holland, Amsterdam, 1972.

5. Gatlin, L. L., Math. Biosci. 13, 213 (1972).

6. Gatlin, L. L., "Information Theory and the Living System," Columbia University Press, New York, 1972.

7. Goel, N. S., Rao, G. S., Ycas, M., Bremermann, H. J., and King, L., J. Theor. Biol. 35, 399 (1972).

8. Goldberg, A. L., and Wittes, R. E., Science 153, 420 (1966).

9. Hasegawa, M., and Yano, T., "Tampakushitsu Kakusan Koso Bessatsu," p. 134, Kyoritsu, Tokyo, 1972.

10. Hasegawa, M., and Yano, T., submitted to Math. Biosci.

11. Hasegawa, M., and Yano, T., Fourth International Conference on the Origin of Life, Barcelona, 1973.

12. Josse, J., Kaiser, A. D., and Kornberg, A., J. Biol. Chem. 236, 864 (1961).

13. Kimura, M., Nature 217, 624 (1968).

14. King, J. L., and Jukes, T. H., Science 164, 788 (1969).

15. King, J. L., Proc. Sixth Berkeley Symp. Math. Stat. Prob., p. 69, University of California Press, Berkeley, 1972.

16. Labadie, M., Ducastaing, S., and Breton, J. C., Bull. Soc. Pharm. Bordeaux 107, 61 (1968).

17. Lind, S. C., Bardwell, D. C., and Perry, J. H., J. Am. Chem. Soc. 48, 1556 (1926).

18. Matthews, C. N., and Moser, R. E., Nature 215, 1230 (1967).

19. Mizutani, H., and Noda, H., "Tampakushitsu Kakusan Koso Bessatsu," p. 89, Kyoritsu, Tokyo, 1972.

20. Mizutani, H., Mikuni, H., and Takahasi, M., Chem. Letters 573 (1972).

21. Okihana, H., in preparation.

22. Oparin, A. I., "The Origin of Life on Earth," Macmillan, New York, 1938.

23. Smith, T. F., Math. Biosci. 4, 179 (1969).

24. Sueoka, N., "Evolving Genes and Proteins" (Bryson, V., and Vogel, H. J., eds.), p. 480, Academic Press, New York, 1965.

25. Swartz, M. N., Trautner, T. A., and Kornberg, A., J. Biol. Chem. 237, 1961 (1962).

26. Veis, A., and Cohen, J., J. Am. Chem. Soc. 78, 6238 (1956).

27. Veis, A., "Biological Polyelectrolytes" (Veis, A., ed.), p. 211, Marcel Dekker, New York, 1970.

28. Villars, D. S., J. Am. Chem. Soc. 52, 61 (1930).

29. Vogel, H., and Zuckerkandl, E., "Biochemical Evolution and the Origin of Life" (Schoffeniels, E., ed.), p. 352, North-Holland, Amsterdam, 1971.

30. Völker, T., Angew. Chem. 72, 379 (1960).

31. Voorn, M. J., Rec. Trav. Chim. 75, 405 (1956).

32. Yano, T., and Hasegawa, M., submitted to J. Mol. Evol.

33. Zuckerkandl, E., Derancourt, J., and Vogel, H., J. Mol. Biol. 59, 473 (1971).

THE COACERVATE-IN-COACERVATE THEORY OF THE ORIGIN OF LIFE

Vladimír J.A. Novák

Czechoslovak Academy of Sciences

Prague, Czechoslovakia

INTRODUCTION

The chief merit of Oparin's coacervate theory of the origin of life on the Earth was that it opened to modern science a field of research which up to then had been a subject of mere speculation. This was even more the case after the theory had been enriched by new experimental findings on the chemical synthesis of more complicated organic matter, including protein-like substances and nucleotides, in line with the constructionist approach (see Fox, 9). Many prominent biologists and biochemists and Oparin (22,23, etc.) himself developed this concept further and supported it by new findings and arguments of principal importance, e.g. Calvin (4), Fox and Dose (8), Bernal (2), Ponnamperuma (25), Buvet (3), Florkin (7), Dose (5), Lipmann (15), Krasnovsky (14), Eigen (6), and Oró (24).

Many of these views differ in their details, e.g. as to how, from what substances, in which sequence, and under what conditions the process of the formation of the first living organisms from non-living organic matter began and how it continued. The majority of the relevant authors, however, as distinct from creationists, vitalists, panspermists, etc., are agreed that life as we know it started, and developed continuously, on the surface of this planet, without any intervention by super- or un-natural forces, that it was not the outcome of a mere isolated chance in the Earth's history, and on various other presumptions and subsidiary working hypotheses stemming both from experimental data and from a thorough consideration of existing concepts and scientific arguments. They are further agreed that primordial proteins preceded primordial nucleic acids, that it was the interaction of proteins and nucleic acids and the determinant role of the latter (coding) which gave rise to the first real organisms, and that this was not a single, isolated occurrence, but that it happened over and over again wherever the essential substances came into contact under the right conditions, e.g. an aqueous medium allowing a given minimum concentration (coacervation), given temperature limits, sterility, i.e. the absence of higher organisms, such as bacteria, which would destroy the developing materials before they could interact, etc.--conditions which were undoubtedly present in the Earth's primordial surface.

Some questions are still under serious debate, however, and more facts are required before it can be definitely decided which of the various eventualities are correct. They include problems connected with the question of the instant from which we can speak of commencement of the life process, the criteria of cell structure, the conditions of self-replication, the instant at which natural selection started, etc. The present paper attempts to answer at least some of these questions and to present a coherent picture of the events of the "coacervate-in-coacervate" theory, a preliminary outline of which was submitted at the Varna Symposium in 1971 (Novák and Liebl, 17).

Before any such synthesis could be attained, reciprocal comparison and verification of all available findings and presumptions, of both constructionists and reductionists (in the sense used by Fox, 9) and of all other authors, irrespective of their classification, was necessary. Another source of findings and bases for presumptions was the structure of various extant viruses constituting an unbroken series of transitions from a replicating nucleic acid molecule to a cell. Further, there were findings on the laws of Darwinian evolution among higher organisms, with special reference to the author's principle of sociability (16) and, last but not least, biochemical findings on the relevant chemical reactions and geological data on the existing conditions to be considered.

The process of the origin and development of life on the Earth can be subdivided into three periods: 1) biochemical evolution, from the origination of the first organic matter on the Earth's surface up to the autoreproductive reactions of primordial nucleic acids, 2) subcellular evolution, from the commencement of NA self-replication up to the origin of the first true, continuously dividing cells, 3) the period of Darwinian evolution from the primordial cell up to present-day human society. The coacervate-in-coacervate theory is predominantly concerned with the second phase of the evolution of life.

BASIC PRESUMPTIONS

It was originally suggested by Bernal (1) that the time span between the origin of the first self-replicating molecules and the first cells was as long as, if not longer than, the period between the first cell and human society. This, however, is untenable if we consider the number of mutations possible in a subcellular organism such as a virus, in a unicellular organism and in the highest multicellular plants or animals. It was clearly refuted by the discovery of well-developed Cyanophyta in the oldest known sediments, like the Onverwacht layers in South Africa, the age of which was estimated to be 3.5 milliard years, i.e. more than two-thirds of the Earth's age. Definitive evidence is provided by recent findings showing that the precursors of cellular structures (minimal cells or protocells) are formed by simple physical processes (9). This does not necessarily mean that the simplest true cells, e.g. the most primitive bacteria, also originated as quickly, but a period of one or a few million years would undoubtedly have been sufficiently long for this process.

PRIMORDIAL PROTEINS AND COACERVATION

All extant organisms, from the simplest viruses upwards, are known to contain both proteins and nucleic acids. It can therefore feasibly be assumed that they were equally necessary for the first organisms in the Earth's history. They could not have appeared at the same time, however, and there is every reason to presume that it was the proteins which appeared first, in an abiotic manner (see Oparin, 22). Several different ways of amino acid polymerization have been described. Of these, the simplest--and therefore the commonest and most completely elaborated--is the heat polymerization of α-amino acids described by Fox (10-12), on which the proteinoid theory was based (9).

The necessary temperature (110-180°C) was surely not uncommon on the early Earth's surface. It is comprehensibly possible that materials produced in this way, together with other originating organic substances, were flushed by rains from the places of their origin (e.g. volcanic regions) and were carried by torrents to the coastal areas, where they accumulated. In contact with water, they swelled up and then occurred in the form of coacervates (microspheres, etc.). From the aspect of the further evolution of living systems, the great advantage of the coacervate state was that it accumulated the substances within a given water container which prevented their dispersion into too low concentrations, while keeping them in the most active aqueous solution state. Moreover, coacervates are known to attract other substances of both low and high molecular weight from their surroundings, thereby facilitating their reciprocal contact and interaction.

THE OLDEST ENZYMES AND NUCLEIC ACIDS

There is reliable evidence showing that the constituents of nucleic acids could also have originated in a chemical, abiotic manner in those times. Their organization into nucleic acid molecules is possible only in the presence of a suitable enzyme (polymerase), however. Most of the extant enzymes are highly specific, but this can hardly have been the case at the time of the primordial proteins. Since enzyme specificity is also essentially the outcome of gradual evolution, however, some enzymatic activity of a general character must likewise presumably have existed in the most primitive proteins, although it would have been much slower and less effective than the activity of maybe any of the known recent enzymes.

It can thus be accepted that abiogenetic nucleotides and bases were attracted into the above protein or proteinoid coacervates and that only then were they polymerized into the most primitive nucleic acid. This is the most feasible explanation, which now looks much less improbable than it did only a few years ago, despite the fact that we do not, at present, know the relevant reaction. If synthesis occurred as a result of the action of the protein, however, it is highly probable that the resultant primordial nucleic acid was determined by the character of the protein, although this does not necessarily mean that its base sequence would then code a similar protein. All that was needed was that the resultant molecule would give rise to a protein at all and that it would be capable of catalyzing its replication under the given conditions. If we accept that this type of reaction took place many times over, possibly with different proteins, wherever conducive conditions were present, it can be regarded as the first step towards the main feature of all living organisms, i.e. their continuous autoreproduction at the expense of the surrounding matter.

THE INCEPTION OF DNA REPLICATION

As soon as such a primitive nucleic acid molecule once originated in a given coacervate droplet or layer (it might have resembled recent RNA transfer, as suggested by Ohno, 20), its further synthesis was catalyzed by itself, that is to say, autoreplication started. At the same time, the molecule started to function as the matrix for synthesis of the type of protein where, again, the original protein might have acted as catalyzer. In this way, the contents of the given coacervate droplets were gradually converted to NA molecules and the corresponding protein. With this, metabolic activity diminished and the coacervate eventually disintegrated. When released from the internal medium, the DNA molecules were incapable of further metabolism, but did not lose their capacity for renewing metabolism if they came across another medium (another coacervate droplet) of a suitable composition. An analogy of this, at a more advanced level of evolution, is the penetration of bacteriophage DNA into the bacterial cell.

In this way, the original NA molecules continued to reproduce both themselves and the specific protein molecule. Sooner or later, however, the necessary components within reach of the molecules were utilized and their metabolism finally stopped. Even then, of course, they might not have been destroyed, but could have persisted for a time in an abiotic (cryptobiotic) state (see 13) to resume reproduction, even after long periods in a dry state, on re-entering a suitable medium. (This is no argument on behalf of Hinton's conclusion that life originated on land, in various freshwater pools, where it is far less likely that all the necessary compounds would have occurred together.)

In such a monomolecular state, the NA molecules were far more exposed to various mutagenic factors than in cellular organisms and this undoubtedly raised their mutability rate. This meant an increase in the rate of evolution, although the possibilities of changes in one short-chain molecule are limited compared with the DNA systems of higher animals. Mutation resulted in permanent changes in some of the reproducing molecules and this led to changes in their replication rate, resistance, and various other characters, which thus became subject to natural selection.

COMMENCEMENT OF THE ACTION OF NATURAL SELECTION

The point in the evolution of life from which we can speak of natural selection is a much debated question. Some authors already speak of the action of natural selection among coacervate droplets (23) or proteinoid microspheres (9). There is a difference between its action before and after autoreproduction has started, however. Before the continuous reproduction of more or less identical individuals commences, as is the case from the inception of NA replication, the stability of different systems (coacervate droplets) causes differences in their persistence, but it does not affect further evolution directly, since each new individual has to originate afresh. In this case, therefore, it seems more appropriate to speak of differential persistence, or differential stability, although, of course, this also has its evolutionary consequences. Actual natural selection would thus be limited to the influence of differential survival and reproduction on subsequent generations.

THE QUESTION OF THE "LIVING MOLECULE" AND MONOMOLECULAR ORGANISMS

NA molecules in the simplest viruses have often been spoken of in the above sense, i.e. as living molecules or monomolecular organisms. This is definitely not correct in the sense that such a molecule would differ from other NA molecules and that it would possess a permanent property (quality) which would justify the epithet "living." If kept in a bottle in the powdered state, it is no more living than any other chemical. If, by the word "living," we understand a reversible state of permanent metabolic activity and if we bear in mind that it is not the molecule alone which is a higher organism, but various molecules of an environmental substance interconnected by continuous metabolism, then the word "living" seems quite appropriate to express the temporary state of the molecule (i.e. for as long as metabolism continues). In this meaning, and within these limits, it is also fully applicable to any higher organism except in quantity (the size of the body and number of constituents).

Similarly, the term "monomolecular" is justified only in its specific sense, which approximates to that of "individual" in chemistry. Actually, each replication makes two molecules from each one, so that the manifestation of life is at the same time the end of its individuality. In all cases like the tobacco mosaic virus, therefore, we practically always have to do with molecule colonies and not with individual molecules, although they remain individuals as far as identical daughter and sister molecules are concerned and as long as no chemical bonds occur between them.

Whether we regard similar autoreproducing NA molecules as organisms, or whether we prefer to keep this term for cellular living beings (not forgetting that there is no strict division between the highest viruses and the lowest cellular organisms, however) is merely a matter of terminology. We should nevertheless bear the following points in mind:

1. A replicating NA molecule of this type possesses, in the form of its replicative capacity, all the main features generally accepted as characteristic for life, though in an undifferentiated state, i.e. a) continuous metabolism (NA replication and proteosynthesis, with all the necessary part reactions); b) growth and reproduction united in the form of NA replication; c) heredity--replication, is in fact, the actual basis of heredity, its further mechanisms at higher levels, like mitosis and sexual reproduction, being simply improvements based on natural selection, adapting it for conditions in higher organisms.

All other features of living organisms can derive from these three main qualities present in replicating NA molecules, as long as the medium is suitable and contains all the necessary constituents.

2. NA molecule replication initiated the continuous process of evolution, the culmination of which has so far been the development

of human society. In other words, one (or perhaps several) of the many original NA molecules which developed abiotically within the primordial coacervates started the uninterrupted, continuously branching phylogenetic line leading to every one of us without exception. We can thus rightfully claim that the evolution of man, which has passed through many stages, including the elasmobranch fish and the anthropoid primate, started from a single nucleic acid molecule.

THE SOCIABILITY PRINCIPLE AT SUBCELLULAR LEVEL

The sociability theory, as formulated by the author (16) shows that all extant and extinct organisms belong to one of the five levels or degrees of individuals, starting with monomolecular organisms in the sense described above and terminating with animal and human societies (1), that each higher degree of organisms had to pass, in the course of its evolution, through all the lower degrees (2), and that evolution from a lower to a higher level passed through the same five main stages, i.e. non-division (non-separation), differentiation, development of an internal medium, correlation, and integration (to an individual of a higher order) (3; Fig. 1).

It can be demonstrated that development from the first to the second degree, i.e. from the monomolecular to the unicellular organism, passes through the same five main phases. The analogy with the development of unicellular to multicellular plants and animals and with that of the two higher degrees (vascular plants and metameric animals as the fourth degree and animal and human societies as the fifth degree) was thus a great help for deeper comprehension of the evolution of cellular structure.

The nonseparation of self-replicating NA molecules at the outset of the self-replication period, together with the association of various products of both protein and nucleic acid molecule metabolism with replicating molecules, are the first essential step in evolution towards cell level. They can be observed in various types of simple extant viruses, such as influenza virus or some plant viruses. A similar comparison was at one time refuted on the basis of the "degradation" theory, which nowadays has much less support than it had some fifteen or twenty years ago, however. In the first place, intermediate phylogenetic stages between the molecule and the cell are a technical necessity; secondly, metabolism and growth in higher cellular organism is, in principle, chemical reaction of both high and low molecular weight substances; and thirdly, as emphasized by Oparin (21), proteins and other organic substances which today are found free only within the bodies of cellular organisms (plant and animal) were spread in an active coacervate form on many parts of the primordial Earth's surface, in a sterile state, before the

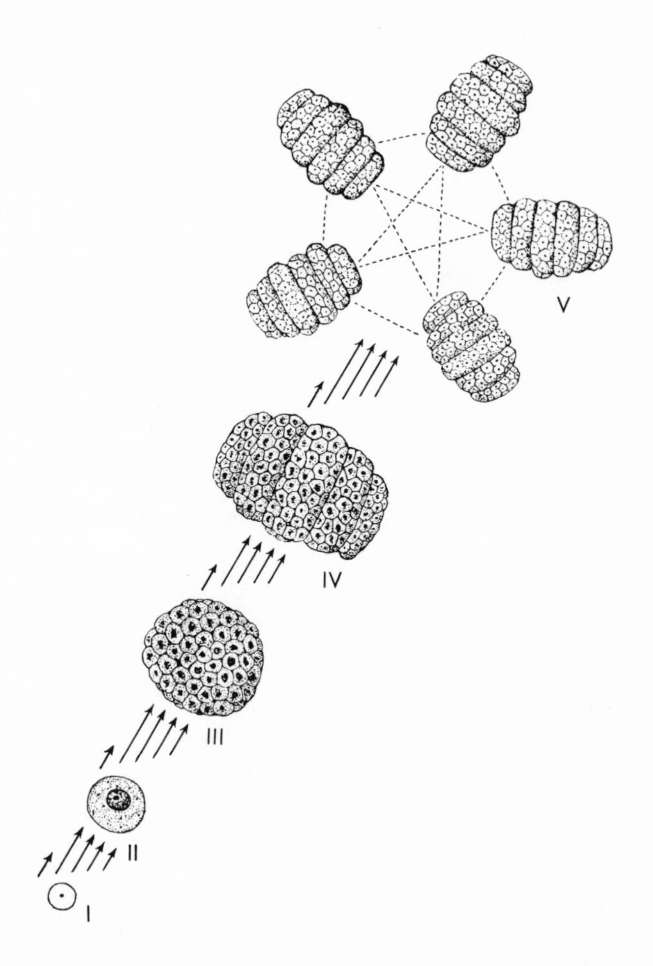

Figure 1. The principle of sociability (scheme). The five degrees of sociability in the evolution of organisms: 1) monomolecular protobionts, first reproducing, self-replicating organisms. 2) Unicellular organisms. 3) Simple multicellular organisms (plants and animals). 4) Cormophyta and metameric animals. 5) Societies of metameric animals. The five arrows between each two degrees: the phases of sociability. The broken lines interconnecting all original units (individuals of 4th degree) inside the individual of 5th degree suggest their mutual interactions and coherence.

first organisms developed. It is further known that some higher parasitic animals lack certain important enzymes, which they obtain from the host's body. As mentioned above, the role of enzymes, under primordial conditions, was played by the original proteins, although they were far less specific and active.

Similarly, the other four phases of sociability can also be found among extant viruses. Bacteriophages, for instance, stopped at phase 2, i.e. differentiation (before formation of the internal medium). Mycoplasmatales are probably the phylogenetically lowest organism in which phase 3 (formation of the internal medium) appears--still without a permanent outer membrane. Phase 4 of sociability, i.e. the development of correlation mechanisms, starts with the formation of a permanent surface membrane, which later becomes the cell wall of most unicellular organisms, especially of a plant character. The first example of this among extant organisms are rickettsiae. This is also the first step towards phase 5--the integration of the original molecule colony to an individual of a higher order, a unicellular organism, i.e. a prokaryotic cell, which progressively evolves, through many intermediate stages, into the highest form at this level, i.e. the eukaryotic cell (Fig. 2).

COACERVATE-IN-COACERVATE

The original coacervate theory propounded by Oparin (21), as accepted and gradually completed by the most serious students of probiology (the later proteinoid theory of Fox, 9, is also linked to it), presumed that the first organisms at the cellular level developed through gradual improvement, by natural selection, of the surface membrane of the original coacervate droplet. This view, however, presents us with the problem of how to explain the transition of these structures to the continuous reproduction of similar individuals, which is one of the main characteristics of all living organisms. As already mentioned (17-19), it would be more feasible to assume that such primordial, primary coacervates simply acted as the most suitable medium for the attraction and concentration of the structural material necessary for the origin of the first organisms and that the latter did not actually originate from them, as a result of their possible changes as a whole, but inside them, through internal chemical synthesis of the first nucleic acid molecules capable of replication at the expense of other components of these primary coacervates (irrespective of whether they existed in the form of micro- or macro-spheres, or of continuous layers formed by the fusion of original droplets). Only within such a medium could the future organisms find the necessary nutrient medium for their further reproduction (replication) and the equally necessary protein or proteinoid protoenzymes (pre-enzymes). If the whole of the coacervate droplet changed into a future organism, outside the coacervate it would have nothing on which to live and build others like

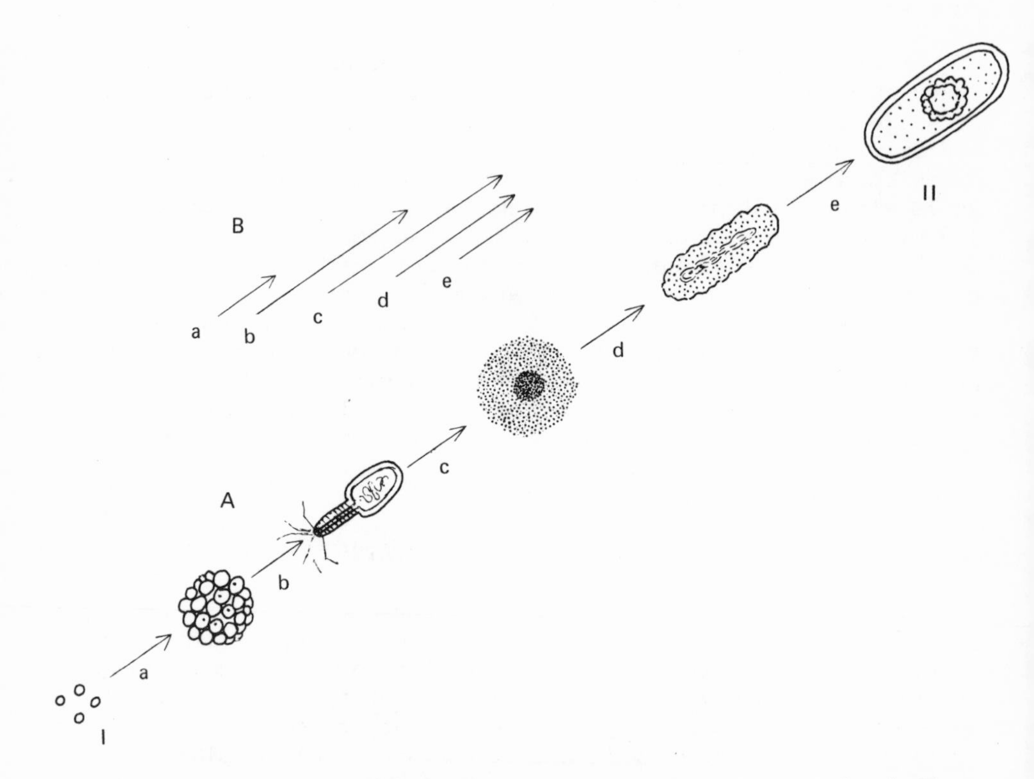

Figure 2. The phases of sociability in the evolution of the individual of 2nd degree (unicellular organism). A) The five gradual phases in the evolution of the cellular structure. 1) Four "monomolecular" protobionts. 2) The simplest cellular procaryote; a) nonseparation (origin of undifferentiated colonies), the stage reached by icosaedric plant viruses and simplest animal viruses; b) differentiation (division of labor), reached on the lowest level by bacteriophages; c) origin of internal medium (Mycoplasms); d) origin of correlations (first structurally permanent membrane) as in rickettsiae; e) integration into an organism of next degree, unicellular procaryote (the simplest bacterium). B) The mutual relation of the five phases in length and time.

itself.

As mentioned above, it can be assumed that the behavior of such primordial NA molecules was a close analogy of the behavior of bacteriophage NA molecules within attacked bacterial cells. The absence of a permanent, solid surface membrane on the coacervate droplets under those archaic, sterile conditions allowed the reproducing NA molecules to penetrate further coacervate droplets without the aid of a special enzyme (unlike extant bacteriophages). If we take into account that similar autoreproducing molecules were capable of cryptobiosis, in the meaning used by Hinton (13, see above), in the original, sterile environments, it is comprehensible that some of these molecules could give rise to the phylogenetic line from which all later organisms, with all their diversity, eventually developed.

Like the extant bacteriophage molecules, their replicating primordial ancestors at first probably produced a specific protein without any permanent chemical bond between the DNA and protein molecules. The products of the metabolism of the NA molecule, including synthesized specific protein molecules, are nevertheless in close contact with the relevant NA molecules, and it may thus feasibly be assumed that a small secondary coacervate originated around each NA molecule. In this way, numbers of new, more condensed, and more specific coacervates appeared in the original, primary coacervate droplets or layers. Their main difference, compared with the simpler primary coacervate, would then also depend on their dynamic character (continuous division and NA molecule replication), all the main components of which (the proteins and the products of their enzymatic activity) are directly determined by the structure (base sequence) of the corresponding DNA molecule. They are therefore hereditary in the true meaning of the term (Fig. 3).

It was undoubtedly such a secondary coacervate-in-coacervate which grew into the phylogenetic starting point of all further biological nonseparation of daughter molecules and their metabolic products, differentiation among the members of the molecule colony, formation of an internal medium (protoplasm), correlation between the original members of the colony starting with the formation of a permanent outer membrane, and integration to an individual of a higher order, i.e. the simplest unicellular organism.

4. The primary coacervates, the main components of which were abiogenic proteinoids or proteins which attracted and concentrated other organic materials, were the most suitable medium for synthesis of the first replicating NA molecules which would, from then onwards, determine the result of further proteosynthesis. The metabolic products of the replicating NA molecules accumulated around them. Unlike the primary coacervate, this secondary coacervate derived directly from the corresponding NA molecule, i.e. its components were all hereditary. It was through progressive changes (mutations) in such

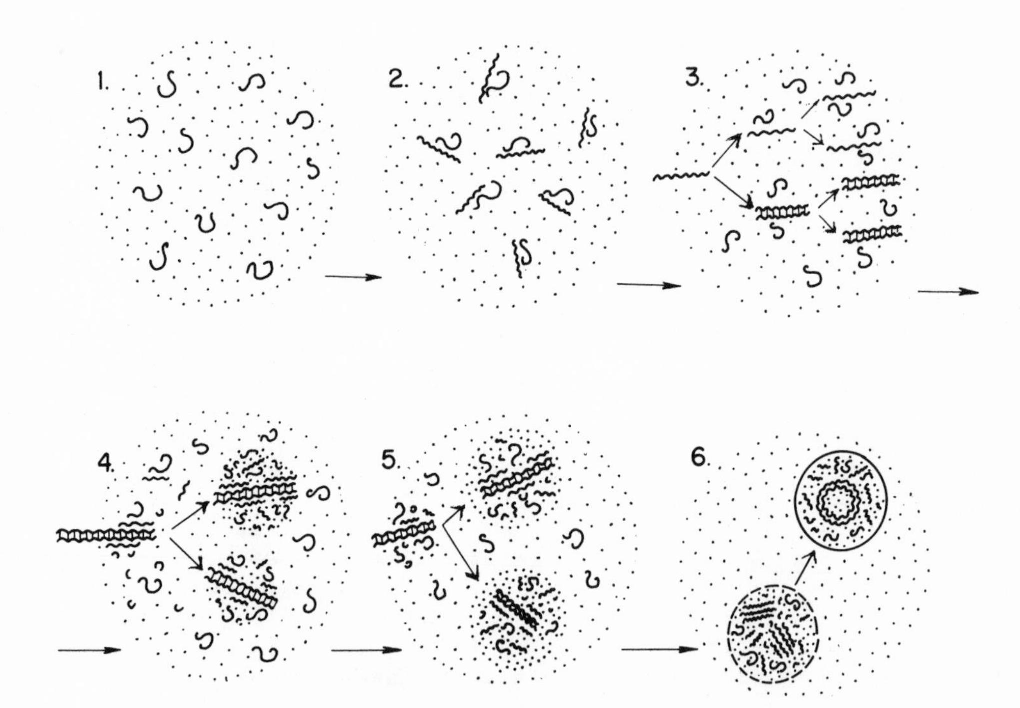

Figure 3. Six gradual stages in the evolution from the self-replicating nucleic acid molecule to the first primitive cell. 1) Primary coacervate of a protein and low molecular substances; 2) origin of first single-stranded NA molecules; 3) origin of a double helix DNA molecule; 4) around each DNA molecule a secondary coacervate originates; 5) reproduction of the secondary, hereditary coacervate; 6) origin of the simplest, primitive cell (from Novák and Liebl; 18).

secondary coacervates, under the action of natural selection, that evolution to the cellular level continued.

5. The question of what constituted the first living organism, in a continuous series, is merely a matter of terminology. It should, however, be borne in mind that such replicating NA molecules possessed all of the main features characteristic for life, i.e. continuous metabolism (NA replication and proteosynthesis), growth and reproduction (united in the form of NA replication), and heredity, from which all the other characteristic features eventually developed.

REFERENCES

1. Bernal, J. D., "The Physical Basis of Life," Routledge and Kegan Paul, London, 1951.

2. Bernal, J. D., "The Origin of Life," p. 345, Weidenfeld and Nicholson, London, 1967.

3. Buvet, R., Third International Symposium on the Origin of Life and Evolutionary Biochemistry, Varna, 1971.

4. Calvin, M., "Chemical Evolution," Oxford University Press, 1969.

5. Dose, K., Theory Exper. Exobiol. 1, 41 (1971).

6. Eigen, M., Naturwissenschaften 58, 465 (1971).

7. Florkin, M., and Mason, H. L., Comparative Biochemistry 1, 1 (1960).

8. Fox, S. W., and Dose, K., "Molecular Evolution and the Origin of Life," W. H. Freeman, San Francisco, 1972.

9. Fox, S. W., Naturwissenschaften 60, 359 (1973).

10. Fox, S. W., and Middlebrook, M., Federation Proc. 13, 211 (1954).

11. Fox, S. W., and Harada, K., J. Am. Chem. Soc. 82, 3745 (1960).

12. Fox, S. W., and Harada, K., "Polyamino Acids, Polypeptides, and Proteins" (Stahmann, M., ed.), p. 47, University of Wisconsin Press, Madison, 1962.

13. Hinton, H. E., Proc. Royl. Soc. B171, 43 (1968).

14. Krasnovsky, A. A., Third International Symposium on the Origin of Life and Evolutionary Biochemistry, Varna, 1971.

15. Lipmann, F., "Molecular Evolution: Prebiological and Biological" (Rohlfing, D. L., and Oparin, A. I., eds.), p. 261, Plenum Press, New York, 1972.

16. Novák, V.J.A., Zurn. Obsc. Biol. 28, 387 (1967).

17. Novák, V.J.A., and Liebl, V., Third International Symposium on the Origin of Life and Evolutionary Biochemistry, Varna, 1971.

18. Novák, V.J.A., and Liebl, V., Biol. Listy 38, 40 (1973).

19. Novák, V.J.A., and Liebl, V., Fourth International Symposium on the Origin of Life, Barcelona, 1973.

20. Ohno, S., "Evolution by Gene Duplication," New York, 1970.

21. Oparin, A. I., "The Origin of Life on the Earth," Oliver and Boyd, Edinburgh, 1957.

22. Oparin, A. I., "The Origin and Initial Development of Life," Med. Publ. House, Moscow, 1966.

23. Oparin, A. I., Scientia 1, 12 (1971).

24. Oró, J., Fourth International Conference on the Origin of Life, Barcelona, 1973.

25. Ponnamperuma, C., Lemmon, R. M., Mariner, R., and Calvin, M., Proc. Nat. Acad. Sci. U. S. 49, 737 (1963).

SEDIMENTARY MINERALS UNDER REDUCING CONDITIONS

L. E. Orgel

The Salk Institute for Biological Studies

San Diego, California 92112

Professor Oparin , in his classical publication in 1924 (1), was the first to emphasize that the primitive atmosphere of the Earth was reducing and that this has important consequences for studies of the origins of life. His proposal that organic compounds were synthesized in a reducing atmosphere provided the conceptual framework for the burst of activity in the field of experimental prebiotic chemistry that was initiated by Urey and Miller (2). It is now widely believed that prebiotic synthesis occurred in a reducing or, at least, a nonoxidizing atmosphere. However, very little has been written about the implications of Oparin's proposal for the discussion of the inorganic chemistry of the prebiotic Earth. Here we indicate some ways in which the inorganic chemistry of oceans and lakes on the primitive Earth is likely to have differed from the chemistry with which we are now familiar.

Our attention was first drawn to this topic by the observation that a mixture of urea and ammonium phosphate in the presence of Mg^{2+} ion forms the basis for an excellent phosphorylating agent (3). When nucleosides are heated with such a mixture, nucleoside - 5' - pyrophosphates are formed in reasonable yield. We wondered whether, on the primitive Earth, magnesium ammonium phosphate (struvite) might have precipitated under certain circumstances in place of hydroxyl apatite, a mineral which is very abundant nowadays. This seemed possible, even though struvite is much more soluble than hydroxyl apatite, because it is known that the Mg^{2+} ion prevents the direct precipitation of hydroxyl apatite from aqueous solution.

It proved easy to show that, in the presence of 0.1 M NH_4^+ ion, struvite does indeed precipitate from a solution containing orthophosphate and equimolar amounts of Mg^{2+} and Ca^{2+} (4). The criticism that this might not be the case in the presence of CO_2 turns out to be ill-founded (5). We have now shown that exposure of the system to CO_2 in the atmosphere does not bring about the precipitation of hydroxyl apatite or interfere with the precipitation of struvite (6).

This volume provides a suitable place to raise a much more general question that is implicit in Professor Oparin's early writing. What other major differences would there have been between the inorganic chemistry of primitive oceans and lakes and the chemistry that we observe today under an oxygen-containing atmosphere?

Hydrogen cyanide is believed by many to be one of the most important precursors of the organic chemicals that formed on the primitive Earth. If current views on prebiotic chemistry are correct, it seems likely that the primitive atmosphere contained substantial amounts of hydrogen cyanide. It is well-known that many of the transition metals form extremely stable, water-soluble complexes with the cyanide ion. Weathering by prebiotic rain may therefore have resulted in the extraction of heavy metals from rock as complex cyanides and the transport of these materials to the oceans. It is possible, therefore, that complex ions such as the ferrocyanide ion, $[Fe(CN)_6]^{4-}$, were abundant components of the primitive ocean.

These arguments raise problems that can only be solved by those with broad experience in geochemistry. Here I wish only to raise the possibility, and remark that if indeed transition-metal cyanide complexes were abundant on the primitive Earth, they could perhaps have been the precursors of the biological complexes of the corresponding metals, for example, the iron-porphyrins.

This is only one of several possibilities that could be explored. Another, which seems of considerable interest, is the role of hydrogen sulphide in the transport and precipitation of metals on the primitive Earth. Again, it is arguable that compounds of metal ions such as Fe^{2+} with sulfide ion were abundant beneath the primitive oceans and were the precursors of ferrodoxin and related enzymes.

These are some of the interesting implications of Professor Oparin's theory of 1924 which, fifty years later, remain to be explored. I hope that geochemists can be persuaded to take an interest in such matters.

REFERENCES

1. Oparin, A.I., "Proiskhozhedenie zhizny," Moscow Izd. Moskovskii Rabochii, 1924. Translation as Appendix I in "The Origin of Life," Bernal, J.D., World Publishing Company, Cleveland and New York, 1967.

2. Miller, S. L., Science 117, 528 (1953).

3. Handschuh, G. J., Lohrmann, R., and Orgel, L.E., J. Mol. Evol., in press (1973).

4. Handschuh, G. J., and Orgel, L. E., Science 179, 483 (1973).

5. McConnell, D., Science 181, 582 (1973).

6. Handschuh, G. J., unpublished results.

MICELLES AND SOLID SURFACES AS AMINO ACID POLYMERIZATION PROPAGATORS

Mella Paecht-Horowitz

Polymer Department, The Weizmann Institute of Science

Rehovot, Israel

INTRODUCTION

In many of his papers and treatises, Oparin advanced the hypothesis that the formation of biomacromolecules from biomonomers might have taken place inside coacervates (1), droplets consisting of polymers separating out from a solution in which a monomer polymerizes in the presence of another polymer. When the new polymer reaches a certain size, phase separation, accompanied by a sharp shift towards synthesis (2), takes place. Evreinova *et al.* have shown that monomers of the surrounding medium can concentrate inside such coacervates up to a hundred-fold, depending on the type of the polymer forming the coacervate and on the nature of the monomers in the solution (3). These experiments led Oparin to the idea that coacervates were the means of concentration of monomers from dilute solutions and that polymerization takes place inside them.

This hypothesis was expanded by Calvin, who pointed out that if, in addition to the concentration of the monomers, also polymerization proceeded more quickly inside the coacervates than in the surrounding solution, the coacervates would contain increasing concentrations of substances which would act as catalysts for further polymerization. The mass of the droplets would then increase and the reaction would accelerate until the droplets would split into fragments, each of identical consistency with the parent droplet. This process would resemble in a way a cellular division (4). Calvin pointed out that the development of biological catalysts from such simple catalytic systems of coacervates could have emerged by natural selection. With regard to the mechanisms of selection and "survival" in the prebiological era, Oparin recently expanded this theory (5) with the hypothesis that multimolecular systems would meet, form coacervates, decompose, and interact again in other combinations and compositions until such combinations would be formed which adapted better to the exterior environment. These would grow and develop further--in other words, they would "survive." The adaptation would depend strongly on the "fitness" between the polymers. Once such "fitting" combinations were formed, they would produce new, better-fitting polymers, the coacervates would decompose again, and the new polymers would find even better fitting partners with which they would form new coacervates, consisting of a further step in advancing the system. This process would be repeated until contemporary self-reproducing systems would evolve.

Bernal, aware that in the natural water reservoirs the concentration of biomonomers is much too low to allow polymerization, suggested that the level of concentration could be raised by adsorption of the monomers on very fine clay particles which exist on the shores of the oceans or rivers (6).

Another form of mineral as support for the polymerization of amino acids was used by Fox and his collaborators (7) who produced proteinoids by thermal polymerization of amino acids on pieces of lava. Hanafusa, in Akabori's laboratory, obtained di- and triglycine by heating amino-acetonitrile with kaolinite (8).

Whatever the reasoning, it was clear to all investigators concerned with the problem of prebiotic synthesis that if polymerization was deemed to occur in a natural aqueous medium (as one would expect from biological or pseudobiological reactions) the concentration of the monomers would have been one of the first criteria, though not the only one. Some sort of activation would then have to have taken place and finally polymerization, preferably a somehow oriented and nonrandom one.

We have tried to elucidate these problems by separating and studying each step in the postulated process.

NONENZYMATIC ACTIVATION OF AMINO ACIDS

The term "activated amino acid" is used to denote an amino acid in a state in which it can undergo polymerization, transpeptidation, esterification, or any other type of reaction. In our system, the activation is accomplished by binding the amino acid's carboxyl group to another acid, that is, through the formation of an anhydride bond (9), using standard organic chemistry procedures. Nature performs this task using enzymes, but how can we envisage the formation of such activated amino acids in the prebiotic period? Kenyon *et al.* have shown that when dicyanamide is added at a constant rate to an acid solution of glycine, peptides up to tetraglycine are formed, each one in time reaching a constant concentration (10). We have obtained the same pattern of peptide formation in a slightly alkaline medium, starting with amino acid adenylates added to the aqueous solution at a steady rate (11).

The pattern as such is typical of any steady state reaction. However, while in Kenyon's system an acid medium was desirable, in our system an alkaline medium gave better results, since the amino acid was already activated. Under these conditions, polymerization proceeds much better when the amino group of the amino acid or the peptide to be condensed is in its free, rather than in the ammonium, form (12). In Kenyon's experiment, activation prior to polymerization was necessary, and was produced by the action of dicyanamide, an acid pH being required for the reaction to take place.

As in nature the pH of solutions is usually more or less neutral, we looked for a way to produce acid spots on which activation of amino acids might take place in an otherwise neutral medium. Suitable ion exchangers might provide the answer. As synthetic ion exchangers do not exist in nature, we started to investigate various minerals with regard to their ion exchange properties.

One class of minerals, zeolites, swell only very slightly in water and are composed of alumina silicates arranged in such a way that they form canals leading through the whole body of the mineral. These canals are of varying width, according to the original mode of formation of the zeolites; this structural feature allows the zeolites to serve as molecular sieves. In addition to these properties in common, some zeolites exhibit strong membrane hydrolysis, that is, the salts they form are very readily hydrolyzed, and their cations pass quickly into the surrounding aqueous solution. These zeolites naturally will be also very good ion exchangers, and might cause the free carboxyl group of the amino acid to react with ATP to give a mixed anhydride. This idea has been tested with an artificial zeolite, having canals wide enough for large molecules to pass through them; it was indeed found that mixed anhydrides were formed under these conditions. Yet it should be stressed that not all batches of presumably the same zeolite gave the mixed anhydrides and the mechanism of their formation might be more complicated than it seemed at first. Anyhow, even when formed, the anhydrides hydrolyzed quickly as the zeolites are not polymerization catalysts (13).

HETEROGENEOUS POLYMERIZATION

As mentioned previously, the concentration of biomonomers in nature is not sufficiently high for meaningful reactions to take place. Even in the laboratory, our attempts to produce homogeneous polymerization by dissolving amino acid adenylates at concentrations of 10^{-2} - 10^{-1} M in water, at pHs 7-8, either by dissolving the whole substance all at one time (12) or by performing the reaction under steady state conditions (11), did not yield high molecular weight substances; the reaction usually stopped at the tetra- or penta-mers. The polymerization yields, even for the formation of these small peptides, were rather low (30-40% at the most). We therefore started a search for nonbiological catalysts that would enable us to obtain peptides of higher degrees of polymerization, without having to form also the whole spectrum of the lower ones.

We considered three pathways: the first was to take as the phosphate part of the anhydride a long chain carbohydrate ester which would be nonmiscible with water and would provide a monomolecular layer surface on which the polymerization could take place. Using decanol phosphate, dodecanol phosphate, and docosanol phosphate (10, 12, and 20 carbon atoms) we obtained peptides up to 10-12 mers, but still in yields of only 8-10% (14).

A second pathway was to test various metals, in colloidal or sponge form, as surface catalysts. Optimal results were obtained with metals of the transition groups, cobalt and copper being the most active ones. Discrete molecular weight spectra were obtained, showing peptides even up to 30-40 mers, but these in yields not exceeding 2-3%, while most of the peptides were in the 4-6 mer range (15).

Various ion exchangers tested as surface catalysts were found not to influence the polymerization process at all.

The most promising type of surface tested so far for catalytic properties with regard to polymerization is that of clays. And here we return to Bernal's hypothesis, which we applied to Oparin's coacervate conception. After some preliminary tests, we found that certain clays, namely those which are able to swell, are excellent catalysts for the kind of polymerization in which we are interested. This would imply that the polymerization is a pure surface phenomenon, but as will be shown later, this is not the case. The clays we have studied so far are of the montmorillonite group.

The montmorillonites are composed of three unit layers, two of silicon oxide, with a metal oxide-hydroxide layer between them. The major part of our experiments were carried out on aluminium montmorillonite, and some were on iron montmorillonite (nontronite). The experiments with the latter showed that as long as the clay is able to swell, the nature of the metallic component is of secondary importance, producing only minor changes in the general pattern. Clays of the same configuration but without the ability to swell (like kaolinite) have very little effect on the polymerization.

The general pattern of polymerization in the presence of montmorillonite is the following: practically no hydrolysis of the monomers occurs, and the yield of polymer is nearly 100%. A discrete spectrum of degrees of polymerization is obtained, such as 8, 15, 22, 30, 36 mers, in more or less equal amounts. The higher peptides, even after prolonged washings or chromatographic separations, are still linked to one molecule of adenylic acid, but in the final stage this is no longer by an anhydride bond, but by an ester bond between the carboxyl groups of the peptide and the hydroxyl group of the sugar part of the adenylic acid. It is this esterification which terminates the reaction (16-18).

As to the mechanism of the polymerization on clay though we do not have any definitive answer, the following suggestions are plausible:

On measuring the adsorption of amino acids, peptides, and adenylic acid on montmorillonite, at pH 7-9, we found that none of these substances is adsorbed and their liberation from the clay occurs only after polymerization has taken place. What is desorbed is the free adenylic acid and the peptides formed by the polymerization process. The desorption process proceeds at a much lower rate than the polymerization reaction; hence it is clear that the polymerization is not a continuous process but must stop for various intervals of time, until the peptides formed and the adenylic acid liberated leave the clay, and new monomers can polymerize.

Another factor tending to produce an oscillating process of polymerization is the following: when adenylic acid is liberated, it produces a sharp local drop in pH which stops polymerization; only when the pH returns to a neutral or slightly alkaline level can polymerization again take place. The superposition of these factors would give peptides of a discrete spectrum of degrees of polymerization (18).

COPOLYMERIZATION OF AMINO ACIDS IN THE PRESENCE OF MONTMORILLONITE

In order to investigate the preferences of the different amino acids for mutual interaction and the influence of montmorillonite on these preferences, we studied the copolymerization of adenylates of several pairs of amino acids in the presence and in the absence of montmorillonite. A suspension of montmorillonite of particle size ∿150Å, at a concentration of 500 mg/l, was maintained by a pH stat at a constant pH of 7.8. The experiment was carried out at room temperature. The mixture of the adenylates of the two amino acids was added in very small portions until the concentration of the amino acids was about 0.05 M. A portion of the montmorillonite suspension containing montmorillonite of a larger particle size was then added, and subsequently several portions of the amino acid adenylates. This procedure was continued until about 2 mM of adenylates (1 mM of each amino acid) and fifteen portions of montmorillonite (up to particle sizes of 5000 Å) had been added. The mixture was left overnight with stirring, still at controlled pH. The suspension was then centrifuged, the precipitate washed numerous times, and the combined supernatants passed through membranes of various pore sizes. Each filtrate was concentrated by lyophilization and then passed consecutively through columns of Sephadex of various mesh sizes with water and 0.01 N HCl as eluants. The separations were repeated until the analysis of each fraction was constant. When it could be reasonably assumed that a fraction contained only one single substance, the amino acid sequence in every such fraction was determined (19). Tables I and II show the results of these experiments. As can be seen, the affinities between the amino acids differ according to the circumstances in which the polymerization has been carried out--whether in the presence or absence of montmorillonite. So far, the only pair which gave practically the same results in the presence as in the absence of clay is aspartic acid-histidine. The reasons for this will be discussed presently.

The degrees of polymerization and the amount of copolymerization depend also on the interacting partners. It can be seen from Table I that in some cases DL-alanine gives more homopeptides than in others and these homopeptides differ in their degree of polymerization from one case to another. For comparison, Table III gives the mean degrees of polymerization of the amino acid adenylates when each one is polymerized separately, in the presence of montmorillonite.

As to the various fractions obtained as the result of copolymerization in the presence of montmorillonite, generally, in these cases where a strong interaction between the two amino acids occurs, the degrees of polymerization obtained are higher than when the interaction is weak. The exception is glycine, which interacts very well with other amino acids, but the resultant peptides are of low molecular weights. This is probably due to the low solubility of glycine peptides; a peptide cannot continue to grow once it has precipitated. This is probably also the reason why L-alanine practically does not polymerize at all, while DL-alanine does so very well in the presence of montmorillonite.

In the absence of montmorillonite, a large amount of the adenylates hydrolyzes, while the part which does polymerize gives only small peptides; the relative degree of affinity between the amino acids can be calculated also for these. Naturally, it is possible that the affinity may change with the length of the chain, a problem which has not been investigated yet.

It is curious that although aspartyl adenylate polymerizes rather well by itself in the presence of montmorillonite (it does

Table I

Results of Copolymerization of Adenylates of Pairs of Amino Acids in the Presence of Montmorillonite

Interacting Substances	Type of Bond	Relative Yields of Bonds* (%)	% of Oligo-peptides	% of Homo-peptides	Mean Degree of Polymerization of Homopeptides	Mean Degree of Polymerization of Oligopeptides
Alanyl Adenylate	Ala-Ala	40		$(Ala)_n$-8	10	
	Gly-Gly	32				
Glycyl Adenylate	Ala-Gly	15	81	$(Gly)_n$-11	11	9
	Gly-Ala	13				
Alanyl Adenylate	Ala-Ala	23		$(Ala)_n$-19	22	
	Val-Val	52				
Valyl Adenylate	Ala-Val	12	81	$(Val)_n$-0	-	31
	Val-Ala	13				
Alanyl Adenylate	Ala-Ala	47		$(Ala)_n$-39	21	
	Asp-Asp	49	18			
Aspartyl Adenylate	Ala-Asp	2		$(Asp)_n$-43	27	17
	Asp-Ala	2				
Alanyl Adenylate	Ala-Ala	37		$(Ala)_n$-10	8	
	Ser-Ser	37				
Seryl Adenylate	Ala-Ser	12	69	$(Ser)_n$-21	12	40
	Ser-Ala	14				
Aspartyl Adenylate	Asp-Asp	55		$(Asp)_n$-32	10	
	Gly-Gly	21				
Glycyl Adenylate	Asp-Gly	9	43	$(Gly)_n$-25	2	9
	Gly-Asp	15				

Table I (cont'd)

Interacting Substances	Type of Bond	Relative Yields of Bonds* (%)	% of Oligo-peptides	% of Homo-peptides	Mean Degree of Polymerization of Homopeptides	Mean Degree of Polymerization of Oligopeptides
Aspartyl Adenylate	Asp-Asp	59		$(Asp)_n$-32	29	
	Ser-Ser	22				
Seryl Adenylate	Asp-Ser	10	65	$(Ser)_n$-13	10	10
	Ser-Asp	9				
Aspartyl Adenylate	Asp-Asp	36		$(Asp)_n$-37	5	
	His-His	44				
Histidyl Adenylate	Asp-His	8	36	$(His)_n$-27	3	10
	His-Asp	12				

*The numbers given are the yields as related to the overall amount of bonds. In some of the homogeneous polymerizations, this amount is very low as most of the substance simply hydrolyzes, giving the free amino acid.

Table II

Results of Copolymerization of Adenylates of Pairs of Amino Acids in the Absence of Clays

Interacting Substances	Type of Bond	Relative Yields of Bonds* (%)	% of Oligo-peptides	% of Homo-peptides	Mean Degree of Polymerization of Homopeptides	Mean Degree of Polymerization of Oligopeptides
Alanyl Adenylate	Ala-Ala	29		$(Ala)_n$-26	1	4
	Gly-Gly	15				
Glycyl Adenylate	Ala-Gly	26	42	$(Gly)_n$-32	1	
	Gly-Ala	30				
Alanyl Adenylate	Ala-Ala	13		$(Ala)_n$-10	1	
	Val-Val	15				
Valyl Adenylate	Ala-Val	44	79	$(Val)_n$-11	1	4
	Val-Ala	28				
Alanyl Adenylate	Ala-Ala	18		$(Ala)_n$-15	2	
	Asp-Asp	-				
Aspartyl Adenylate	Ala-Asp	43	74	$(Asp)_n$-11	1	3
	Asp-Ala	39				
Aspartyl Adenylate	Asp-Asp	42		$(Asp)_n$-45	1	
	His-His	42				
Histidyl Adenylate	Asp-His	5	10	$(His)_n$-45	2	6
	His-Asp	11				

*The numbers given are the yields as related to the overall amount of bonds. In some of the homogeneous polymerizations, this amount is very low as most of the substance simply hydrolyzes, giving the free amino acid.

Table III

Results of the Homopolymerization of Amino Acid Adenylates in the Presence of Montmorillonite

Type of Amino Acid	Mean Degree of Polymerization
Alanine	12
Glycine	5
Histidine	6
Serine	18
Valine	14
Aspartic Acid	12

not polymerize without montmorillonite, due to its charge repulsion), it gives very little copolymerization. Surprisingly, not only does it not interact well with the basic histidyl adenylate, but in the presence of histidyl adenylate even the homopolymerization of aspartyl adenylate is much reduced. When the homopolymerization of histidyl adenylate in the presence of montmorillonite was investigated, it was found that the mode of polymerization is the same as when no clay is present; a high degree of hydrolysis occurs and those low molecular weight (not more than 5-6 mers) peptides which are formed are of consecutive degrees of polymerization.

These data have caused us to revise our previous assumption that the absorption of amino acid adenylates on clays is a surface phenomenon and hence the greater the surface, the more absorption of substance and correspondingly the greater the extent of polymerization. If this assumption were true, histidine would behave like any other amino acid, which it obviously does not. As the main difference between histidine and the other amino acids investigated so far is its bulk, we have tried to envisage a mechanism in which the bulk of the molecule might be decisive. We constructed a molecular model on which two sheets of clay were separated by only the breadth of a water molecule and checked what part of the amino acid adenylate molecule would fit in between the clay sheets. We based this model on the fact that although montmorillonite expands in pure water practically _ad infinitum_, its expanded layers collapse and precipitate in the presence of salts and approach a width which depends on the precipitating agent. However, no matter what the agent, the layers never become so close that a molecule of water cannot come between them. With this model, we could observe that in the case of alanine, glycine, serine, valine, and aspartic acid, the amino acid as well as the sugar phosphate part of the molecule fitted in rather well between the layers, while the adenine protruded. Thus we could envisage the molecules as arranging themselves around the clay particle and the polymerization proceeding along the edges of the clay, with the peptide inside and the adenine portion protruding.

According to this model, the bulkier amino acids could not polymerize on montmorillonite as they are unable to penetrate between the layers of the clay; indeed, histidine also is much too bulky to penetrate between the layers of the clay (20).

This model agrees rather well with the Oparin conception that polymerization is supposed to take place inside the coacervates. Assuming that the clay replaces the coacervates, the polymerization in this case takes place inside the clay and not on its surface.

From the negative polymerization results with histidyl adenylate, it would appear that not only is histidine unable to penetrate the layers of the clay, but that it blocks them, so much so that aspartyl adenylate which, when alone, can penetrate the layers, in the presence of histidine is unable to do so and hence does not polymerize either.

DISCUSSION AND CONCLUSIONS

Summing up the role of clays in polymerization of amino acids, one may say that their main functions seem to be the concentration of the reactants and the scattering of some of the molecular charges, thus allowing the interaction of molecules which cannot interact without clays due to the repellent effect of their charges. Thus adenylates are able to interact in the presence of montmorillonite, but not in its absence. In the latter case, a stepwise reaction proceeds between one molecule of adenylate and one molecule of free amino acid or peptide which gives only low peptides of consecutive order. In the presence of montmorillonite, the adenylates can interact directly with each other and produce active peptides which interact further, resulting in the formation of relatively high peptides.

In the presence of montmorillonite (Table I) independent of the ratio of interaction between the two amino acids, the preference is always for interaction with themselves to interaction with one another.

This preference might be the reason also for the formation of repeating units in a polypeptide; once a certain small peptide is formed, the preference is for interaction with its own kind rather than with another. However, the overall proportion of such repeating units is not very high.

It would be interesting to learn whether the nucleotide part of the molecule could also polymerize in the presence of clays, giving the respective polynucleotides (21); so far, no such polymerization has been detected.

If our assumption that polymerization could take place only between the layers of clay is valid--and, as we have seen, in our model, adenylic acid could not penetrate--it is not surprising that no nucleotide polymers are obtained. Possibly if a way could be found to introduce the nucleotide between the layers, polymerization would take place. Future experiments should include the testing of various other clays, changes in the buffer to allow the layers to stay a fairly large distance apart or even fixing the distances between layers initially by introducing large molecules between them. However, in the latter case another problem could arise, namely, fixing molecules would block the entrance to those molecules in the polymerization of which we are interested.

REFERENCES

1. Oparin, A. I., "Life, Its Nature, Origin and Development," p. 47, Oliver and Boyd, Edinburgh, 1961.

2. Oparin, A. I., and Serebrovskaya, K. B., Dokl. Akad. Nauk. SSSR 148, 943 (1963).

3. Evreinova, T. N., Pogosova, P., Tsukawara, T., and Lapinova, T., Nauch, Dokl. Vyssh. Shk. 1, 159 (1962).

4. Calvin, M., Science 130, 1170 (1959).

5. Oparin, A. I., Abstr. of the 4th Int. Symp. of the Society for the Study of the Origin of Life, Coloqu. IV, 22, Barcelona, 1973.

6. Bernal, J. D., "The Physical Basis of Life," Routledge and Kegan Paul, London, 1951.

7. Fox, S. W., "The Origins of Prebiological Systems" (Fox, S. W., ed.), p. 363, Academic Press, New York, 1965.

8. Akabori, S., "Aspects of the Origin of Life" (Florkin, M., ed.), p. 117, Pergamon Press, London, 1960.

9. Katchalsky, A., and Paecht-Horowitz, M., J. Am. Chem. Soc. 76, 6042 (1954).

10. Kenyon, D. H., Steinman, G., and Calvin, M., Biochim. Biophys. Acta 124, 339 (1966).

11. Paecht-Horowitz, M., and Katchalsky, A., Biochim. Biophys. Acta 140, 14 (1967).

12. Lewinsohn, R., Paecht-Horowitz, M., and Katchalsky, A., Biochim. Biophys. Acta 140, 24 (1967).

13. Paecht-Horowitz, M., and Katchalsky, A., J. Mol. Evol. 2, 91 (1973).

14. Paecht-Horowitz, M., unpublished results.

15. Lewinsohn, R., and Paecht-Horowitz, M., unpublished results.

16. Paecht-Horowitz, M., Abstr. Fifth Internat. Symp. Chem. Natural Products, p. 232, London, 1968.

17. Paecht-Horowitz, M., "Molecular Evolution, Vol. 1, Chemical Evolution and the Origin of Life" (Buvet, R., and Ponnamperuma, C., eds.), p. 245, North-Holland, Amsterdam, 1971.

18. Paecht-Horowitz, M., Berger, J., and Katchalsky, A., Nature 228, 636 (1970).

19. Paecht-Horowitz, M., Israel J. Chem. 11, 369 (1973).

20. Paecht-Horowitz, M., Proc. Fourth Internat. Symp. Soc. Study Origin Life, Barcelona, 1973, in press.

21. Paecht-Horowitz, M., Angew. Chem. Int. Ed. (English) 12, 349 (1973).

POSSIBLE WAYS OF IDENTIFYING THE ABIOGENESIS OF BIOCHEMICALLY IMPORTANT COMPOUNDS

T. E. Pavlovskaya

A. N. Bakh Institute of Biochemistry

Academy of Sciences of the USSR, Moscow

Five decades lie between our time and the time when A. I. Oparin put forward his hypothesis of the origin of life on Earth (1). Oparin's hypothesis has become a theoretical stimulation all over the world for experimental modelling of life's origin; this research is one of the main trends in modern biology. A wealth of investigations on theoretical and experimental problems, particularly on prebiological chemical evolution, on the main indications for this stage of evolution, and on the ways inorganic matter is converted to organic matter have been carried out during these five decades.

Whereas the first years of developing the theory of life's origin were devoted mostly to study of individual stages of chemical evolution, connected with abiogenic synthesis of biochemically important compounds, later on the greatest attention was focussed on the prebiological multimolecular systems as a path to appearance of primordial cells and to metabolism, on finding out the most ancient stages of biogenesis by comparative study of actual organisms, or by paleobiochemical investigations.

The conversion of simple inorganic and organic molecules in the gas phase and in aqueous media under the action of various types of energy seems to represent the problem most often tackled in theoretical and experimental research in the field of chemical evolution.

This paper deals with certain possible approaches to the abiogenesis of biochemically important compounds. It reviews our results on photochemical formation of biochemically important compounds involving lower aldehydes, and it also considers certain properties of macromolecules relative to the problem in question.

The theory implied that the formation and regular evolution of carbonaceous compounds, of hydrocarbons and their derivatives, under the reducing conditions of the primitive Earth, were the basis of the origin-of-life specificity (2). This was the starting point in creating experimental models for study of chemical evolution. The possibility of abiogenic formation of many compounds inherent to present organisms has been established as a result of extensive research carried out in many countries. An important implementation was the evidence in favour of possible independent synthesis of polypeptides and polynucleotides under primitive Earth conditions. These results have been enlarged by new data obtained recently (3).

Research in cosmic organic chemistry is of great importance for the problem of life's origin (4). The discovery of polyatomic molecules, such as CH_2O (5), NH_3 (6), HCN (7), HNCO (8), and many others, in the interstellar space is a convincing argument in favour of the key part played by these molecules in the abiogenesis of biochemically important compounds.

Photochemical conversion of these molecules both in the atmosphere and in the hydrosphere of the primitive Earth might have had great significance in the stage of prebiological evolution. It is just these processes, characteristically involving the elementary steps of formation of free radicals, atoms, and energy-rich molecular intermediates, that could lead to appearance of diverse compounds in simple starting systems.

It has been established in our laboratory that photochemical conversions of lower aldehydes (formaldehyde and acetaldehyde) in aqueous solutions of ammonium salts yielded various compounds different in composition and structure and belonging to the class of amino acids, peptide-like compounds, imidazoles, indoles, amines, amides, etc. (9,10). A set of amino acids relatively stable in composition: glycine, alanine, serine, threonine, and lysine were detected in CH_2O solutions irradiated with 254 nm UV rays (11). The yield of amino acids increased with the exposure time (up to 4 x 10^{10} erg/cm^2 in 355 hrs) approaching the maximum value. Amino acids with a quantitative predominance of glycine were synthesized at the highest rate (12). Imidazole compounds were synthesized simultaneously, with similar dynamics of accumulation in time and with quantitative predominance of imidazole (13).

We suggest that the synthesis of amino acids and imidazole compounds under experimental conditions follows mechanisms involving common initial stages. These are the formation of free radicals by CH_2O photolysis and their dimerization to form molecular products of a high chemical activity. ESR study of individual steps of the photochemical synthesis of amino acids in formaldehyde and ammonium nitrate (14) has shown that formyl radicals, CHO, were formed by photolysis of aqueous formaldehyde glass (Fig. 1, spectrum 1). The ESR spectrum

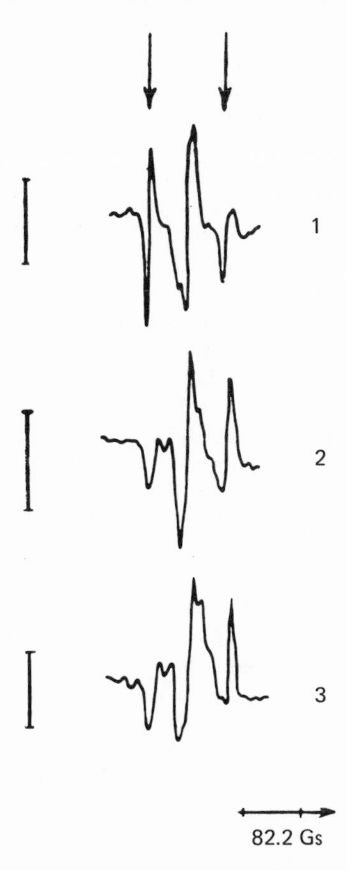

Figure 1. ESR spectra of frozen and UV-irradiated solutions of formaldehyde (1), ammonium nitrate (2), formaldehyde and ammonium nitrate mixture (3). Vertical arrows show the position of formyl radical components. The vertical lines at the left show the intensity of the standard line.

of the formaldehyde and ammonium nitrate glass solution represents a superposition of ESR spectra of individual components in a mixture (Fig. 1, spectrum 3). Additional identification of formyl radicals was made in studying their thermal properties. The pattern of ESR signal decay for frozen UV-irradiated CH_2O solutions at various temperatures (103 to 193°K) coincided with similar data reported in the literature (15). Glyoxal detected in UV-irradiated aqueous solutions of CH_2O could be due to dimerization of formyl radicals: $CHO + CHO \rightarrow (CHO)_2$. The highly reactive dialdehyde, glyoxal, seems to be the molecular product responsible for glycine and imidazole synthesis in the irradiated CH_2O and ammonia salt system. This was confirmed by experiments in which glyoxal and its derivative, methylglyoxal, were left to stand in aqueous solutions of ammonium nitrate or ammonium chloride, or ammonia. Glycine was predominant (80-90%) in glyoxal-containing solutions, and alanine in methylglyoxal solutions. Compounds showing a positive reaction with diazotized sulphanilic acid (imidazole and its derivatives) were also detected in the reaction products (14).

Predominant synthesis of glycine, along with alanine, in UV-irradiated (254 nm) fog containing formaldehyde and ammonium nitrate seems to involve the same active intermediates (16).

The specific mechanism of acetaldehyde photolysis with accumulation of energy mostly in the intermediate complex molecules (17) might be responsible for synthesis of a wider range of compounds in UV-irradiated aqueous solutions of this aldehyde and of ammonium nitrate. Fractions of amino acids and peptide-like compounds (10), imidazoles (9), indoles (10), were isolated from the diverse reaction products by separation on Dowex-50 (H^+ form) and Dowex-2 (OH^- form). The amino acids and peptides were separated in turn by ligand-exchange chromatography using Sephadex G-25-Cu (OH_2), and this permitted obtaining a fraction of compounds resembling peptides in their chromatographic and electrophoretic properties and in their reactions with specific reagents to the peptide bond (using an iodine-benzidinic reagent [18]). These were called peptide-like compounds. The IR spectra of these compounds exhibited bands characteristic of linear polymers at 1400 cm^{-1} and 1580 cm^{-1} that coincided in position and form with the peaks displayed by glycylglycine taken as reference compound (Fig. 2). Amino acid analysis for the summary fraction of amino acids and peptides in such compounds before and after hydrolysis (6 N HCl, 110°C, 24 hrs) has shown that all amino acids identified in the free state: alanine, glycine, lysine, glutamic acid, leucine, serine, valine, aspartic acid, isoleucine, were contained in the peptide-like compounds. Taking the amount of lysine in peptide-like compounds as unity (the content of this amino acid in the chromatographic sample was 0.440 μM), the fraction of glutamic acid will be 0.19, of valine 0.15, of glycine and alanine 0.13 each, of aspartic acid 0.08, of leucine and isoleucine 0.06 each, of serine 0.03. Almost the whole amount of lysine, glutamic acid, valine and aspartic acid (90 and 80%), 40% of serine and glycine, and 16% of alanine in the peptide-like compounds were found in the bound state. More than half of the nitrogen fixed in amino acids and peptide-like compounds under experimental conditions was found in peptide-like compounds, total amount about 3 mg/ml. [Identification was made by the Kjeldahl technique with subsequent reaction with ninhydrin according to Moore and Stein (19)]. Up to four amino acid residues compose the peptide-like compounds.

The formation (along with amino acids and peptide-like compounds) of an indole ring that represents a component of tryptophan and other metabolites is also of interest in studying the primordial synthesis of biochemically important compounds (10).

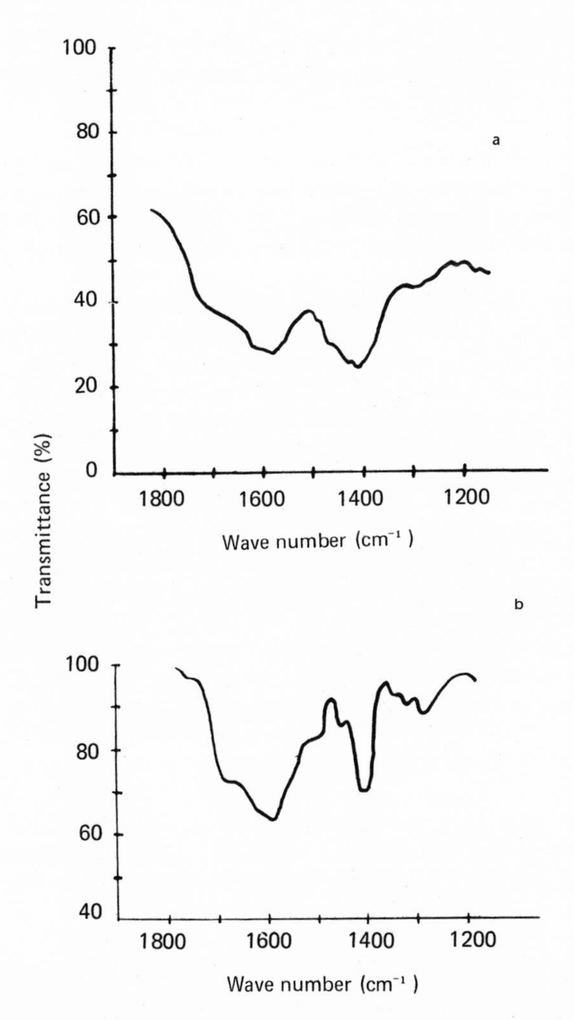

Figure 2. IR spectra of peptide-like compounds synthesized in UV-irradiated aqueous solutions of CH_3CHO and NH_4NO_3 (a) and of glycylglycine (standard) (b).

The evidence for abiogenic photochemical formation of imidazoles (4-hydromethylimidazole, 2-methylimidazole, 4,5-dimethylimidazole, 4-oxomethylimidazole, amino derivatives of imidazole) and of the imidazole, as such, in irradiated aqueous solutions of acetaldehyde and ammonium nitrate (9) and in similar systems containing formaldehyde (13) is particularly helpful in understanding the appearance of catalytically active polypeptides in the presence of histidine, as well as the prebiological activation processes in which these compounds could play the part of organic catalysts (20,21).

A salient feature of the abiogenesis of biochemically important compounds is their radiation and photochemical resistance. The theoretical aspect of this feature established by quantum-mechanics calculation of electron structures of the biochemically important molecules was considered by Pullman (22) relative to primordial biochemical evolution. In its turn, analysis of numerous experimental data on photochemical and radiation-chemical changes of biochemically important compounds would permit establishing the most resistant monomer molecules (as, for instance, is the case for adenine), as well as finding certain model approaches to possible ways of the evolution of macromolecular structures, specifically those of proteins.

We shall deal here only in brief with some observations made by us relative to radiation-chemical alterations of globular proteins under the action of ionizing radiation. It will be emphasized that radiation-chemical study of molecules such as those of proteins are bound to be limited by their exceptional complexity. The number of possible chemical and structural alterations of such molecules by irradiation is very large.

However, comparison of certain results shows a very important specific feature of proteins, namely the extreme variation in their radioresistance at various sites of the molecule. In studying the radiation-chemical alterations of proteins with respect to their chromophoric groups (from their IR and UV spectra), to their functional sulfhydryl groups, to conformational alterations (from absorption in shortwave UV and the Amide-I band), and to aggregation we have obtained results showing that the radiation-chemical yield relative to the different indices may differ by several powers of ten.

Measurement of IR absorption spectra (human serum albumin, equine serum globulin, etc.) within the range of 4000-1800 cm^{-1} upon irradiation up to 1.5-2.0 million roentgen showed only a minor decrease of absorption, without appearance of other characteristic frequencies. This corresponds to usual denaturational changes accounted for by cleavage of weak intramolecular bonds in protein molecules, without essential chemical changes. As to G determination, it will be noted that its value is undoubtedly very low ($G < 0.01$ [23]). Measurement of IR spectra of human serum albumin solutions and of dry preparations in the range of characteristic Amide-I absorption frequencies (1500-1750 cm^{-1}), upon Co^{60} irradiation in doses close to one ionization per molecule, showed that disturbance of the secondary protein structure was only about 5% (24). The insignificant changes of this protein absorption spectra at 190 nm irradiated in solution with X- and gamma-rays in doses up to 200 kr also are in favour of the above (25).

More marked changes are observed in UV absorption spectra of proteins resembling serum albumin (at a ratio of tyrosine to tryptophan >1) in the range of 250-335 nm. Taking into account the Rayleigh correction for UV rays scattering by aggregated protein fractions (26), the increase in absorption (at λ = 280 nm) by a protein solution (0.1%) irradiated at 500 kr was about 25%, which corresponds, relative to this index, to G $\simeq$ 0.01-0.03 (27).

Unlike the low G values observed on changes in chromophoric protein groups, oxidation of the protein-SH group displays high yields. The quantum yield of SH group oxidation $\phi = 2\ 10^{-3}$ and the ionic yield of this reaction upon X-ray irradiation, I = 1, were found in determining the change in SH group content in egg albumin upon X-ray and UV irradiation of the protein in an aqueous solution. This corresponds to G = 3 per 100 ev of spent energy (27).

The aggregation of proteins on irradiation of their solutions is a firmly established fact. This phenomenon was confirmed by the increasing opacity of protein solutions (0.1%) irradiated at 500 kr in air and in vacuum, by sedimentation curves, and by osmotic measurements revealing that the molecular weight of human serum albumin irradiated at 300 kr (2% solution, pH 7.1) increased from 69,000 to 117,000 (28). The aggregated product, dimer, yield in gamma-irradiated bovine serum albumin in solid state (irradiations were carried out in air in a dose corresponding to an average of one hit per mole -8.5 Mrads) was G = 0.8 (29).

Radiation-chemical conversions of globular proteins by the aggregation mechanism are predominant over those occurring by unfolding of the peptide chains.

Thus, even these results provide evidence of the extremely high inhomogeneity of the radioresistance of protein molecules relative to their different properties. The characteristic differences in radiation-chemical yields amount almost to two powers of ten (from G = 3.0 to 0.01-0.03).

This remarkable variety of the radiationchemistry of proteins suggests that also in the prebiotic environment a selection of macromolecules on the basis of their radio-sensitivity has played a role in the control of the evolution of proteins and their evolutionary predecessors.

Experiments on the photosynthesis of monomers and the study of the radiationchemistry of macromolecules help to set guidelines for the understanding of prebiotic evolution of biologically significant compounds.

REFERENCES

1. Oparin, A. I., "Proiskhozhdenie zhizny," Izd. Moskovskii Rabochii, Moscow, 1924.

2. Oparin, A. I., "Vozniknovenie Zhizni na Zemle" (Akad. Nauk SSSR, ed.), Moscow, 1957.

3. Proceedings of the Fourth International Conference on the Origin of Life, Barcelona, June 25-28, 1973. Origins Life 1 (1974), various papers.

4. Sagan, C., Nature 238, 77 (1972).

5. Snyder, L. E., Buhl, D., Zuckerman, B., and Palmer, P., Phys. Rev. Letters 22, 679 (1969).

6. Cheung, A. C., Rank, D. M., Townes, C. H., Thornton, D., and Welch, W. I., Phys. Rev. Letters 21, 1701 (1969).

7. Snyder, L. E., and Buhl, D., Astrophys. J. 163, L47 (1971).

8. Buhl, D., Snyder, L. E., Schwartz, P., and Edrich, J., Nature 243, 513 (1973).

9. Pavlovskaya, T. E., Pasynskii, A. G., Sidorov, V. S., and Ladyzhenskaya, A. I., "Abiogenez i Nachal'nii Stadii Evolyitsii Zhizni," pp. 41-48, Izd. "Nauka," Moscow, 1968.

10. Telegina, T. A., Pavlovskaya, T. E., and Ladyzhenskaya, A. I., Zhur. Evol. Biokhim. Fiziol. 10, 129 (1974).

11. Pavlovskaya, T. E., and Pasynskii, A. G., in "Problemy Evolyutsionnoi i Tekhnicheskoi Biokhimii," pp. 70-79, Izd. "Nauka," Moscow, 1964.

12. Sidorov, V. S., Dokl. Akad. Nauk SSSR 164, 692 (1965).

13. Sidorov, V. S., Pavlovskaya, T. E., and Pasynskii, A. G., Zhur. Evol. Biokhim. Fiziol. 2, 293 (1966).

14. Niskanen, R. A., Pavlovskaya, T. E., Telegina, T. A., Sidorov, V. S., and Sharpatyi, V. A., Izv. Akad. Nauk SSSR, ser. biol. 238 (1971).

15. Smirnova, V. I., Zhuravleva, T. S., Yanova, K. G., and Shigorin, D. N., Zhur. Fiz. Khim 38, 742 (1964).

16. Pavlovskaya, T. E., Telegina, T. A., Sokol'skaya, A. V., and El'piner, I. E., Izv. Akad. Nauk SSSR, ser. biol., No. 6, 922 (1971).

17. De Groot, M. S., Emeis, C. A., Hesselman, I. D., Drent, E., and Farenhorst, E., Chem. Phys. Letters 17, 332 (1972).

18. Reindel, F., and Hoppe, W., Chem. Ber. 87, 1103 (1954).

19. Moore, S., and Stein, W. H., J. Biol. Chem. 211, 907 (1954).

20. Yuasa, S., Ishigami, M., Honda, Y., and Imahori, K., Sci.Repts. (Japan) 19 (No. 2), 7 (1970).

21. Lohrmann, R., and Orgel, L. E., Nature 244, 418 (1973).

22. Pullman, B., in: "Biogenèse" (Colloque sur les systèmes biologiques élémentaires et la biogenèse)(Gavaudan, P., ed.), pp. 162-179, Paris, 1967.

23. Pavlovskaya, T. E., and Pasynskii, A. G., Kolloid. Zhur. 27, 305 (1955).

24. Kharchenko, L. I., and Pavlovskaya, T. E., Radiobiologyia 13, 669 (1973).

25. Pavlovskaya, T. E., Tongur, A. M., Volkova, M. S., Slobodskaya, V. P., and Kharchenko, L. I., "Radiobiologyia," inf. bull. No. 14, 35 (1972).

26. Pavlovskaya, T. E., and Pasynskii, A. G., Biokhimiya 22, 266 (1957).

27. Pavlovskaya, T. E., and Pasynskii, A. G., "Trudy Pervogo Soveshchaniya po Radiatsionnoi Khimii" (Akad. Nauk SSSR, ed.), p. 177, Moscow, 1958.

28. Komarova, L. V., and Pasynskii, A. G., Ukrain. Biokhim. Zhur. 31, 5 (1959).

29. Rosen, C. G., Int. J. Radiat. Biol. 19, 587 (1971).

EVOLUTION OF MODELS
FOR EVOLUTION*

Duane L. Rohlfing

Department of Biology
University of South Carolina
Columbia, South Carolina 29208 USA

INTRODUCTION

Of the many accomplishments of Professor Oparin, two are outstanding. His experimental investigations with coacervate droplets, designed and interpreted in the context of the origin of life, have provided much insight into the origins and evolution of protocells. Of even greater overall impact is his pioneering hypothesis concerning the origin of life that suggested conditions, chemical species, and processes of prebiotic evolution. He has presented us with a theoretical model for molecular and cellular evolution.

Opportunity is taken here to discuss models and their value, critiques of models and the value of critiques, and the evolution of models as mediated by critiques. Emphasis is placed on models for evolution, although the concepts may be applicable to other models.

Richard Levins states (1) that a model is "The basic unit of theoretical investigation . . . [and] is a reconstruction of nature for the purpose of study." A schematic phylogenetic tree, for example, is a model for biological evolution that shows (reconstructs) what is known, or what is believed, about present and past relationships of living things; the reconstruction emphasizes the results of biotic evolution and provides a guide to investigations for missing links. Fig. 1 is one representation of Oparin's model for prebiotic evolution. This depiction reconstructs one concept of the

*Dedicated in memory of Mark William Dean

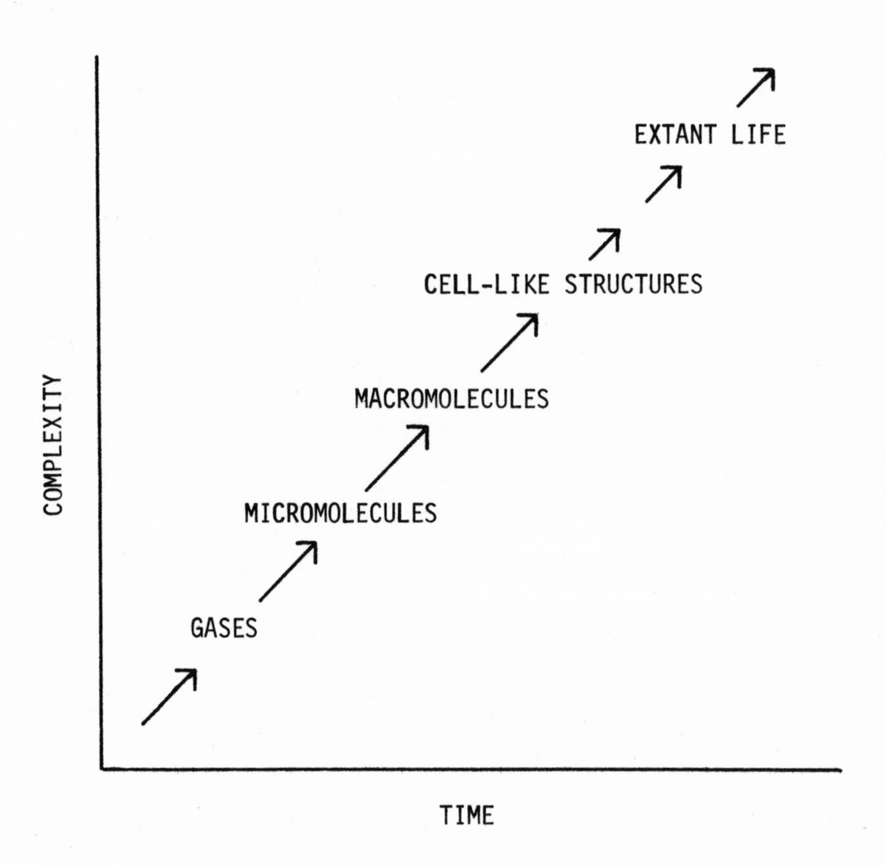

Figure 1. A model for molecular and cellular evolution, emphasizing complexity and time. Energy, not shown, is needed throughout.

relationship between complexity and time and indicates a few levels of complexity that are easily recognizable. (The level, "first life," is accordingly omitted from the figure.) His model has served as a guide to areas to be studied, and the studies have converted it from a largely theoretical (2) to a largely experimental model. The experimental studies not only have added details to the model but have resulted in "submodels" germane to the various transitions of Fig. 1.

The value of a model or submodel depends in part on its purpose. A phylogenetic tree may be valuable to emphasize the results of biotic evolution, but it provides little information about the mechanisms of evolution. This latter purpose would be better served by a different model, one that considered factors such as reproduction, mutations, environmental pressures, and changes in environments. Oparin's model, as depicted in Fig. 1, emphasizes at an easily understandable level gradual changes in complexity, a feature of considerable value when contrasted to the miraculous jumps in complexity (e.g., mud being transformed into frogs) that were associated with concepts of spontaneous generation that preceded his early writings. The depiction, however, does little to indicate the details of the various transitions. Also, it is partially erroneous; a staircase presentation of increases in complexity would be more realistic (cf. 3) and also the time scale is inaccurate. These are, however, very minor if not invalid criticisms of the model if its main purpose is to emphasize evolutionary changes. Precision is sacrificed for generality and ease of understanding (cf. 1). Details or mechanisms are of greater consequence at the submodel level; precision and realism become more important than generality, for the purpose is different, and, incidentally, the overall model becomes more complex and thus meaningful to smaller numbers of people.

The value of a model is interrelated with its validity, and the criteria for validity may vary both with purpose and type of model. However, a model for prebiotic molecular evolution at any level or stage has two criteria against which it must be evaluated (4); these are evolutionary continuity and geological (or, better, environmental) relevance. These criteria are expressed collectively by the term "evolutionary relevance." A model that does not explain the prebiotic source of reactant materials lacks evolutionary continuity; a model based on the use of unrealistic conditions lacks environmental relevance.

Models are evaluated against these (or any) criteria by means of critiques, published or unpublished. ('Critique' as used here is broadly defined as the cognizance of potential flaws in a model.) If based on theoretical grounds only, critiques do little more than alert to possible weaknesses in the model, or merely place doubt on the model. To prove that a model is faulty, or to show that an alternative model is more valid, the theoretical basis of the critique must itself be tested and verified by experiment or by observation.

There are many more critiques than tested critiques.

EXPERIMENTS

The objectives of the experimental section of this paper are 1) to illustrate the experimental testing of possible weaknesses (i.e., critiques) of a model for one phase of molecular evolution and 2) to suggest, by way of the experiments, that some aspects of the model should be changed. The model investigated concerns the formation of prebiotic protein from amino acids. S. W. Fox has shown that amino acids copolymerize thermally to form polyamino acids that resemble contemporary protein in many ways (5). The model has dual aspects; the process of forming polyamino acids is regarded as a model for the formation of protoprotein,* while the product polyamino acids represent a model for protoprotein.* Both aspects of the model are considered in four of the five experiments below, which test alternative conditions that may be evolutionarily more relevant than previously considered and that in theory could cause the model to be modified.

Concerning the Composition of Prebiotic Protein

Contemporary protein is comprised of twenty different coded (proteinous) amino acids, glutamic and aspartic acids (inclusive of their amides) occurring most frequently (6). The development of the model for protoprotein was guided by this information. Sets of proteinous amino acids enriched with dicarboxylic amino acids were used as reactants, and the resulting polyamino acids, termed proteinoids (5), were similarly constituted. [Subsequently, lysine-rich (7) and neutral (ref. 5, p. 148) proteinoids, comprised of proteinous amino acids, have been described.]

Contemporary protein, however, may not be an accurate guide to the composition of prebiotic protein, nor even to that of early biotic protein. For example, some proteinous amino acids may be late-comers to biotic protein (8), and the nonproteinous amino acid, ornithine, may once have been a constituent (9). The composition of prebiotic protein, if synthesized from free amino acids, had to be influenced by the kinds of amino acid that were available prebiotically.

---

*The term "protoprotein," as used in association with thermally synthesized polyamino acids (Rohlfing, D. L., Abst. of Papers, Annual Meeting, American Association for the Advancement of Science, Philadelphia, 26-31 December, 1971), connoted prebiotically synthesized polyamino acids with functional capabilities and evolutionary potential, for which the thermal polymers served as models. In a recent discussion, S. W. Fox mentioned that this usage may be outdated and that the word perhaps should denote only early ribosomally synthesized polyamino acids. The word without precise definition obviously can be ambiguous, and its earlier connotation may now be inapporpriate. In the absence of an unambiguous alternative, however, "protoprotein" is yet used here in the sense employed earlier.

Many amino acids and other compounds have been formed (sometimes after hydrolytic procedures) in numerous experiments conducted under a variety of simulated prebiotic conditions (e.g., ref. 5, chap. 4). Although differences in conditions lead to different rosters of amino acids, some generalities are evident (10). Glycine, rather than dicarboxylic amino acids or lysine, is usually the most prevalent amino acid, and alanine is often next plentiful. Also, several nonproteinous amino acids are formed, often in appreciable quantities. This general pattern is augmented by analyses of hydrolyzates of meteoritic and lunar samples (e.g., 11-14), although nonproteinous amino acids have not been reported in the latter case.

If such sets of amino acids are indicative of prebiotic amino acids, it is environmentally unrealistic to use reactants comprised (solely) of proteinous amino acids enriched with aspartic and glutamic acids or lysine (15; cf. ref. 16 and note 2 therein). If such sets would not polymerize thermally, then the model for the formation of protoprotein would lack evolutionary continuity, and an alternative for the model (or for the sets of amino acids; 15) would be mandatory. If such sets would polymerize to yield polyamino acids of different composition than that described for proteinoid, then the current proteinoid model for protoprotein would lack environmental relevance and should be considered for revision. These possibilities have been tested (16).

Eight sets of amino acids reported by others as resulting from simulated prebiotic syntheses or as constituents of hydrolyzates of extraterrestrial samples were used as reactants under conditions typically used in the synthesis of proteinoid. All tested sets polymerized thermally (16). In each case the predominant constituent of the resulting polyamino acid was glycine, not dicarboxylic amino acids, even in one case when it was not the most prevalent in the reactants. Nonproteinous amino acids, when present in the reactants, were constituents of the polymers (17).

The experiments indicate that the thermal model for the formation of protoprotein is applicable to sets of reactant amino acids that may be more realistic environmentally than those previously tested. Because, however, the conditions inherent in the critique had an effect on the compositions of the resulting polymers, an alternative to the model for protoprotein is provided by the experiments, and a modification of the model is suggested for consideration.

Concerning the Atmospheric Pressure for Polymerizing Amino Acids

The contemporary pressure of 1.0 atm has routinely been used in preparing thermal polyamino acids and in many other simulated prebiotic experiments. The atmospheric pressure of the early Earth, however, may not have been 1.0 atm (18,19), and certainly the pressure of some extraterrestrial bodies is not 1.0 atm. Alternative pressures could have an influence on simulated abiotic experiments, as has been suggested by Haldane (20).

We recently tested (21) whether amino acids would polymerize at low pressures (0.7, 0.5, 0.3, 0.2, and 0.001 atm) and, if so, whether such pressures would influence the resulting polyamino acids. The reactants were sets of amino acids reported as resulting from the hydrolysis of lunar and meteoritic extracts (11,13). [The conditions are interpreted to represent two simultaneous steps toward extraterrestrial relevance: sets of amino acids and low pressure (22).] As controls, the same sets of amino acids were heated under nitrogen at 1.0 atm.

Polyamino acids formed at all pressures tested. Some data are presented in Table I. Accordingly, the model for the formation of protoprotein is concluded to be applicable to previously untested pressures that may be evolutionarily more relevant than 1.0 atm. [Given water for the hydrolytic generation of free amino acids from precursors, polyamino acids could exist extraterrestrially--*cf*. ref. 16.] However, yields, one property (color intensity), and amino acid compositions were influenced by the low pressures (0.02, 0.0001 atm) relevant extraterrestrially. (Whether the intermediate pressures perhaps germane to the prebiotic Earth influenced the results is currently unresolved.) The experiments provide alternatives to the model for protoprotein (or for extraterrestrial protoprotein; ref. 16) and suggest that close estimates of and attention to terrestrial and extraterrestrial pressures may be essential in other simulated abiotic experiments (23).

Concerning the Type of Atmosphere Relevant to the Preparation of Proteinoids

Proteinoids have been prepared typically under a flow of nitrogen gas, which provides a nonoxidizing atmosphere (5). The primitive atmosphere, however, probably would have contained additional gases, more reactive than the relatively inert nitrogen. Although not primary reactants, atmospheric constituents could influence the polymerization reactions, as is suggested by the law of mass action and by reports in the literature (24,25).

Table I

Polyamino Acids Prepared at Different Pressures[a]

Parameter	Polymer and Pressure: From "lunar" amino acids[b] 0.02 atm	From "lunar" amino acids[b] 1.0 atm	From "meteoritic" amino acids[c] 0.02 atm	From "meteoritic" amino acids[c] 1.0 atm
Yield, wt. %	4.5 ± 0.6	2.9 ± 0.4	5.0 ± 1.2	1.5 ± 0.4
Color intensity, $A \cdot g^{-1} \cdot l^{-1}$	3.9 ± 0.5	2.7 ± 0.3	2.2 ± 0.2	1.1 ± 0.1
Biuret color, $A \cdot g^{-1} \cdot l^{-1}$	1.3 ± 0.4	1.4 ± 0.2	1.5 ± 0.1	1.8 ± 0.4.
Partial composition, moles/1000 moles amino acid				
Asp	85 ± 1	121 ± 2	74 ± 25	60 ± 3
Glu	54 ± 4	32 ± 5	66 ± 3	49 ± 2
Gly	704 ± 12	678 ± 7	519 ± 27	560 ± 8
Ala	120 7	120 ± 15	136 ± 8	136 ± 4
β-Ala	-	-	31 ± 3	30 ± 2

[a]Synthesized at 175°C, 2 ("lunar") or 3 ("meteoritic") hrs. Each value is the mean and standard deviation of analyses of quadruplicate polymers. Yields are of water-soluble, nondiffusible material after dialysis. Color intensity, measured at 400 nm, refers to the coloration of the polymers. Biuret color was measured at 300 nm; serum albumin under these conditions gave a value of 1.6.

[b]Reactant amino acids proportioned according to ref. 13.

[c]Reactant amino acids proportioned according to ref. 11.

We recently tested whether amino acids would copolymerize under various atmospheres (21). The set of amino acids obtained by Miller (26) from methane, ammonia, water, and hydrogen was heated under a continual flow of these gases, proportioned according to Miller's initial (27) proportions. Also, the set formed by Harada and Fox (28) from methane, ammonia, and water was heated under a flow of these gases proportioned according to the original report (28). (These conditions are interpreted to represent two simultaneous steps toward evolutionary relevance: sets of amino acids and atmospheres.)

Also tested were pure gases (oxygen, ammonia, hydrogen, carbon dioxide), used separately over the set of amino acids reported by Lowe et al. (29). In each case, nitrogen was used as a control atmosphere.

Each tested set of amino acids polymerized under the various atmospheres (Table II). The applicability of the model for the formation of protoprotein is thus expanded to atmospheres perhaps more realistic than formerly studied. The yields, tested properties, and amino acid compositions of the resulting polyamino acids were not influenced by the nature of the atmosphere, even in cases in which pure gases of extremes of oxidation or reduction capability (oxygen vs. hydrogen) were involved. The model for protoprotein itself therefore cannot be regarded as being subject to modification, for a choice between alternative models (in terms of composition, properties) is not provided by the experiments. The tested alternate conditions may (or may not) be environmentally more realistic, but they are not influential (cf. 30).

Concerning the Copresence of Geological Materials with Amino Acids

In synthesizing proteinoids, mixtures of pure amino acids are almost [25,31,5 (p. 150)] invariably used. In many natural situations, amino acids would be greatly contaminated with other organic compounds and with inorganic, geological matter. Such materials could influence the polymerization process [cf. the role of clays in other model syntheses (32,33)].

We recently tested (21) whether finely ground basalt [a constituent of the lunar surface (34)] would influence the polymerization of a "lunar" set (13) of amino acids. The proportions of basalt:amino acids tested were 0:1, 1:1, 2:1, 4:1, and 8:1 (w:w); the syntheses were conducted at 0.0004 atm. [The experimental design is interpreted to represent three simultaneous steps toward extraterrestrial realism: set of amino acids (22), low pressure, and presence of inorganic material.] Also tested in similar admixture with lava was a set of amino acids (29) synthesized under simulated terrestrial conditions.

The two sets of amino acids polymerized in the presence of all tested proportions of added inorganic material, in yields that ranged from 1.6 to 8.0 wt. %. The applicability of the model for the formation of protoprotein is thus regarded as being further expanded (ref. 5, p. 150). Whether the product polyamino acids were influenced by the proportion of additives is yet unresolved, due to difficulty in obtaining reproducible quantitative results. Accordingly, the model for protoprotein is not yet subject to

Table II

Polyamino Acids Prepared under Different Atmospheres[a]

Parameter	Polymer and Atmosphere								
	From "Miller" amino acids[b]			From "Harada-Fox" amino acids[c]			From "Lowe" amino acids[d]		
	Gases	Air	N_2	Gases	Air	N_2	H_2	O_2	N_2
Yield, wt. %	2.0 1.9	1.2[e] 1.7	1.7 1.7	1.0 1.2	1.0 1.2	1.0 0.8	4.4 4.6	5.1 4.2	5.6 4.5
Color intensity, $A \cdot g^{-1} \cdot l^{-1}$	0.9 0.9	1.0 1.0	0.9 0.8	1.7 1.6	1.8 1.7	1.8 1.7	1.7 1.7	1.6 1.7	1.7 1.6
Biuret color, $A \cdot g^{-1} \cdot l^{-1}$	1.7 1.5	1.9 1.7	1.5 1.6	1.4 1.3	1.6 1.6	1.6 1.6	2.0 1.9	1.9 1.7	1.8 2.3
Partial composition, moles/1000 moles amino acid									
Asp	6 6	6 5	9 5	55 52	56 55	58 57	47 44	47 51	49 50
Glu	8 8	9 8	31 11	44 44	46 44	44 43	-	-	-
Ala	99 108	100 107	102 105	104 105	113 104	110 109	102 79	84 73	79 77
Gly	678 699	685 692	670 673	690 701	679 698	681 683	641 665	657 634	664 667
β-Ala	143 104	128 117	121 137	-	-	-	138 151	158 168	149 154
α-Abu	40 40	43 39	38 40	14 13	14 13	15 14	4 7	6 5	6 5

[a]Synthesized in duplicate at 175°C, 1.5-2.5 hr. Yields are of water-soluble, nondiffusible material after dialysis. Color intensity, measured at 400 nm, refers to the coloration of the polymers. Biuret color was measured at 300 nm; serum albumin under these conditions gave a value of 1.6.

[b]Reactant amino acids and gases (CH_4, H_2, NH_3, H_2O) proportioned according to ref. 26.

[c]Reactant amino acids and gases (CH_4, NH_3, H_2O) proportioned according to ref. 28.

[d]Reactant amino acids proportioned according to ref. 29.

[e]Known loss during processing.

modification, because of insufficient testing of the critique.*

Concerning Coacervate-like Microspheres

The final experimental example pertains to protocell models and concerns not so much the potential refinement of a model but rather the possibility for the merger of different and perhaps parallel models.

The importance of microlocalities was early recognized by Oparin, and his studies with coacervate droplets have provided an experimental basis for extrapolations toward even higher levels of complexity (35). However, polymers of biotic origin have been used to form coacervate droplets, and in a prebiotic context the coacervate model does not explain the source of the constituent material. Evolutionary continuity is lacking. [At the time of Oparin's experiments, means were not widely evident (35) for obtaining constituent material of simulated prebiotic origin.]

An alternate model for protocells is afforded by Fox' microspheres (5), particles that form from proteinoid rich in dicarboxylic amino acids and that differ considerably from coacervate droplets (36). Because proteinoids are of simulated prebiotic origin, the microsphere model has the necessary evolutionary continuity.

It has been shown recently (37) that microparticles can result from proteinoids rich in lysine due to a process different than that used with proteinoid rich in dicarboxylic amino acids. The two types of microparticle differ especially in their range of pH stability. Those made from lysine-rich proteinoid resemble coacervate droplets in several ways (isothermal formation involving charge neutralization, collapse at surfaces, stabilization by quinone, uptake of dyes, concentrating of constituent material).

These results expand the known range of conditions (type of polymer, mode of formation) applicable to the proteinoid model for protocell formation. Perhaps more importantly, if the microparticles from lysine-rich proteinoid prove to be very similar to coacervate droplets, the experiments would suggest a source of prebiotic material for coacervate formation and thereby provide the previously lacking evolutionary continuity (38). The results can also be interpreted as providing the potential for the merging of two models for protocell formation, as illustrated in Fig. 2.

---

*More recent experiments indicate that the presence of sea sand results in a statistically significant increase in yield of polymer.

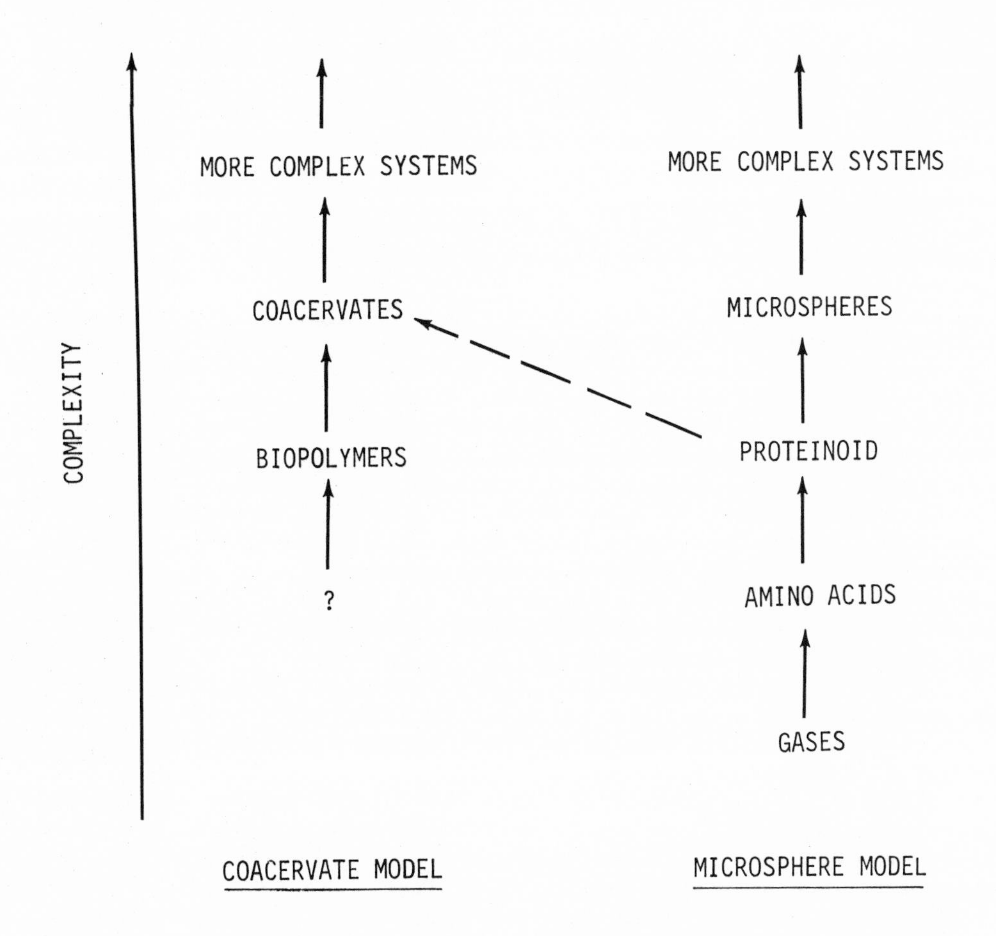

Figure 2. Two models for protocells and their evolutionary continuity. The similarities between coacervates and microparticles made from lysine-rich proteinoid suggest that the lacking evolutionary continuity for coacervation phenomena can be provided and that the models may merge.

Other Experiments--Performed and Proposed

Other theoretical critiques, of course, have been tested. For example, opinions differ concerning the composition of the primitive atmosphere and the relative contribution of various energy sources; several atmospheres and energy sources have been tested (ref. 5, chaps. 3 and 4), and the results of such experiments document alternate conditions and rosters of products as meriting consideration. The resulting alternatives have in a sense provided a broad general model for especially the formation of micromolecules. Another example concerns one model for the formation of protoprotein (39) that has been suggested by experiment to be faulty in that peptide-bond containing polymers are not formed (40). As a third example, the proteinoid microsphere model for protocell formation can be cited as an experimentally tested alternative to the coacervate model.

Many critiques or suggestions have been advanced that, to my knowledge, have not been tested experimentally. One critique concerns the use of closed systems in simulated prebiotic syntheses. Closed systems could have occurred inside the primitive Earth but not on it, and experiments that use both closed systems and energies that could occur only on the surface of the Earth (e.g., simulated lightning, ultraviolet light) lack environmental realism. Reasons have been advanced (41) to indicate that reactions would proceed differently, if at all, in open or steady-state systems, especially where the reactants and intermediate products are gaseous. Doubt is thus cast on closed-system models by the critique, but the issue probably will not be resolved until the critique is tested experimentally. (The Apollo program can be regarded as an observational testing, through analyses of natural experiments; *cf.* 41.)

Another aspect concerns the experimental use of single-energy sources, whereas more than one (e.g., ultraviolet light plus lightning or heat) probably would have been acting simultaneously in a natural situation. Different energy sources acting separately lead to different results, and the effect of combined energy sources remains largely (42) untested. As another example, the study of the formation and properties of protocell models under the present-day atmosphere is environmentally unrealistic (*cf.* Keosian, this volume); the use of various proposed prebiotic atmospheres could lead to different concepts of protocellular phenomena. The proposed photosynthetic protocell (43), using ultraviolet energy, is of interest in this context. The proposals, however, remain to be tested.

A final example pertains to the confines of laboratory glassware. Reactions that take place in 20 ml. test tubes or 5 liter flasks cannot closely mimic the geologically realistic situation of almost limitless boundaries (43,44). Enough experience in scaling-up various chemical processes is available to indicate that results are influenced by the size of the apparatus; by connotation, simulated prebiological syntheses could similarly be influenced. The suggestion (45) of a large environmental chamber, with dimensions of many cubic meters, seems attractive, but remains untested.

DISCUSSION

Evaluation of Critiques

In the previous section, doubt was cast on a model for protoprotein and its formation by four theoretical critiques that suggested alternative environmental conditions (sets of reactant amino acids, various atmospheres, low pressure, presence of inorganic additives). When tested, none of the theoretically more realistic conditions prevented polyamino acids from forming. Any merits in the criticisms as directed to the model for the formation of protoprotein are thus shown by experiment to be inapplicable, or at least noncrucial, and the critiques no longer cast doubt on the model. The model is expanded by the experiments, and it approaches more than before a status of general applicability.

Some of the critiques, however, appear to have merit when applied to the model for protoprotein. The alternative conditions that were the foundations of some of the critiques (e.g., theoretically more realistic sets of reactant amino acids) resulted, when tested, in product polyamino acids with different compositions and properties than previously described.* The theoretical bases of the critiques are shown by experiment to be worthy of further consideration, and any doubts cast on the model for protoprotein by the critiques are augmented by experiment (e.g., 16). The results of the tested critiques offer alternatives to the model.

Evolution of Models

R. A. Daly has stated (46) that "The best model is the one that works best. The perfect model, working infinitely well, is not for men now living."

At any given time, a particular model may be regarded as the best model, but the best existing model is not necessarily a valid model and it certainly is not a perfect one. To have continuing value, any model must evolve; a static model has only historical interest. The availability of tested alternatives is prerequisite to the evolution of models.

---

*The model for protoprotein appears to be uninfluenced by one of the critiques (type of atmosphere), within the range of conditions tested.

The tested alternatives to the current model for protoprotein, however, may not prove to be improvements. Further evaluations may be necessary before one can decide which alternative is the best.* In a broader context, alternatives also exist to the thermal (proteinoid) model for protoprotein and its formation. In my opinion, the proteinoid model is the best of the existing models. It works best, for it provides polyamino acids of high molecular weight that show a variety of protein-like properties envisionable as being important prebiotically (e.g., catalytic activities of several types, selective interactions, morphogenesis); by comparison, other models (ref. 5, chap. 5) are scarcely characterized and in some cases yield at most small peptides, if even peptides (40). However, the proteinoid model in the long run may not prove to be the best of the available models. Before this could be evaluated, data comparable to the array documented for proteinoid would be required of the other models; the evaluation would rely in part on existing concepts of "good" and "best" (i.e., concepts of evolutionary relevance), which also are subject to change. Conceivably, none of the current models is valid. All existing models may in time become obsolete, one important function of a model, but this almost certainly would require the genesis, development, and acceptance of a better model (47).

A model is forced to evolve when sufficient information is available to dictate a judicious choice among alternatives (*cf.* 48).

Projections

The process of refining models, or forcing them to evolve, is itself largely evolutionary, not revolutionary. For example, at most only three improvements (if they prove to be that) in experimental design were simultaneously incorporated into the experiments cited above. Many improvements would be required to closely approximate the real conditions, and such approximation would likely be a slow process, involving modifications that incorporate earlier modifications. The modifications themselves may have to await increased knowledge of evolutionarily relevant conditions. On this planet, prebiological conditions may never be perfectly mimicked, although this is a goal that should be approached.

---

*At this time I personally believe that the model for the composition--and hence properties--of protoprotein should be modified to include higher proportions of glycine and of nonproteinous amino acids.

The evolution of experiments toward more realistic conditions is accompanied by increasing difficulties. More simultaneously employed variables lead to greater complexities in experimental design and analytical procedures, possibly more complex results, and probably more difficulties in interpreting results (49). Interdisciplinary approaches may become more and more mandatory. Greater complexities and difficulties, however, are unavoidable if a closely realistic model for molecular and cellular evolution is to evolve.

ACKNOWLEDGMENTS

Gratitude is expressed to numerous persons, particularly S. W. Fox, J. Keosian, and M. A. Saunders, for their many helpful suggestions. Additional thanks are due the former two for providing me with preprints of their papers for this volume. The research discussed in this paper was supported by grants from the National Aeronautics and Space Administration, currently NGR 41-002-034 (Supplement 1).

NOTES AND REFERENCES

1. Levins, R., "Evolution in Changing Environments," Chapter 1, Princeton University Press, Princeton, 1968.

2. Early studies of coacervation phenomena provided early experimental support for part of the model.

3. Fox, T. O., in: "Molecular Evolution: Prebiological and Biological" (Rohlfing, D. L., and Oparin, A. I., eds.), p. 35, Plenum Press, New York, 1972.

4. Fox, S. W., Am. Biol. Teacher 36, 161 (1974).

5. Fox, S. W., and Dose, K., "Molecular Evolution and the Origin of Life," W. H. Freeman and Co., San Francisco, 1972.

6. Vegotsky, A., and Fox, S. W., in: "Comparative Biochemistry" (Florkin, M., and Mason, H. S., eds.), vol. IV, p. 185, Academic Press, New York, 1962.

7. Fox, S. W., Harada, K., and Rohlfing, D. L., in: "Polyamino Acids, Polypeptides, and Proteins" (Stahmann, M., ed.), p. 47, University of Wisconsin Press, Madison, 1962.

8. Jukes, T. H., "Molecules and Evolution," p. 65ff, Columbia University Press, New York, 1966.

9. Jukes, T., Abstr. of Paper 34, Fourth International Conference on the Origin of Life, Barcelona, 1973.

10. The evolutionary relevance and analytical thoroughness of these and other experiments may be, of course, subject to question. The results, however, have been used as reported, for the purpose of additional experimental study of other parameters; errors may thereby be propagated.

11. Cronin, J. R., and Moore, C. B., Science 172, 1327 (1971).

12. Kvenvolden, K., Lawless, J., Pering, K., Peterson, E., Flores, J., Ponnamperuma, C., Kaplan, I. R., and Moore, C., Nature 228, 923 (1970).

13. Harada, K., Hare, P. E., Windsor, C. R., and Fox, S. W., Science 173, 433 (1971).

14. Fox, S. W., Harada, K., and Hare, P. E., Space Life Sciences 3, 425 (1972); Gehrke, C. W., Zumwalt, R. W., Kuo, K., Rash, J. J., Aue, W. A., Stalling, D. L., Kvenvolden, K. A., and Ponnamperuma, C., ibid., p. 439.

15. Although natural fractionations to yield mixtures of amino acids rich in glutamic and aspartic acids can be envisioned, means for fractionating all proteinous from all nonproteinous amino acids are not apparent; cf. Rohlfing, D. L., and Saunders, M. A., Abstr. of Paper Bc4, Ninth International Congress of Biochemistry, Stockholm, 1973.

16. Saunders, M. A., and Rohlfing, D. L., Science 176, 172 (1972).

17. α-Aminoisobutyric acid was not incorporated in all cases.

18. Sagan, C., in: "The Origins of Prebiological Systems" (Fox, S. W., ed.), p. 200, Academic Press, New York, 1965.

19. Holland, H. D., in: "Petrologic Studies: A Volume to Honor A. F. Buddington" (Engel, A. E., James, H. L., and Leonard, B. F., eds.), p. 466, Geological Society of America, New York, 1962.

20. Haldane, J.B.S., in: "The Origins of Prebiological Systems" (Fox, S. W., ed.), p. 201, Academic Press, New York, 1965.

21. Fouche, C. E., and Rohlfing, D. L., Federation Proc. 32, 640 Abs (1973); Fouche, C. E., Ph.D. dissertation, University of South Carolina, 1973.

22. The amino acids occur in precursor form, water being needed to generate free amino acids. The amino acids are regarded as being indicative of extraterrestrial sets formed (given water) in natural experiments; the question of whether correct conditions were used in simulated prebiotic experiments is thus avoided. The experiments reported here attempt to mimic general extraterrestrial conditions but not specifically lunar or meteoritic conditions. Lunar thermal energy, however, has been documented [Keays, R. R., Ganapathy, R., Laul, J. C., Anders, E., Herzog, G. F., and Jeffery, P. M., Science 167, 490 (1970); Skinner, B. J., ibid., p. 652].

23. Signs of molecular evolution on other planets are currently being sought. A potential costly error is inherent in expectations and analytical devices that are based on simulated experiments conducted at pressures not germane to the planet in question. Cf. Fox, S. W., Bull. Atomic Scientists 29 (10), 46 (1973).

24. Harada, K., and Fox, S. W., J. Am. Chem. Soc. 80, 2694 (1958).

25. Fox, S. W., Nature 201, 336 (1964).

26. Miller, S. L., Biochim. Biophys. Acta 23, 480 (1957).

27. The proportions of gases we used mimicked Miller's proportion only briefly, because the latter continually changed throughout the reaction due to the use of a closed system.

28. Harada, K., and Fox, S. W., in: "The Origins of Prebiological Systems" (Fox, S. W., ed.), p. 187, Academic Press, New York, 1965.

29. Lowe, C. U., Rees, M. W., and Markham, R., Nature 199, 219 (1963).

30. Miller, S. L., and Orgel, L. E., "The Origins of Life on the Earth," p. 145, Prentice-Hall, Englewood Cliffs, N.J., 1974.

31. Snyder, W. D., and Fox, S. W., Federation Proc. 32, 640Abs (1973).

32. Hanafusa, H., and Akabori, S., Bull. Chem. Soc. Japan 32, 626 (1959).

33. Paecht-Horowitz, M., Angew. Chem. 12, 349 (1973).

34. Apollo 15 Preliminary Examination Team, Science 175, 363 (1972).

35. The experiments of Herrera are of interest in this context [e.g., Herrera, A. L., in: "Colloid Chemistry" (Alexander, J., ed.), vol. II, p. 81, The Chemical Catalog Company, New York, 1928].

36. Fox, S. W., J. Evolut. Biochem. Physiol. 6, 131 (1970); cf. Fox, this volume.

37. Rohlfing, D. L., Origins of Life, in press (1974).

38. On appropriate treatment, microspheres from acidic proteinoid are reported to resemble coacervate droplets in some ways [Smith, A. E., and Bellware, F. T., Science 152, 362 (1966); cf. also Young, R. S., in: "The Origins of Prebiological Systems" (Fox, S. W., ed.), p. 347, Academic Press, New York, 1965].

39. Matthews, C. N., and Moser, R. E., Nature 215, 1230 (1967).

40. Ferris, J. P., Donner, D. B., and Lobo, A. P., J. Mol. Biol. 74, 499 (1973).

41. Fox, S. W., Bull. Atomic Scientists 29(10), 46 (1973).

42. Both thermal and electrical energies are involved in spark discharge experiments; electrical energy alone gives different results (ref. 26).

43. Keosian, J., in: "Molecular Evolution: Prebiological and Biological" (Rohlfing, D. L., and Oparin, A. I., eds.), p. 9, Plenum Press, New York, 1972; cf. Keosian, this volume.

44. Some experiments have been conducted on a small scale on geological material, rather than in glassware; e.g., ref. 25.

45. Pattee, H. H., in: "Molecular Evolution" (Buvet, R., and Ponnamperuma, C., eds.), p. 42, North-Holland, Amsterdam, 1971.

46. Daly, R. A., "Architecture of the Earth," p. 1, Appleton-Century Co., New York, 1938.

47. R. Levins [Am. Scientist 54, 421 (1966)] has pointed out that a model is discarded when the current issues are no longer those for which it was designed.

48. Platt, J. R., Science 146, 347 (1964).

49. Interpretations of mechanisms might become very difficult; fortunately, results are more important (*cf*. Keosian, this volume) than details of mechanism in constructionistic approaches to molecular evolution.

PHOTOSYNTHESIS AND SOME PROBLEMS OF EVOLUTION OF BIOLOGICAL SYSTEMS

A. B. Rubin

Moscow State University

Moscow, USSR

According to the Oparin evolutionary theory of the origin and development of life (1), the appearance of photosynthesis on the Earth was the result of preceding biological evolution and was a turning point in the transition from anaerobic to aerobic life. In this connection, it would seem of interest to analyze some features of the photoautotrophic mode of metabolism as a stage in the evolution of bioenergetic systems, which included the structures and complexes of the catalytic mechanisms of evolutionally earlier heterotrophic organisms. On the other hand, a study of present-day forms of autotrophic organisms may cast light on the historical development of the phototrophic mode of metabolism that passed through a number of interconnected stages during biological evolution.

In this respect, it would seem profitable to compare the biochemical nature of metabolism in higher plants and in various photoautotrophic bacteria; a comparison of the oxygen-sensitivity of the latter would also seem to be desirable. In our opinion, an analysis of the properties of photoautotrophic types of metabolism from the viewpoint of the nature of their organization is of no less interest.

It has become increasingly clear during recent years that the primary photosynthetic processes, which are the most specific trait of photosynthesis, are characterized by a definite functional, structure-morphological, and genetic autonomy.

The primary processes, therefore, comprise a sufficiently complex biological system. This, evidently, is due to the important evolutionary position that photosynthesis occupies as an independent stage in the development of bioenergetic systems. The formation of such an integral system during evolution was of course a gradual process and was inseparably linked with the development of the photo-autotrophic type of metabolism.

It is well known that one of the laws of evolution of living matter is that new organisms make use of the catalytic mechanisms and structures inherited from their evolutionary predecessors, but adapt and develop them under new conditions. The main forms of dynamic organization characteristic of biological systems, and which originated at earlier evolutionary stages, evidently continued to improve and develop during the evolution of photoautotrophic organisms. Naturally, the set of photosynthetic processes, at least in their modern form, cannot be considered as a complete biological system with all its attributes, capable of existing independently. Therefore only the more essential characteristics of biological organization inherent in the bioenergetic system of photosynthesis which underlie the functioning of the system will be examined.

Life is connected with the existence of open systems in which uptake and removal of matter and energy continuously take place.

The kinetic peculiarities of living systems is due to the intricate set of reactions underlying metabolism, and which is organized in a definite manner in space and time. Such features of the dynamic properties of open systems as equifinality of the final state, the bottleneck principle or feedback regulation, have long attracted the attention of investigators. Characteristic in this respect are the experimental and theoretical results obtained recently on the appearance in open biological systems of oscillatory processes due to the nonlinear nature of the interaction between the components (2). Energetically such processes can occur only in open systems which are far from thermodynamic equilibrium and this is one of the most important properties of biological systems.

In contrast to simple chemical open systems which do not possess an excess of free energy and whose state is completely determined by the relative rates of the "input" and "output" processes, biological systems possess special intrinsic regulatory mechanisms. A characteristic feature of the latter is that small amounts of matter or energy transferred in the control process itself may evoke a response in the regulated system which involves the utilization of much larger amounts of energy and matter (3). In essence, this is the energy basis for regulation of the state of open living systems by the "feedback" principle. Thermodynamically, it is clear that such "active self-regulation" processes are possible only in the presence of free energy reserves produced in systems far removed from equilibrium. On change of internal state of the system this energy should be immediately available.

These properties are consistent with the principle of maximal thermodynamic potential (free energy) attainable under the given conditions as a result of the high degree of energy coupling and organization of various matabolic processes. This principle is the result of evolution of biological systems (4).

It would of course be wrong to ascribe the causes and direction of biological evolution exclusively to thermodynamics but nevertheless energy factors undoubtedly played an enormous role in the formation and development of highly organized biological systems capable of self-regulation. Indeed, on a world scale life requires a constant inflow of free energy from an outside source, which at present is solar energy harnessed during photosynthesis. On the other hand, an analysis of the factors which determined the rate and path of formation of living systems reveals the important energy role which solar energy played at all stages of preceding evolution.

Thus, at the dawn of evolution, long before the existence of photosynthesis in its present form, sunlight and especially ultraviolet rays supplied energy for abiogenic formation of those compounds which subsequently served as substrates for further evolution on the Earth.

The role of light, apparently, did not only consist in this but also in the acceleration of further transformations of complex compounds produced abiogenetically. Indeed, in abiogenetic synthesis light served as a free energy source, the energy being required for the very formation of complex molecules from simpler one; however, for further chemical evolution of the complex compounds, which could proceed spontaneously under favorable temperature conditions of the primeval atmosphere of the Earth, no additional inflow of energy was required in principle.

Under these conditions the absorption of light by complex molecules could result in the appearance of photochemical reactions in which high energy barriers could be surmounted at the expense of light energy at ordinary temperatures. This, in turn, was the cause of many transformations being practically irreversible during evolution of the complex compounds. In this respect light excitation of the acceptor molecules could also have played a catalytic role in regulating the direction and rate of the chain of complex reactions; this could be due to preferential acceleration of certain of the photochemical steps at a period preceding the formation of specific enzymes. Fluorescing complex molecules abiogenetically synthesized by light could probably have been a factor which hastened further evolution at the expense of subsequent photochemical transformations.

The life activity of some of the earliest living systems could also be decisively affected by the photochemical reactions of compounds which they contained. It is not a coincidence that many biologically important substances contained in modern organisms and which do not have any direct relation to photosynthesis and are not illuminated under ordinary conditions, nevertheless change their activity on illumination and undergo serious alterations which affect the general metabolism (5).

It is possible, moreover, that the appearance of pigmented organisms on the Earth initially involved the utilization of the photochemical reactions in porphyrins for acceleration of some metabolic processes, but not for direct photosynthesis of organic substances. Later on, these properties of the photoreactions of porphyrins became the physicochemical basis of photosynthesis.

Initially, therefore, the inflow of light free energy led to an increase of the total thermodynamic potential of the initial simple compounds in relatively simple photochemical processes, resulting in their transformation into more complicated molecules. All further evolution proceeded at the high energy level thus attained, ensuring in this way the ultimate appearance of heterotrophic living systems.

Under conditions in which penetration of ultraviolet rays to the Earth's surface was impeded, the free energy sources in the form of ready-made organic substances required for the functioning of open living systems produced during the development of heterotrophic life, gradually became exhausted. At this stage the disappearance of life from the Earth as a result of the energy deficiency produced by the ecological conditions on the planet could be avoided only by invoking photosynthesis. The latter could fulfil the role of "rescuing" developed forms of heterotrophic life only by creating a complex biological system of light energy storage processes. The biological organization of living matter attained at this time as a result of evolution was used and then developed further.

Contemporary photosynthesis theory, which certainly cannot be reduced to the occurrence of a single photochemical reaction, however specific it may be, is based on an analysis of intricate space-time relations between interacting photosynthetic components.

Just as the essence of life cannot be elucidated without a study of its origin and evolution, the investigation of the nature of photosynthesis must be based on an analysis of its major evolutionarily conditioned traits of which the nature of organization of photosynthetic processes is one of the most important.

The major evolutionary attainment underlying photosynthesis as a qualitatively new stage in the development of bioenergetic systems on the Earth, is the organization of the electron transport chain. As is well known, the photochemical reactions of the pigments induce electron transport along a chain of intermediate compounds and also a coupled ion (proton) transport. However it is not the products produced as a result of direct reduction in these photoreactions which are employed, as is the case for simple photochemical processes involving dye molecules, but rather the final products of a complex chain of electron and hydrogen transfer, viz. ATP and NADPH which are necessary and sufficient for dark reduction of carbon dioxide.

The main significance of the electron flow is not that these compounds are formed, since they were synthesized by preceding heterotrophic anaerobes, but that it turned out to be the most efficient means of storing electron excitation energy in chemical bonds. Efficiency here is not meant to signify energy efficiency, which for overall photosynthesis is not very high, but the ability to regulate the direction and general state of processes under various conditions. This, in turn, is determined by the general regulatory properties at the level of structure-functional organization and kinetic peculiar ities of the integral electron transfer system. As an example, we cite some results obtained in our laboratory on investigation of the regulatory mechanisms in the system of primary photosynthetic processes.

The primary processes of photosynthesis consist essentially of a noncyclic electron flow through a number of intermediate carriers from water, in the case of higher plants, or from external donor, in the case of photosynthesizing bacteria, and also of a cyclic flow along the closed chain: pigment → cofactors → cytochrome → pigment.

An investigation of the kinetics of these flows has shown that cyclic transport cannot be regarded as an independent cycle since the intermediate components involved in it exchange electrons with the noncyclic system during a period of time which greatly exceeds the mean turnover time of a cycle (5).

In essence the topographical difference between cyclic and noncyclic electron transport systems is that the former involves feedback from photoreduced products to components oxidized by the pigment.

Thus, cyclic flow plays the role of a feedback loop in the unified open system of electron transfer. As in any complex system, the regulatory significance of the latter is great and in the given case is connected with possibility of excessive light energy being stored under conditions of saturation of noncyclic flow. Regulation according to the feedback principle is also conditioned by the nonlinear nature of interaction between the photosynthetic carriers, which means that the electron transport rate not only depends on concentration of the reduced molecules but also on concentration of the oxidized acceptors.

An important feature of kinetic regulation of electron flow is that the latter involves various processes whose characteristic times differ by several orders of magnitude. In bacterial photosynthesis, for example, cytochrome oxidation induced by a laser pulse is over after 1-3 msec, whereas transition processes may occur over a period of several seconds, and these also involve molecules of the cytochrome complex (6). Generally speaking, the simultaneous occurrence of processes with different rates is an important criterion of stability of complex systems and particularly of biological systems. In photosynthesis, this property is ensured by the existence of electron pools in the chain, distributed in a specific manner. These pools also exert a stabilizing effect (6).

From the foregoing, it can be concluded that the electron transfer chain in photosynthesis is based on general principles of organization of open biological systems. However, a comparison of bacterial and plant photosynthesis shows that during evolution of photoautotrophic nutrition these principles gradually changed.

The principle of alternativeness of electron flow is particularly pronounced in the chloroplasts of higher plants. This means that attainment of final results of the light stage can be ensured by various optimal paths depending on the conditions of operation of the photosynthetic apparatus. An example is the participation of various components of the electron transfer chain depending on the type and physiological state of the organism. The absence of the Emerson effect in many cases also signifies that coupling between the two photochemical reactions in photosynthesis of higher plants is not necessarily rigid, and may vary depending on the rates of electron exchange between intermediate carriers or between the carriers and other cellular metabolites.

Generally speaking, the electron transport system in higher plants contains a more varied set of electron carriers than that observed in bacteria. Consequently the relations between the carriers is more diverse in plants and, for example, additional electron transfer cycles may arise (7). This is an important property and is a result of further development of the regulatory pattern inherent in the cyclic system of bacterial photosynthesis.

The peculiarity and high degree of organization of electron flow is closely linked to another important function of the flow, which is its ability to induce the synthesis of macroergic bonds.

Perhaps the most widely accepted is the chemiosmotic hypothesis proposed by Mitchell. It is assumed that phosphorylation is caused by a transmembrane electrochemical potential produced by the electron flow. We shall not stop to discuss the controversial question regarding the cause and effect relations between the processes of electron transfer, pH gradient formation and phosphorylation; it will suffice here to note that in an evolutionary sense this mechanism is a great step forward compared to the substrate phosphorylation mechanism encountered in earlier heterotrophic anaerobes. This leap did not consist in the appearance of a new physical mechanism for direct storage of electron-vibrational energy in macroergic bonds; this mechanism is apparently the same in substrate and photosynthetic phosphorylation and nothing about its nature is contained in the Mitchell theory.

The point here is that the interaction between electron transport, ion transfer, and ATP synthesis opens new opportunities for the energy factors which regulate the general state of the photosynthetic system.

It is precisely in the photosynthetic system that such regulatory mechanisms as photosynthetic control by ADP/ATP ratio or energy-dependent NADP reduction have found a wide application. Moreover, coupling between the electron flow and transmembrane potential production opens new possibilities for a space-time regulation under conditions when the potential produced at one part of the membrane may affect electron transfer and ATP formation at other parts without direct electron exchange occurring between them.

The electron pools may be important for generation of transmembrane potentials. They charge the membrane surface and play a regulatory role in the intermediate kinetic link between the rapid electron transport reactions and the slower processes of ATP synthesis and ATP utilization in cellular metabolism (6).

A study of the regulatory factors determining the behavior of the photosynthetic system led us to the conclusion that electron flow does not only induce a redistribution of electrons in the photosynthetic chain, but also alters the conformational state of the carriers as a result of electron-vibrational interactions (6). As a consequence, the carrier coupling constants are altered and this in itself affects electron transfer in the chain.

The system of primary photosynthetic processes is therefore a bioenergetic system with a unique ability of closely coupling energy and kinetic regulatory factors. This was the basis for the appearance and development during evolution of the aerobic respiration system in which the regulatory properties mentioned above are particularly pronounced.

The appearance of photosynthesis was also connected with a number of peculiar physiochemical mechanisms being incorporated into the initial system; these new mechanisms were closely related to the general structural features of the photosynthetic apparatus. For example, the electron-vibrational interactions, which are manifest under these conditions in the photosynthetic structures, make possible the oxidation-reduction transformations of chloroplasts which can occur at temperatures as low as that of liquid nitrogen(8).

An important difference between higher plants and photosynthesizing bacteria is, as is well known, that plants carry out a specific reaction, photolysis of water, which bacteria do not carry out.

In this connection it might be mentioned that our measurements of the temperature-dependence of the mean fluorescence times showed that in cells of higher plants, but not of bacteria, there are two different physical processes of photosynthetic deactivation of chlorophyll-excited states. One of the processes, also operative in bacteria, is temperature-independent (down to -196°), whereas the efficiency of the second is manifest only at temperatures lower than approximately -80°. The first process occurs in the bacterial photosystem and in the long-wave system in plants; the second is specific to plants. This difference between photosynthesizing bacterial and plant cells can evidently be ascribed to the evolutionary position of the plant and bacterial types of photosynthesis.

It is conceivable that the appearance of green plants on the Earth and the utilization of water as hydrogen donor resulted not only in alteration of the biochemical aspect of the photosynthesis reactions but also in a change in the functional structure of the pigment apparatus. It is quite possible that the specific reaction of water photolysis, for which no analogues previously existed, required the creation of a novel physicochemical mechanism, a result being the appearance of a new type of reaction center responsible for the photolysis of water.

The evolutionary development of the photoautotrophic mode of metabolism therefore included the mobilization of new physicochemical mechanisms whose action, however, was determined by the general structure-functional peculiarities of the photosynthetic systems.

As a whole, the material presented in this paper confirms the conclusion that the appearance of photosynthesizing organisms on the Earth was a new stage in the evolution of bioenergetic systems and their regulatory mechanisms, and signified a radical change in biological organization.

REFERENCES

1. Oparin, A.I., "Life, its Nature, Origin and Development "(Russ.) Moscow, Izd. AN SSSR, 1960.

2. "Oscillatory Processes in Biological and Chemical Systems" (Russ.) Izd. Nauka, Moscow, 1967.

3. Lyapunov, A.A., Problems cybernetics (Russ.) 10, 179 (1963).

4. Dechev, G., and Maskona A., Dok AN SSSR 162, 202, 1965.

5. Rubin, L.B., Yeremeyeva, O .V., and Akhobadze, V.V., Uspekh. sov. bio. 71, 220 (1971).

6. Rubin, A.B., in: "200-year Anniversary of the Discovery of Photosynthesis" (Russ.), Moscow State University, 1973.

7. Korshunova, V.A., Krendelyeva, T.E., and Rubin A.B., Biokhimiya 32, 980 (1967).

8. Chance B., and Nishimura M., Proc. Nat. Acad. Sci. US 46, 19 (1960).

THE OPARIN THEORY AND SOME PROBLEMS OF EVOLUTION OF BIOENERGETIC SYSTEMS

B. A. Rubin
Faculty of Biology
Moscow State University
Moscow, USSR

One of the most important achievements of modern experimental biology is the establishment of similarity of the principles of ultrastructural and functional organization of all forms of living things on the Earth.

This signifies that perfection of the catalytic mechanisms and functionally active structures during biological evolution did not primarily occur as a result of replacement of the preceding mechanisms or structures but rather as a result of their modification and increasing complexity. The appearance of additional mechanisms was necessary to permit the organisms to adapt their metabolism to the changing conditions of life on our planet. This is what Linus Pauling implied (1) when he said that living matter, as opposed to other forms of matter, retains in its organization information on its own history.

This proposition is completely justified by cellular mechanisms which are the basis of the major bioenergetic functions such as photosynthesis, respiration, and fermentation. Molecular biological data show that, despite the disparity of these functions with respect to their biological role, they are in fact genetically and evolutionarily closely interconnected.

We would like to emphasize at the very beginning that the usual meaning of the word photosynthesis is too restricted and narrow. The initial mode of utilization of solar energy was the photochemical formation of organic compounds from inorganic ones, and this occurred before the appearance of primeval forms of life. The abiogenetically produced compounds were the initial substrate, both material and energetic, used in subsequent formation of the primordial organisms.

This is the basic idea of Oparin's theory formulated over fifty years ago. It has obtained experimental confirmation in various laboratories.

According to the Oparin theory, heterotrophs were the earliest forms of life on our planet and their energy metabolism proceeded under anaerobic conditions. This idea is also universally accepted.

Contemporary anaerobe-heterotrophs possess reasonably well developed membranes and other biologically active structures, as well as catalytically active systems of energy metabolism including dehydrogenases, isomerases, transferases, ATP synthesis systems, etc.

Photoautotrophy was the next stage in development of bioenergetics; "The time arrived when all, or almost all dissolved organic substances, which hitherto had served as food for the primordial organisms, were exhausted. Only those organisms could continue to exist which were able to adapt themselves to the new conditions of life. There are two possibilities; if the old mode of nutrition were to be retained the organic substances required for life could be acquired only by consuming the weaker organisms; an alternative would be to move along a new path and create an apparatus which would make possible the consumption of inorganic compounds." This was written by Oparin in 1924 (2).

This latter line of development was inevitable, due to a number of causes and primarily due to the changes in the ecological conditions on the Earth (decrease of intensity of hard ultraviolet rays, screening effect of ozone derived from the oxygen produced as a result of photodissociation of water vapor in the atmosphere, etc.).

Photoautotrophy was inseparably connected with the formation of catalytic systems of a principally new type. These were the pigments which possessed the wonderful property of utilizing light energy for transition to an electronically excited state. The energy of the free electron which thus appeared and its transformation into chemical energy involved a new system of catalytic mechanisms which comprised the electron transport chain (ETC). As a highly complex mode of energy metabolism, photoautotrophy underwent a number of interconnected changes in the course of evolution.

It is widely recognized that the first in this chain of events were the phototrophic bacteria (3), which carried out a form of bacterial photosynthesis sometimes called photoreduction. A confirmation of this is the fact that the majority of the phototrophic bacteria are anaerobes, i.e. organisms which evolutionarily are closest to the most ancient (with respect to energy metabolism) group of organisms. A property common to all contemporary photosynthesizing bacteria is that the latter are able to utilize only highly reduced compounds with a low oxidative potential as electron donors. This also seems to show convincingly that these organisms are the closest to the anaerobic heterotrophs. The photosynthetic apparatus of autotrophic bacteria is quite complex. It includes a photosynthetic unit, an electron transport system with cytochromes, flavin enzymes, quinones that are also found in later forms of autotrophs, the higher green plants included. The electron flow is coupled to phosphorylation and, along with ATP, one of the products of photosynthesis is NADPH.

The cyclic and noncyclic electron flows are quite distinct in bacterial photosynthesis and in this respect they resemble higher plants and algae. In bacteria, however, the two systems interact with a single photoactive center, whereas in green plants there are at least two photochemical centers. Photosynthesizing bacteria do not possess specialized structures like chloroplasts and the simpler chromatophores fulfil their function.

Although the overwhelming majority of bacteria are anaerobes, there are some which are of a transition form. Some anaerobic purple bacteria, for example, can thrive in the presence of oxygen and even grow at the expense of respiratory energy. Such forms evidently appeared as a result of repeated adaptation occurring at a later stage of this period of evolution.

The phototrophic bacteria differ with respect to other properties as well.

The appearance of the phototrophic mode of nutrition was a powerful stimulus for further evolution of the biochemical and biological properties of organisms.

A certain role in this process could have been played by the photosynthetic bacteria but, as a whole, this role was quite restricted. This is due to the paucity of compounds assimilated by the organisms and also to the fact that the latter mainly utilize long waves of the solar spectrum. Further progress in the evolution of autotrophic energy metabolism required the removal of "bottlenecks" which restricted the activity of photosynthetic bacteria, and in the capture of free substrates and new spectral regions. In the course of this evolution such organisms arose as possessed mechanisms permitting them to mobilize electrons from the most oxidized form of hydrogen, which is the water molecule.

The resulting organisms were green plants. Their main advantages over the photosynthetic bacteria consisted in the following: a) the presence of chloroplasts with an ultrastructure which impedes the back reaction between the active oxidant (oxygen) and strongly reduced carbon compounds; b) capability of absorbing high intensity visible rays of the spectrum; c) the presence of two (or perhaps even three) photochemical centers; and d) ability of directly reducing NADP.

With these properties, plants were not only able to inhabit almost all parts of the Earth where water and carbon dioxide were available, but also to produce a broad spectrum of new forms.

An enormous influence on the evolution of bioenergetic and biological processes was exerted by oxygen evolved during photosynthesis and possessing a maximal oxidative potential. With the appearance of oxygen, it was inevitable that mechanisms were created and selected which were capable of deactivating oxygen under mild biological conditions by reducing it. But this signified the beginning of a qualitatively new stage in bioenergetic evolution, the transition from the anaerobic to aerobic type of metabolism.

The appearance of oxygen was a critical event for the anaerobic organisms which existed on the Earth at that time. Their first problem was how to defend themselves from the lethal effect of this powerful oxidizer, to which the anaerobes were not adapted. In connection with this, the time of appearance of organisms capable of liberating oxygen was separated by a long period from the time of appearance of organisms capable of efficiently using oxygen as an electron acceptor.

It was during this time that various transitive forms of organisms originated. Initially the oxygen evolved could not be utilized, but gradually it began to be involved in energy metabolism.

The catalytic systems which these forms possess are a reflection of the various modes of utilization of electron donors and of the movement towards the use of oxygen. As an example, we might cite the obligate autotrophic blue-green algae which evolve oxygen in photosynthesis but cannot use it, as they exist anaerobically.

Another type of organism changing over to aerobic metabolism are the chemosynthetic bacteria which also comprise an inhomogeneous group. The latter includes both autotrophs and (to a less extent) heterotrophs; similarly, along with the aerobes, which are predominant, aerobic species can be encountered.

Some chemosynthetic organisms (such as the iron bacteria) possess a practically complete set of components of the ETC (NAD and FAD dehydrogenase, cytochromes, ferredoxin, ubiquinones, vitamin K, etc.). Nevertheless, the energy efficiency of oxidative metabolism, as a whole, is not high in these organisms. In particular, coupling between respiration and phosphorylation is weak and there is no respiratory control (except for the iron bacteria; 4). One of the causes of this is the absence of specialized morphological structure such as mitochondria. In chemosynthetic organisms the function of these organelles are fulfilled by a single-layer, chtoplasmic membrane devoid of cristae.

The evolution of the ability of plants to utilize oxygen terminated with the appearance of aerobic respiration. The latter was a result of improvement of existent enzyme systems and of the formation of some new ones involved in the transfer of electrons to activated oxygen. Efficient coupling of the functions of this complicated set of enzymes was conditioned by the formation of mitochondria. The ultrastructure of the latter was such that oxidation not only extracted energy from the substrates but also produced physiologically active intermediate metabolites.

It should be emphasized that the ETC mechanisms existing in photosynthetic organisms were incorporated into the catalytic systems of respiration during their formation. The respiratory electron transport chain may be regarded as the photosynthetic ETC uncoupled from the pigments. Some differences in the nature of the components of the two chains are not significant and are not of a principal character. The energy mechanisms of anaerobes are also widely used in the oxygen respiratory systems. Glycolysis, for example, is the initial stage of the so-called dichotomous respiration path. Some principal features of the ultrastructure of the organoids and of the molecular organization of membranes, etc. are also used.

Other evidences of a close evolutionary relation between respiration and photosynthesis is the chemical similarity of a number of paths of carbohydrate biosynthesis in photosynthesis and carbohydrate degradation in respiration (the reductive and oxidative pentose phosphate cycles), the identity of the intermediate products, and many other facts.

It should be emphasized that the formation of the ETC of photosynthesis and respiration is based on the same principle. Namely, the length of the chain should be long enough for the drop of energy in each step to be sufficiently small, since in this way a higher efficiency of utilization of the electron energy can be attained. As Szent-Gyoergyi has so colorfully remarked (5), the living cell is able to avoid large energy losses as it pays in small change.

Another important piece of evidence in favor of the genetic relation between respiration and photosynthesis is the existence of alternative mechanisms and paths of each of the two processes.

Thus, for example, molecular oxygen can become involved in metabolism by three different types of mechanisms: 1) reduction by uni- or bi-valent donors; 2) bivalent reduction and subsequent incorporation of the reduction product into the oxidized molecule (hydroxyl formation); 3) direct incorporation of oxygen into the oxidized molecule (oxygenation).

All of these possibilities were postulated in their time by A. N. Bakh and V. I. Palladin. Important work on their experimental verification was carried out by Oparin (6). Each of the types of interaction with oxygen is catalyzed by a large number of enzymes differing with respect to chemical structure, nature of cofactors, kinetic properties, electron source and behavior toward temperature, partial pressure of oxygen, etc.

A factor which amplifies this alternative nature of the respiratory catalytic mechanisms of plants is that many components of the oxidative apparatus are polyfunctional systems.

A striking example of this is peroxidase, which Bakh considered an obligatory component of the biological oxidation system. It is well known that for many dozens of years this idea did not attract attention or was virtually rejected. It has been found that, besides oxidation at the expense of the activated oxygen of peroxides, peroxidase can function as a typical oxidase mediating the oxidation of a large number of compounds of various chemical nature by free molecular oxygen.

The principally important role of the oxidase functions of peroxidase is due to the fact that in some cases they may be coupled to oxidation of biologically important components (reduced forms of NAD and NADP). On this basis, peroxidase may be regarded as a full-fledged alternative component of the oxidation apparatus of plants. One may also term as alternative those enzymes which participate in the terminal stages of oxidation, and those enzyme systems which are involved in photorespiration, etc.

The adaptive significance of the existence of alternative respiratory paths is enormous.

Experimental proof has been obtained recently that the alternativeness principle is at the basis of all bioenergetic systems, including those of photosynthesis.

At present, for example, it is widely recognized that two photosystems (I and II) operate in photosynthesis. On the other hand, some reliable data point to the existence of a system III which carries out so-called "pseudocyclic" electron transport. The same can also be said about the number of photoactive centers in green plants and the nature of interaction between the systems (Z-scheme, parallel or independent schemes, etc.). The quantum yields of photosynthesis, relation between the cyclic and noncyclic photophosphorylation systems, electron transport paths, etc. are also alternative.

Pronounced differences have been detected in the ultrastructure of chloroplasts from various groups of plants and even in different tissues and cells of a single plant. The chemical paths of transformation of carbon assimilated in photosynthesis are also different (C_3 and C_4 paths).

In conclusion, it should be emphasized once again that the wide application of alternative paths was a powerful factor in increasing the efficiency of bioenergetic processes and played a progressive role in evolution as a whole.

The contents of our short survey can serve as a convincing proof of the correctness of the thesis formulated by John Bernal (7) in his book "The Origin of Life." He wrote: "All phenomena studied in biology comprise a continuous chain of events and each subsequent stage cannot be explained if the preceding ones are not taken into account. The unity of life is a consequence of its history and therefore is a reflection of its origin."

It is exactly this idea which lies at the basis of the theory of Academician Aleksander Ivanovich Oparin whose eightieth anniversary is being observed by scientists from over the world with a feeling of deep gratitude to its creator.

REFERENCES

1. Pauling, L., and Zuckerkandl, E., Acta Chem. Scand. 17, 9 (1963).

2. Oparin, A. I., "Proischogdenije gizni," Moskovskij rabotshij, Moskva, 1924; in Bernal, J. D., "The Origin of Life," World Publishing Co., Cleveland, 1967.

3. Kondratjeva, E. K., "Fotosinthesirujuschtshie bacterii," Moskva, 1963; Gaffron, H., "Opredelja juschtchije etapi fotochimitcheskoi evolucii," p. 49, Horizonti Biochimijj, Moskva, 1964; Calvin, M., "Wosmognije puti evolucii fotosintesa i konversii kvantov," p. 24, Horizonti Biochimijj, Moskva, 1964.

4. Oishi, K., Kim, R., Aida, K., and Uemura, T., J. Gen. Appl. Microbiol. 16, 301 (1970).

5. Szent-Gyoergyi, A., "Über den Mechanismus der biologischen Verbrennungen," Nobel Vortrag, 1937.

6. Bakh, A. N., J. Russ. Fis.-Chim. Obschestva 12, 44 (1912); Palladin, W. I., Z. Akad. Nauk 20, 5 (1907); 2, 977 (1908); Oparin, A. I., Biochem. Z. 124, 90 (1921); James, W. O., "Plant Respiration," Oxford, 1953.

7. Bernal, J. D., "Wosniknovenije gizni," p. 15, Moskva, 1969.

AN EVOLUTIONARY MODEL FOR PREBIOTIC PHOSPHORYLATION

Alan W. Schwartz

Department of Exobiology, University of Nijmegen

The Netherlands

INTRODUCTION

Not long ago, in writing on the problem of prebiotic phosphorylation, I concluded: "Clearly a great deal more work is necessary before any definitive model can be proposed to account for the sources of phosphorus on the primitive Earth" (1). It is still premature to speak of a "definitive model" and more work is certainly necessary. It is particularly appropriate, however, that in honoring the 50th anniversary of the publication of "Proiskhozdenie zhizny" and the 80th birthday of Academician Oparin, a tentative outline of such a proposal can now be made. This model is made possible by recent work in several laboratories and unites a number of earlier studies into what may be described as an evolutionary continuum linking phosphorus to the other biogenic elements.

THE REQUIREMENTS OF A MODEL

Any model which attempts to provide a pathway for phosphorylation must take into account the following facts (1-4): the only source of phosphorus now known to exist to any extent on the surface of the Earth (apart from biologically produced oddities) is apatite, $Ca_{10}(PO_4)_6(F,Cl,OH)_2$. Apatite is not only the exclusive source of phosphorus on the Earth at present, where it occurs as an accessory mineral in all important classes of igneous rocks, but very probably also was at the time of the crystallization of the first crustal rocks [a suggestion that whitlockite, $Ca_3(PO_4)_2$, may be a more significant phosphate mineral than was previously thought does not materially alter the problem (5)]. Apatite (primarily fluorapatite) also occurs as well-separated, crystalline masses, formed by differentiation from cooling, basic magma. All of the known forms of apatite are highly insoluble in neutral and basic solution and slightly soluble in solutions of weak acids.

A model, to be considered geologically relevant, should either explain (preferably demonstrate) the origin of a secondary source of phosphorus which is more amenable to reaction, or somehow deal with the problem of the refractory nature of apatite. This might be done, for example, by making a plausible argument for the presence of some acid which would dissolve apatite (4). Another approach to this problem has been the suggestion that elemental phosphorus and phosphine could have been introduced into the atmosphere during catastrophic early degassing of the Earth (1). Although the possibility of such a process is in agreement with the reduced state of phosphorus in the iron meteorites as well as in lunar basalt (5), this does not necessarily constitute evidence for the occurrence of this process on the Earth. (It is widely accepted that iron meteorites are of secondary origin while the carbonaceous chondrites, which are much closer to "primitive" material, contain oxidized phosphorus). Recently the mineral struvite, $Mg(NH_4)(PO_4)6H_2O$, has been shown to precipitate from ammonia-enriched, artificial seawater under certain conditions and must be regarded as a possible secondary phosphate source on the primitive Earth (6). The model which is the subject of the present chapter, however, is independent of the possibility of precipitation of any other mineral. Indeed, both processes may have operated jointly on the primitive Earth, where, as I hope to show, phosphate could have been incorporated into evolutionary pathways which already involved H, C, N, and O.

THE TIME-SCALE FOR PHOSPHATE INCORPORATION

It is not yet clear when the first liquid water appeared on the Earth. Some suggestions have recently been made that some of the oldest known rocks, which have ages of 3.7×10^9 y, may be derived from sediments (7). However, if liquid water existed prior to this date, a large scale metamorphic event at about 3.7×10^9 y BP, for which there seems to be much evidence, would very likely have caused a reevaporation of any preexisting water (8). The best first approximation is probably to use 3.7×10^9 y BP as the earliest date for the presence of liquid water. Now, if we accept the morphological and chemical evidence found in the various sediments of the Swaziland and Buluwayan Systems as indicating the presence of organisms, then at most 5×10^8 years separate the appearance of liquid water from the appearance of life in the geological record (9,10). Conceivably, the time available was much less. Admittedly this is pretty shaky speculation, but it does point out the possibility, at least, that the origin of life required a period of time which was much shorter than has usually been assumed. However, if the time scale for prebiotic molecular evolution on the surface of the Earth seems to shrink with each new microfossil discovery, it is expanding with respect to that part of the process which could have occurred at an even earlier stage.

The elements of primary biological importance--H, C, N, and O--were certainly at a fairly advanced stage of evolution prior to the formation of the oceans and probably prior to the accretion of the Earth itself, as is evidenced by the organic compounds present in carbonaceous chondrites as well as in interstellar dust clouds (11,12). It is noteworthy that a number of the observed interstellar molecules are also important intermediates in prebiotic simulation experiments (particularly those containing the CN moiety). Although survival of preformed organic material during the accretion of the Earth has been questioned (13), it is well to remember that CN itself is stable enough to exist in the atmospheres of stars (12).

There is no evidence that phosphorus enters into chemical evolution at a preplanetary stage. In carbonaceous chondrites, it is present primarily as whitlockite (1). Chemical evidence for the association of phosphorus with the organic material in these objects is absent, with the possible exception of the Orgueil meteorite (14). No information is yet available on the association of phosphorus with interstellar dust.

In the absence of other indications, therefore, it would appear that the association between phosphorus and the other biogenic elements probably took place at a later stage in the evolutionary process. It is now clear that as soon as liquid water appeared on the Earth, an enormous potential for phosphate uptake could have been provided by the presence of CN.

PHOSPHATE AND THE CN MOIETY

It has been demonstrated that oxalic acid could have played an important role in prebiotic phosphorylation reactions (4). Complexing and precipitation of calcium by oxalate has a strong solubilizing effect on apatite, especially at reduced pH. The effect of evaporating a solution of ammonium oxalate is, in fact, to lower the pH due to a partial loss of ammonia. Continued evaporation leads to a concentration of phosphate in solution (15). These conditions, therefore, favor phosphorylation in aqueous solution in the presence of any of a number of condensing agents which have been utilized for phosphorylation with soluble phosphates (16-20).

Evaporation to dryness results in the deposition of ammonium phosphates, which have previously been suggested as prebiotic phosphorylating agents (21). The heating of ammonium phosphate salts leads to loss of ammonia, the synthesis of condensed phosphate and, in the presence of nucleosides, to the synthesis of nucleotides (22,23). Inorganic polyphosphates have been suggested frequently as primitive phosphorylation agents and sources of free energy (3,24-27). Their activity has been demonstrated in the phosphorylation of nucleosides under anhydrous as well as aqueous circumstances and throughout the pH range from strongly acid to strongly basic conditions (28,29). Under acidic conditions, temperatures in the range 0-20°C suffice (28).

In view of the foregoing considerations, therefore, it should be no surprise that substantial yields of nucleotides have been obtained from the evaporation and heating of dilute solutions of nucleosides and ammonium oxalate in the presence of apatite, with and without the addition of organic condensing agents (in the absence of oxalate no phosphorylation occurs; 30,31). The condensing agent, cyanamide, has a marked effect on the yields of nucleotide obtained. Urea, which also stimulates phosphorylation in the apatite-oxalate system, has previously been shown to promote the thermal conversion of ammonium phosphates to condensed phosphate (6,32) and the thermal phosphorylation of nucleosides (33). Temperatures as low as 65°C have been employed successfully for this reaction (34). [Phosphorylation with apatite-urea mixtures has also been accomplished by establishing acid conditions with concentrated ammonium chloride (33).]

Cyanamide and its hydrolysis product, urea, are produced (together with guanidine) by irradiation of cyanide solutions (35). More significantly, it has been known for some time that cyanogen solutions produce, upon standing, a number of products including oxalate, urea, and cyanate (36,37). Recently it has been demonstrated that both cyanogen and cyanide solutions produce, upon oligomerization and hydrolysis, oxalate and urea as major products (38). These observations posed the question whether cyanogen or cyanide solutions could alone serve as reaction systems for phosphorylation. Subsequently, the oligomerization and evaporation of 0.06 M cyanogen solutions in contact with apatite was shown to produce extensive phosphorylation of nucleosides (31,39). The identity of urea and oxalate as major products under these conditions was confirmed. Oxamate and oxamide were also identified in smaller quantity.

All of the above leads to the conclusion that a direct reaction pathway was available on the primitive Earth, linking the solubilization and activation of mineral phosphate to one of the simplest molecular combinations of the biogenic elements, that is, to CN (see Fig. 1). It would seem that the mere formation of aqueous solutions of compounds generally thought to be key intermediates in

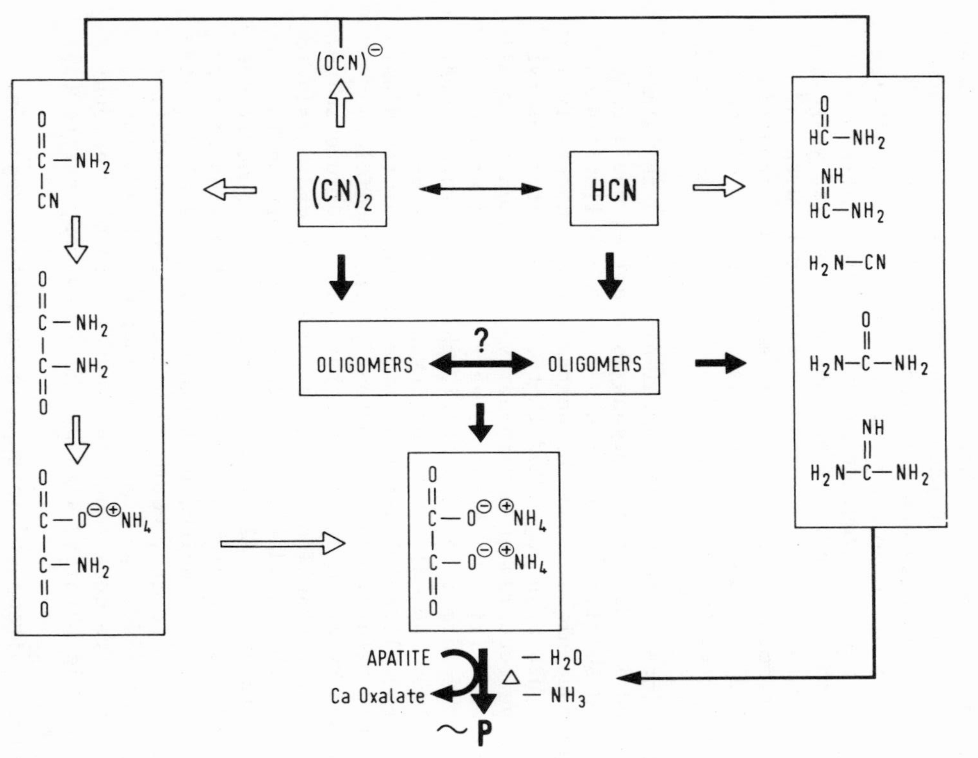

Figure 1. Some aspects of the reaction system relating the CN moiety to phosphorylation. The symbol ∿P represents phosphate, activated either by reaction with one of the organic compounds formed in solution or by thermal condensation. Solid arrows indicate processes which occur primarily in concentrated solution, while hollow arrows indicate dilute aqueous reactions. The double arrow between $(CN)_2$ and HCN indicates the summation of gas-phase as well as aqueous interconversions.

prebiotic molecular evolution produces a powerful reaction system for the activation of phosphate.

DILUTE AQUEOUS CHEMISTRY AND EVAPORATION

Cyanogen and HCN undergo oligomerization when allowed to stand in aqueous solution, at room temperature, in concentrations greater than about 0.01 M (38,40). Under these conditions, a brown-black precipitate appears which has been referred to as "azulmic acid" (41). It does not appear to be known if the material precipitated differs depending on whether cyanogen or HCN solutions have been employed. However, the similarities in the low molecular weight, soluble products formed from both compounds (primarily urea and oxalate) suggest a common mechanism. Under more dilute conditions, where the rate of hydrolysis is greater than the rate of oligomerization, HCN is converted to formamide and ammonium formate.

It has been pointed out that the total nitrogen in the atmosphere, if dissolved in the present oceans as cyanide, would only give rise to a 0.2 M solution (40). Since this upper limit is only one order of magnitude higher than the concentrations necessary for oligomerization, it has been suggested that oligomerization of cyanide could only have occurred at a very early period when the available liquid water was limited. Alternatively, a concentration mechanism such as the formation of eutectic phases in frozen ponds may have been involved (40,42). However, cyanamide has been reported to form in the ultraviolet-irradiation of HCN solutions as dilute as 7.5×10^{-5} M (42). More importantly, the conditions necessary for phosphorylation should be attainable in dilute solutions of cyanogen, without oligomerization, simply by hydrolysis to ammonium oxalate. Although additional work is necessary on the kinetics of the hydrolysis of dilute solutions of cyanogen, it seems plausible that conditions could have been established even in very dilute aqueous environments which, upon evaporation and mild heating in the presence of apatite, would have led to phosphorylation. It is noteworthy that model experiments conducted with a natural, igneous fluorapatite have been found to produce much more extensive phosphorylation than is obtained with synthetic hydroxyapatite (39).

The term "microenvironment" is enjoying a certain vogue in describing such models as the "evaporating pond" for processes on the primitive Earth. That this is an inaccurate description of a massive process is evidenced by the extensive evaporite deposits which have long been known, as well as by more recent discoveries. In Permian time alone, areas of evaporites that have been recognized extend to hundreds of thousands of square kilometers and thousands of meters in thickness (44). Recent discoveries of giant evaporite deposits under the Mediterranean involve millions of square kilometers (45). This preserved evidence is primarily marine in nature and probably represents the formation and evaporation of transient inland seas (46). Not visible in the geological record, but probably of even greater importance in consideration of prebiotic evolution, is the constant process of formation, evaporation, and reformation of inland pools and even films. At the present time, of the order of 10^{17} liters of rainfall evaporate from land masses yearly (47), an amount of evaporation sufficient to equal the total volume of the oceans in 10^4 years. This is a process which can be extrapolated back beyond the formation of substantial oceans, to the first contact of liquid water with igneous rock. In the presence of an atmosphere containing reactive and water-soluble molecular species such as cyanogen, the process illustrated in Fig. 1 may already have begun.

REFERENCES

1. Schwartz, A. W., "Molecular Evolution: Prebiological and Biological" (Rohlfing, D. L., and Oparin, A. I., eds.), p. 129, Plenum Press, New York, 1972.

2. Gulick, A., Am. Scientist 43, 479 (1955).

3. Miller, S. L., and Parris, M., Nature 204, 1248 (1964).

4. Schwartz, A. W., "Chemical Evolution and the Origin of Life" (Buvet, R., and Ponnamperuma, C., eds.), p. 207, North-Holland, Amsterdam, 1971.

5. Nash, W. P., and Hausel, W. D., Earth Planet. Sci. Letters 20, 13 (1973).

6. Handschuh, G. J., and Orgel, L. E., Science 179, 483 (1973).

7. Moorbath, S., O'Nions, R. K., and Pankhurst, R. J., Nature 245, 138 (1973).

8. Cloud, P., Amer. J. Sci. 272, 537 (1972).

9. Kvenvolden, K. A., "24th IGC-Section 1," p. 31.

10. Schopf, J. W., "Exobiology" (Ponnamperuma, C., ed.), p. 16, North-Holland, Amsterdam, 1972.

11. Buhl, D., Nature 234, 332 (1971).

12. Oró, J., Space Life Sci. 3, 507 (1972).

13. Sagan, C., Nature 238, 77 (1972)

14. Rossignol-Strick, M., and Barghoorn, E. S., Space Life Sci. 3, 89 (1971).

15. Schwartz, A. W., and Deuss, H., "Theory and Experiment in Exobiology, Vol. 1" (Schwartz, A. W., ed.), p. 73, Wolters-Noordhoff, Groningen, 1971.

16. Steinman, G., Lemmon, R. M., and Calvin, M., Proc. Nat. Acad. Sci. U.S. 52, 27 (1964).

17. Lohrmann, R., and Orgel, L. E., Science 161, 64 (1968).

18. Ferris, J. P., Science 161, 53 (1968).

19. Halmann, M., Sanchez, R. A., and Orgel, L. E., J. Org. Chem. 34, 3702 (1969).

20. Stillwell, W., Steinman, G., and McCarl, R. L., Biorganic Chem. 2, 1 (1972).

21. Beck, A., Lohrmann, R., and Orgel, L. E., Science 157, 952 (1967).

22. Ponnamperuma, C., and Mack, R., Science 148, 1221 (1965).

23. Rabinowitz, J., Chang, S., and Ponnamperuma, C., Nature 218, 442 (1968).

24. Jones, M. E., and Lipmann, F., Proc. Nat. Acad. Sci. U.S. 46, 1194 (1960).

25. Schwartz, A., and Fox, S. W., Biochim. Biophys. Acta 87, 696 (1964).

26. Kulaev, I. S., "Chemical Evolution and the Origin of Life" (Buvet, R., and Ponnamperuma, C., ed.), p. 458, North-Holland, Amsterdam, 1971.

27. Baltscheffsky, H., "Chemical Evolution and the Origin of Life" (Buvet, R., and Ponnamperuma, C., eds.), p. 466, North-Holland, Amsterdam, 1971.

28. Waehneldt, T. V., and Fox, S. W., Biochim. Biophys. Acta 134, 1 (1967).

29. Schwartz, A., and Ponnamperuma, C., Nature 218, 443 (1968).

30. Schwartz, A. W., Biochim. Biophys. Acta 281, 477 (1972).

31. Schwartz, A. W., van der Veen, M., Bisseling, T., and Chittenden, G.J.F., BioSystems 5, 119 (1973).

32. Osterberg, R., and Orgel, L. E., J. Mol. Evol. 1, 241 (1972).

33. Lohrmann, R., and Orgel, L. E., Science 171, 490 (1971).

34. Bishop, M. J., Lohrmann, R., and Orgel, L. E., Nature 237, 162 (1972).

35. Lohrmann, R., J. Mol. Evol. 1, 263 (1972).

36. Sidgwick, N. V., "The Organic Chemistry of Nitrogen," p. 300, Oxford, London, 1949.

37. Varner, J. E., and Burrell, R. C., Euclides 15, 1 (1955); Chem. Abstr. 49, 11735h.

38. Ferris, J. P., Donner, D. B., and Lobo, A. P., J. Mol. Biol. 74, 499, 511 (1973).

39. Schwartz, A. W., van der Veen, M., Bisseling, T., and Chittenden, G.J.F., Origins of Life, in press.

40. Sanchez, R. A., Ferris, J. P., and Orgel, L. E., J. Mol. Biol. 30, 223 (1967).

41. Sidgwick, N. V., "The Organic Chemistry of Nitrogen," p. 306, Oxford, London, 1949.

42. Sanchez, R., Ferris, J., and Orgel, L. E., Science 153, 72 (1966).

43. Schimpl, A., Lemmon, R. M., and Calvin, M., Science 147, 149 (1965).

44. Hatch, F. H., and Rastall, R. H., "Petrology of the Sedimentary Rocks," p. 310, Thomas Murby & Co., London, 1965.

45. Hsü, K. J , Ryan, W.B.F., and Cita, M. B., Nature 242, 240 (1973).

46. Hsü, K. J., Earth-Sci. Rev. 8, 371 (1972).

47. Mason, B., "Principles of Geochemistry," p. 198, John Wiley & Sons, New York, 1966.

THE ROLE OF SHOCK WAVES IN THE CHEMICAL EVOLUTION OF EARTH'S PRIMITIVE ATMOSPHERE

A. Shaviv and A. Bar-Nun

Department of Physical Chemistry
The Hebrew University
Jerusalem, Israel

INTRODUCTION

Oparin and Haldane's extension into the prebiological era of Darwin's concepts on evolution have given us all a new insight into the complex phenomena of life on Earth, and opened new horizons for a search for life elsewhere (1,2). Recently, we have witnessed the merging of the disciplines of both theory on the origin of life and mankind's technical capability of evaluating it, by searching for life on other planets. Thus, the contribution of Oparin and Haldane to the field of prebiological evolution cannot be overestimated. The cornerstone of the Oparin-Haldane theory is the reducing character of Earth's primitive atmosphere, in which the primary chemical processes took place. To date, despite the enormous progress in the field of synthesis of complex compounds in a reducing atmosphere (25),relatively little is known about the dynamics of the evolution of Earth's atmosphere. The often used assumption of a state of thermodynamic equilibrium throughout the atmosphere's evolutionary track (30), is not fully justified for a system with a continuous influx of material and energy. Instead, a dynamic approach should be adopted, taking into account the rate of exhalation from the interior of the planet and the effects of the various sources of energy.

The major energy sources which shaped the chemical character of the primitive atmosphere were solar radiation and electrical discharges (6). However, if one accepts the major role played by lightning, one should also consider the thunder which follows as one of the major sources. In the familiar process of lightning, the sudden release of electrical energy raises the temperature and pressure in a narrow region of atmospheric gas along the path of the stroke. The hot, high-pressure gas expands outward from the core in a very short time, forms at its front a supersonic blast wave (shock wave), i.e., a sharp wave-front, across which pressure, temperature, and density rise discontinuously. Consequently, each slab of gas traversed by the shock wave is subjected to a very rapid compression and heating to temperatures ranging from several hundred to several thousand degrees K, depending upon its distance from the core, followed by rapid cooling at a rate of 0.5 to 5×10^7 ^{o}K/sec. A detailed discussion of thunder shock waves is given by Bar-Nun and Tauber (29). Two additional phenomena giving rise to shock waves are meteorites entering the atmosphere at a supersonic speed, sending shock waves in front of them (8), and cavitations in the water due to the action of ocean surface waves (21).

An estimate of the energy put into the primitive atmosphere through thunder shock waves is based upon present day lightning frequency and power. The rate of accumulation of atmospheric electricity, during the build-up of a thunderstorm, depends mainly upon the flux of solar energy reaching the surface (12). According to Cameron and Ezer (18), the Sun was already in the main sequence 4.5 billion years ago; its luminosity was somewhat lower than today. The atmospheric constituents absorbed only a small fraction of light in the UV region and did not absorb visible light. Hence, there is reason to believe that the energy flux due to lightning discharges on the surface of the primitive Earth was somewhat lower than today's rate of 1 cal/cm^2/yr. This is probably an understimate since 3 cal/cm^2/yr of atmospheric electricity is discharged today through corona discharges (6,11), which occur mainly at the top of trees. On the bare earth, during prebiological times, this mode of discharge was much less frequent and most of the atmospheric electricity was discharged through lightning. This could counterbalance the effect of lower luminosity of the Sun and a flux figure of about 1 cal/cm^2/yr will be adopted here. Of the energy released by a lightning stroke, about 40% is available to chemical reactions in the thunder shock wave (29). Therefore, the estimated energy flux through thunder in the lower 5 kilometers of the primitive atmosphere was around 0.4 cal/cm^2/yr. The energy flux due to meteorites is estimated to have been around 0.1 cal/cm^2/yr (26). One has, however, to bear in mind that, while thunder operates in the lower troposphere where about 60% of the atmosphere's mass is concentrated, most of the meteorites affect the upper atmosphere where the density is very low. An exception is a very small number of large meteorites which penetrate the troposphere with their accompanying shock waves (8). Although no estimate is available for the energy in cavitations, it seems to be fairly high, since surface waves are always present in great abundance.

The initial composition of the primitive atmosphere is still under debate with no definite conclusions. According to Rubey (3), the excess volatile material in the present atmosphere, hydrosphere and buried sedimentary rocks contains 1.6 x 10^{24} g of water, 2.5 x 10^{22} g of carbon, 3.0 x 10^{22} g of chlorine, 4.2 x 10^{21} g of nitrogen, 2.2 x 10^{21} g of sulphur and some other minor elements. These excess volatiles were gradually accumulated at the surface by outgassing of the Earth's interior. According to Holland (7), the exhaled materials were reduced on their way to the surface by prolonged contact with hot magma containing large amounts of metallic iron and hydrogen. He maintains that for a period of about 5 x 10^8 years, until most of the iron was segregated into the molten core, the atmosphere consisted mainly of CH_4 and smaller amounts of H_2, H_2O, N_2, H_2S, NH_3 and Ar. As for the nitrogen, it cannot be established with certainty in which form it was exhaled. Ammonia is the more reduced form but molecular nitrogen is by far the more stable. If the exhaled gases were in contact with hot magma at temperatures above 300°C, even in the presence of a large amount of hydrogen, it seems that molecular nitrogen would dominate, since it is favored by the equilibrium constant (9). It will be assumed here that most of the exhaled nitrogen was in molecular form but some emerged as ammonia, which was immediately dissolved in the oceans.

The effect of shock waves upon this atmosphere was studied both experimentally and theoretically. Obviously, the effects of solar radiation and lightning should be considered as well in describing atmospheric evolution. Electrical discharges affect the atmosphere in a manner similar to shock waves (10), but the volume of gas affected by the narrow stroke is quite small. The effect of UV radiation is in some cases different. The combined effects of shock waves and UV radiation will be evaluated towards the end.

EXPERIMENTAL

Of the reactions considered to be the major ones, six were studied in our laboratory by the single-pulse shock tube technique and the seventh was studied by Rao et al.(20). The single pulse shock tube technique was described in detail elsewhere (19). Because of space limitation, the experimental conditions and results are merely summarized in Table I. The data are not treated in a rigorous kinetic fashion, since only the overall results are pertinent to the evolution of the primitive atmosphere. Some detailed accounts of these studies will be published later.

Table I

Shock Wave Experiments

Reaction	Range of compositions of the reaction mixtures	Temperature range covered. °K	Temperature at which conversion and energy efficiency were calculated. °K	Conversion: %(products)/(initial reactants)	Energy-efficiency: (molecules converted)/(erg input)
1) $CH_4 \rightarrow C_2H_6$ [(a)]	1.0-2.0% CH_4 in Ar	1500-4500	2500 3600	52 93	2.4×10^{11} 1.5×10^{11}
2) $C/H + H_2O \rightarrow RCHO$ [(b)]	5.0% C_2H_6 + 5.0% H_2O in Ar	700-1000	1000	∿5	$\sim 5 \times 10^{9}$
3) $C/H + H_2O \rightarrow CO$ [(c)]	2.0-4.5% C/H + 5.0% H_2O in Ar	1500-4600	2500 3600	22 90	1.2×10^{11} 1.4×10^{11}
	4.5% CH_4 + 5.0% H_2O + 16.0% H_2 in Ar	1500-4600	2500 3600	19 80	1.0×10^{11} 1.3×10^{11}
	2.3% CH_4 + 10.0% H_2O in Ar	1500-4600	2500 3600	90 90	1.4×10^{11} 1.4×10^{11}
4) $C/H + H_2O \rightarrow CO_2$ [(d)]	2.3% CH_4 + 10.0% H_2O in Ar	1500-4600	2500 3600	4 9	4.0×10^{9} 1.3×10^{9}
5) $CO + H_2 \rightarrow C/H$ [(e)]	1.0% CO + 9.0% H_2 in Ar	1500-3600	2500 3600	3 9	2.1×10^{11} 4.1×10^{11}
	0.1% CO + 20.0% H_2 in Ar	1500-3600	2500 3600	40 90	1.7×10^{11} 4.1×10^{11}
6) $C/H + N_2 \rightarrow HCN$ [(f)]	5.0% C/H + 5.0% N_2 in Ar	1500-6000	4000 6000	13 21	2.0×10^{10} 1.9×10^{10}
7) C/H/N/O → amino acids [(g)]	4.0% CH_4 + 10.0% C_2H_6 + 20.0% NH_3 in Ar + H_2O	1300-2600	1900-2600	1	$\sim 1.5 \times 10^{9}$
	10.0% C/H + 10.0% NH_3 in Ar + H_2O	1900-2600	1900-2600	0.3	$\sim 5 \times 10^{8}$

LEGEND OF TABLE I.

a) The soot formation and lack of mass balance in the high-temperature runs show that some of the acetylene formed is further pyrolyzed to carbon and polymers.

b) Formaldehyde and acetaldehyde were identified by gas chromatography and by precipitate formation with 2,4-dinitrophenylhydrazine. The quantitative gas chromatographic analysis was inaccurate because the aldehydes condensed on the walls and the yields quoted are approximate.

c) Methane, ethane, or ethylene gave similar yields of carbon monoxide.

d) Carbon dioxide was detected only in this series, with a high water/hydrocarbon ratio.

e) The product consisted mainly of acetylene with smaller amounts of ethylene and methane.

f) This system was studied in a single-pulse shock tube by Rao _et al_. (20). HCN is formed at temperatures above 2500°K. Similar HCN yields are obtained from methane, ethane, or acetylene.

g) The production of amino acids occurs in a nonhomogeneous gas mixture, with a gradient of water vapor and consequently of temperature along the shock tube. Aldehydes are formed in the low temperature region and HCN in the high temperature region. They recombine with ammonia to form α-amino nitriles, of which a fraction is hydrolyzed during the quench period (26). In the second series, amino acids were produced from hydrocarbons such as methane, ethane, ethylene, acetylene, or isobutane.

DISCUSSION

As mentioned earlier, the initial atmospheric composition suggested by Holland will be adopted here. A steady rate of exhalation during 5 x 10^8 years will be assumed. This would imply exhalation rates of 3.2 x 10^{15}, 6.6 x 10^{13}, 8.4 x 10^{12}, and 4.4 x 10^{12} g/yr of water, methane, nitrogen with ammonia, and hydrogen sulphide, respectively. A crucial parameter is the troposphere's humidity, which is determined by its temperature. It will be shown later to govern atmospheric evolution to a great extent. The average present-day surface temperature is 15°C with a linear tropospheric temperature gradient of -0.0065° K/m (17). The average temperature of the atmosphere over an altitude of 5 km is about -1.3°C, corresponding to an average water vapor pressure of 4.1 torr. However, a strong greenhouse effect (23,27) can be inferred in the primitive atmosphere, as some of the atmospheric components were good absorbers in the infrared, implying a higher atmospheric temperature. A value of 24°C for the surface temperature, with an average of 8°C, was adopted, leading to an average water vapor pressure of 8 torr. With a rate of 3.2 x 10^{15} g/yr for water exhalation, water started condensing after about 1.5 x 10^4 years. From then on, an average water vapor pressure of 8 torr prevailed in the lower 5 km of the atmosphere. Consequently, most of the ammonia, hydrogen chloride, and hydrogen sulphide coming out through exhalation were dissolved in the water, leaving only very small concentration of these gases in the atmosphere.

A radius of 11.8 cm was chosen as the effective range of the thunder shock wave (29). The present-day lightning and thunder frequency of 100 per second over the whole Earth's surface (11) was adopted here although, as mentioned earlier, this value might have been somewhat different. Consequently, each year an area of 1.4 x $10^{12}cm^2$ was covered by shock waves. This is 2.8 x 10^{-7} of the whole Earth's surface. Since about 60% of the atmosphere's mass is concentrated in the lower 5 km, a fraction of 1.66 x 10^{-7} of the atmospheric mass was subjected to thunder shock waves each year. The atmosphere's composition was assumed to be homogeneous at all times since diffusion and convection can take care of nonhomogeneities produced at such low frequency. The energy in each lightning stroke is approximately 2.5 x 10^9 cal (17), of which 40% are available for chemical reactions in the thunder shock wave (29). Hence, each column of atmospheric gas, 23.6 cm in diameter and 5 km high, is subjected to an energy of 1 x 10^9 cal. This would imply, for example, 106 kcal/mole for methane at a pressure of 1 bar, whereas the thermodynamic requirement for the conversion of methane to acetylene is only 46 kcal/mole. Thus, there is sufficient energy for the conversion of the gases in each thunder column into products. The shock tube experiments, on the other hand, give the rates of these reactions, namely, to what extent the reactions will proceed during the period of high temperature in the thunder.

The major chemical reactions which occurred in the primitive atmosphere as a result of thunder shock waves are those listed in Table I. The rate constant for each reaction was calculated from the experimental conversion reported in Table I. The rate of methane accumulation in the atmosphere was:

$$(a)\quad d[CH_4]/dt = k_o - k_1\,[CH_4] - k_2\,[H_2O] - k_3\,[N_2].$$

Where $k_o = 4.1 \times 10^{12}$ mole/yr, is the rate of methane exhalation, and $k_1 = 1.66 \times 10^{-7}$ is the fraction of methane pyrolyzed to acetylene each year; $k_2\,[H_2O] = 5.0 \times 10^{11}$ mole/yr is the rate of methane oxidation to carbon monoxide, by the constant amount of water in the troposphere, where $k_2 = 1.66 \times 10^{-7} yr^{-1}$; $k_3[N_2]$ is the rate of methane conversion to HCN. This rate is small compared to the first two terms in equation (a), because of the low nitrogen mole fraction in the exhaled gas and the small experimental conversion to HCN. Integration of equation (a), neglecting the last term, gives the amount of methane in the atmosphere at each time:

$$(b)\quad [CH_4]_t = (k_o - k_2\,[H_2O])\,k_1^{-1}\,[1-\exp(-k_1 t)]$$

When the acetylene is further subjected to thunder shock waves, graphite is formed via a chain reaction, with polymers such as C_4H_2, C_6H_2, C_8H_2, etc., as intermediates (14). The rate of acetylene accumulation in the atmosphere was then:

$$(c)\quad d[C_2H_2]/dt = k'_o - k_4\,[C_2H_2] - k_2[H_2O] - k_3[N_2]$$

After 2×10^7 years the methane reached a steady state and k'_o, the rate of acetylene production, became equal to the rate of methane exhalation. $k_4 = 1.66 \times 10^{-8}$ is the fraction of acetylene converted each year to graphite and polymers, with a yield of $\sim$10%. This is a reasonable estimate since thunder temperatures, high enough for this process to take place, occur only in a small region close to the core. The amount of acetylene in the atmosphere at each time is given by:

$$(d)\quad [C_2H_2]_t = 1/2(k_o - k_2[H_2O])\,k_4^{-1}\,[1-\exp(-k_4 t)]$$

and the amount of graphite and polymers is obtained from:

$$(e)\quad [\text{graphite + polymers}]_t = k_o t - [C_2H_2]_t$$

Similarly, for the nitrogen concentration:

$$(f)\quad [N_2]_t = k_5 k_6^{-1}\,[1-\exp(-k_6 t)]$$

Where $k_5 = 3.0 \times 10^{11}$ mole/yr is the rate of nitrogen exhalation. Since the experimental conversion of nitrogen to HCN is about 12%, the fraction of nitrogen converted each year to HCN was: $k_6 = 2.0 \times 10^{-8}$. The amount of HCN at each time is given by:

$$\text{(g)} \quad [HCN]_t = 2(k_5 t - [N_2]_t)$$

Carbon monoxide and aldehydes accumulated according to:

$$\text{(h)} \quad [CO]_t = k_2[H_2O]t \quad \text{and (i)} \quad [RCHO]_t = k_7 [H_2O]t.$$

Where $k_2 = 1.66 \times 10^{-7}\ yr^{-1}$ and $k_2[H_2O] = 5.0 \times 10^{11}$ mole/yr for a surface temperature of 24^oC. Because of the ~5% yield for the formation of aldehydes, $k_7 = 8.3 \times 10^{-9}\ yr^{-1}$. It was shown experimentally that carbon dioxide is not formed in large amounts in the presence of hydrocarbons and then only when water vapor is in large excess. As long as acetylene and methane were present in the atmosphere, the water vapor oxidized them rather than the carbon monoxide.

As seen from Fig. 1, acetylene dominated the atmosphere only for 1.5×10^8 years, after which carbon monoxide dominated until 7.5×10^8 years, when the carbon dioxide surpassed it. The aldehydes and HCN were washed down and dissolved in the oceans, shortly after their formation, by the rain during the thunderstorms. They formed solutions of 0.007 and 0.038 molar concentration respectively. If all the aldehydes reacted with HCN and ammonia to give α-amino nitriles and ultimately amino acids, a 0.007 molar solution of amino acids would have resulted, i.e., 1×10^{19} moles of amino acids would have accumulated in the oceans. However, a certain fraction of the amino acids was decomposed in the oceans by UV light and a somewhat lower concentration could be expected. Some α-amino nitriles or amino acids could be formed directly in the thunder shock wave, provided that some ammonia were present, but obviously their final quantity would have been similar to the one calculated here. The remaining HCN was available for the synthesis of various complex nitrogen-bearing compounds of biological importance. The hydrogen liberated in all of these processes, up to 3.2×10^{21} moles, was added to the exhaled hydrogen and escaped through the exosphere. McGovern (22) shows that there was sufficient hydrogen in the atmosphere for 5 to 10×10^8 years to keep it in a reduced state. Hydrogen was prevented from escaping by a low exospheric temperature, which was regulated by energy transfer from hot hydrogen in the exosphere to hydrocarbons above the mesopause. These, in turn, radiated it out in the infrared. The presence of large amounts of acetylene in the atmosphere, as a result of thunder shock waves, augmented this energy transfer process and allowed a longer period under reducing conditions.

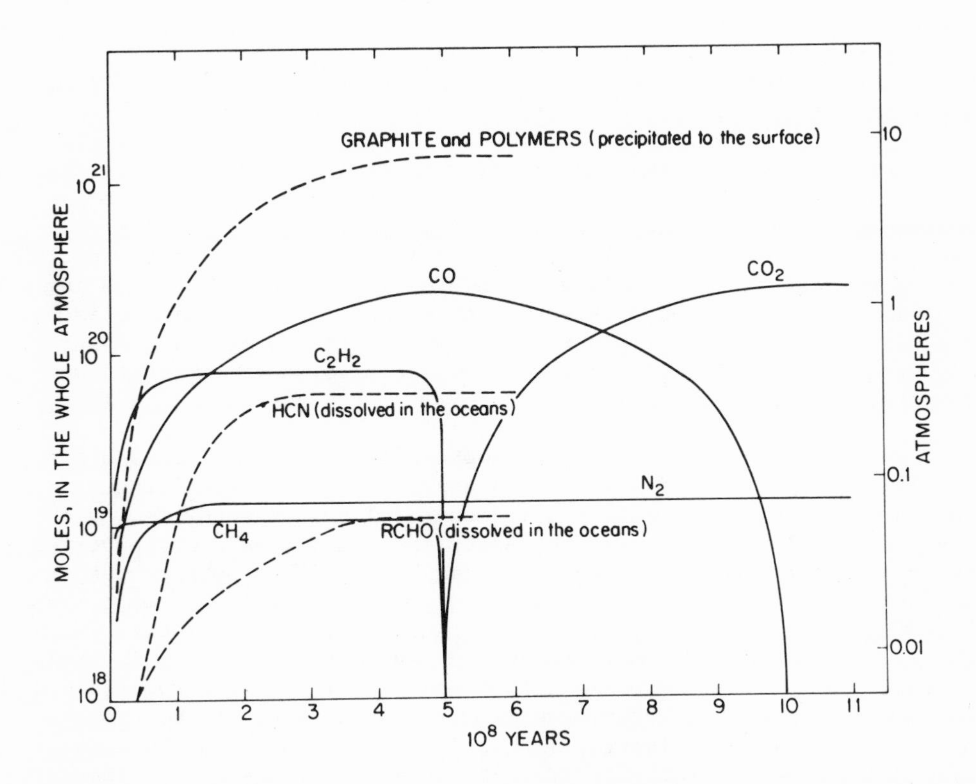

Figure 1. Variation with time of the concentration profiles of the more abundant species in the reducing atmosphere. A steady rate of exhalation during 5 x 10^8 years and a surface temperature of 24°C were assumed.

As stressed in the introduction, the assumption of a state of thermodynamic equilibrium in the atmosphere, as used by McGovern, is not fully justified. Nevertheless, his conclusions regarding the rate of hydrogen escape will not differ by much if a steady state be assumed for the concentrations of ammonia and hydrocarbons in the atmosphere.

Molecular nitrogen at a pressure of 0.08 bar together with 1.3 bar of carbon dioxide dominated the atmosphere after 1 x 10^9 years. It is interesting to note that the ratio N_2/CO_2 = 0.06 in Fig. 1 is quite similar to the ratio of 0.04 in the Cytherean atmosphere (27). Most of the carbon dioxide, however, was precipitated in minerals by reactions such a $CaMgSi_2O_6 + CO_2 \rightleftarrows MgSiO_3 + CaCO_3 + SiO_2$ (27) and a much lower atmospheric pressure of carbon dioxide prevailed.

The most significant feature of this evolutionary track of the primitive atmosphere is that changing the frequency and power of thunderstorms or the rate of exhalation, within reasonable limits, expands or contracts its time scale, while only small changes result in the concentrations of HCN and aldehydes. It should be noted in Fig. 1 that most of the exhaled carbon ended up on the surface as graphite and polymers. This result was obtained by assuming a surface temperature of 24°C; it would have been worse if the contemporary value of 15°C was used.

The great importance of the trophosphere's temperature and its resulting water content now becomes apparent. If the temperature and humidity were low, only small amounts of carbon monoxide would have accumulated and most of the acetylene would have been converted to graphite and polymers. On the other hand, if the surface temperature were 60°C, because of the strong greenhouse effect from acetylene and water vapor (23,27), the average water vapor pressure in the lower 5 km of the atmosphere would have been 60 torr. This would be sufficient, with the present frequency and power of thunderstorms, to convert most of the acetylene to carbon monoxide, at the expense of the graphite and polymers, with only small changes in the aldehyde and HCN concentrations. This is indeed very desirable since most of the carbon on Earth is now in the form of carbonate rocks, with only relatively small amounts of petroleum of possible abiogenic origin. (24) Very large quantities of solid and liquid material were hard to oxidize to carbon dioxide under the mild conditions on the surface of the Earth. A higher surface temperature, around 100°C, might have resulted in a situation similar to that of the Cytherian atmosphere; according to Rasool and De Bergh (27) Venus, because of its proximity to the Sun, was probably too hot for maintaining liquid water at the surface. Consequently, the very large water content of the atmosphere speeded up the oxidation to carbon dioxide which, in the absence of liquid water, did not precipitate as carbonate minerals. The carbon dioxide and water vapor in the atmosphere further increased the temperature, because of a strong greenhouse effect, preventing water condensation even under high atmospheric pressure. Venus was left, therefore, with a massive carbon dioxide atmosphere and a high surface temperature.

Up till now the evolution of the reducing atmosphere was treated as if thunderstorms were the sole energy source operating. To this, however, should be coupled the powerful solar radiation. Of all the species capable of UV absorption in the reducing atmosphere, only methane and acetylene were present in large concentrations. The water vapor concentration above the cloud layer was, like today, in the ppm range. Most of the ammonia was tied up in the oceans, forming NH_4^+ ions with the dissolved hydrogen chloride and hydrogen sulphide. The aldehydes and HCN were washed into the oceans, shortly after their formation, by the rain during the thunderstorms. Thus, the concentrations of all these species in the atmosphere were extremely low. The photochemical reactions of these minor atmospheric constituents certainly played an important role in the production of molecules of biological importance. Their influence on the gross properties of the reducing atmosphere, however, was only marginal, until a large amount of oxygen could be generated through water photodissociation in the upper atmosphere. This, however, could not have occurred during the first 5-10 x 10^8 years (22). Acetylene still has an appreciable absorption coefficient ($\sim$2 atm^{-1} cm^{-1}) around 2000 Å, while methane absorbs only up to 1440 Å (13). Since the black body curve of solar radiation falls off very steeply after 2000 Å, the rate of acetylene photolysis was much higher than that of methane. The products of methane photolysis, such as ethane (and acetylene) and smaller amounts of higher hydrocarbons (13), were also converted to acetylene by thunder shock waves and only added to the large acetylene concentration in the atmosphere. With no thunder shock waves to pyrolyze the higher hydrocarbons in the troposphere, their concentrations might have built up to quite high levels, leading to possible deposition of large quantities of oil on the surface (28). Photolysis of acetylene at 1845 Å yields ethylene, diacetylene, vinylacetylene, benzene, and polymers. The quantum yields, at a pressure of 1.67 torr, are 0.049, 0.106, 0.009, 0.068, and $\sim$2, respectively (4). With a UV flux, at λ <2000 Å, of 2 x 10^{13} photons/cm^2/sec (15), the acetylene atmosphere could have polymerized within around 10^5 years. This, however, did not occur. An acetylene layer about 1.5 km thick, at an altitude of 75 km, was optically thick to radiation at wave lengths below 2000 Å and shielded all the acetylene below it. At this altitude, the acetylene pressure was about 0.01 torr. Extrapolation of the quantum yields of acetylene polymerization to low pressures leads to a zero quantum yield below 1 torr (4). This is probably a result of efficient termination of the polymerization reaction by transfer or disproportionation (5). Ionizing radiation was also incapable of polymerizing acetylene at this low pressure. The products of ion-molecule reactions in acetylene at 0.01 torr are only C_3H_x, C_4H_x, C_5H_x, and some C_6H_x (16). The UV and ionizing radiation, then, did not polymerize the acetylene but produced only its other photoproducts which, in turn, reacted with nitrogen and water in a manner similar to acetylene, thus not altering to a great extent the course of evolution of the reducing atmosphere. It should be emphasized that this evolutionary track could not be followed in the Jovian atmosphere, since the large hydrogen concentration prevented the oxidation of hydrocarbons by water vapor. On the contrary, as shown in Table I, carbon monoxide is converted to methane and acetylene when shock-heated in the presence of large excess of hydrogen.

CONCLUSIONS

(1) Solar radiation hitting the surface of the primitive Earth created thunderstorms at a rate similar to the one on the contemporary Earth. The chemical reactions occurring at the high temperature and thermal quench periods behind thunder shock waves played a major role in the chemical evolution of the primitive reducing atmosphere.

(2) According to calculations, varying the frequency and power of thunderstorms or the rate of exhalation, within reasonable limits, resulted in expansion or contraction of the time scale of atmospheric evolution, with only small changes in the concentrations of the species important to further chemical evolution in the oceans. If, together with methane, some carbon monoxide and carbon dioxide were exhaled, the time required for reaching a carbon dioxide-nitrogen atmosphere was proportionally shorter and lower HCN and aldehyde concentrations were reached. Nevertheless, the overall picture was similar to the one described here.

(3) The surface and troposphere's temperatures were of prime importance to the course and outcome of the atmospheric evolution. Low temperature would have resulted in the formation of large carbon and polymer deposits on the Earth's surface and low aldehyde concentration in the oceans. Very high temperature would have led to a situation similar to that of the Cytherean atmosphere.

(4) The composition of the atmosphere was also of great importance. If high hydrogen concentration was maintained at all times, hydrocarbon oxidation products could not have accumulated, as in the Jovian atmosphere.

REFERENCES

1. Oparin, A.I., "Proiskhozhdenie zhizny," Izd. Moskovskii Rabochii, Moscow, 1924; "The Origin of Life", The Macmillan Company, New York, 1938.

2. Haldane, J.B.S., Rationalist Annual, 148 (1929).

3. Rubey, W.W., Geol. Soc. Am. Spec. Paper No. 62, p. 531 (1955).

4. Zelikoff, M. and Aschenbrand, M., J. Chem. Phys. 24, 1034 (1956).

5. Billmeyer, F.W., "Textbook of Polymer Chemistry", pp. 219-220, Interscience, New York, 1957.

6. Miller, S.L. and Urey, H.C., Science 130, 245 (1959).

7. Holland, H.D., in "Petrological Studies" (Engel, A.E.J., James, H.L., and Leonard, B.F., eds.), p. 447, Geological Society of America, 1962.

8. Hochstim, A.R., Proc. Nat. Acad. Sci. U.S. 51, 200 (1963); ibid., Nature 197, 624 (1963).

9. Bonnie, J., Heimel, S., Ehlers, J.G., and Gordon, S., "Thermodynamic Properties to 6000°K for 210 Substances Involving the First 18 Elements", NASA Report SP - 3001, 1963.

10. Kondrat'ev, V.N., "Chemical Kinetics of Gas Reactions", pp. 525-540, Pergamon Press, Oxford, 1964.

11. Schonland, B.F.J., "The Flight of Thunderbolts", 2nd ed., Clarendon Press, Oxford, 1964.

12. Battan, L.J.,"The Thunderstorm", New American Library, New York, 1964.

13. McNesby, J.R., and Okabe, H., Advances in Photochem. 3, 157 (1964).

14. Gay, I.D., Kistiakowsky, G.B., Michael, J.V., and Niki, H., J. Chem. Phys. 43, 1720 (1965).

15. Sagan, C., in "The Origins of Prebiological Systems " (Fox, S.W., ed.) , pp. 207, 238, Academic Press, New York, 1965.

16. Wexler, S., Lifshitz, A., and Quattrochi, A., Advances in Chem. Series 58, 193 (1966).

17. Handbook of Chemistry and Physics, 47th ed., Chemical Rubber Co., Cleveland, 1966.

18. Cited in: Shklovskii, I.S., and Sagan, C., "Intelligent Life in the Universe", p. 77, Holden-Day, San Francisco, 1966.

19. Bar-Nun, A., and Lifshitz, A., J. Chem. Phys. 47 2878 (1967).

20. Rao, V.V., Mackay, D., and Trass, O., Canad. J. Chem. Eng. 45, 61 (1967).

21. Anbar, M., Science 161, 1343 (1968).

22. McGovern, W.E., J. Atmos. Sci. 26, 623 (1969).

23. Ohring, G., Icarus 11, 171 (1969).

24. Calvin, M., "Chemical Evolution", p. 58, Oxford University Press, New York, 1969.

25. West, M.W., and Ponnamperuma, C., Space Life Sci. 2, 225 (1970); Ibid., 3, 293 (1972).

26. Bar-Nun, A., Bar-Nun, N., Bauer, S.H., and Sagan, C., Science 168, 470 (1970); Bar-Nun, A., Space Life Sci., in press (1974).

27. Rasool, S.I., and De Bergh, C., Nature 226, 1037 (1970).

28. Lasaga, A.C., and Holland, H.D., Science 174, 53 (1971).

29. Bar-Nun, A., and Tauber, M., Space Life Sci. 3, 254 (1972).

30. Miller, S.L., and Orgel, L.E., "The Origins of Life on Earth", Prentice-Hall, Inc., Engelwood Cliffs, New Jersey, 1974.

ACKNOWLEDGEMENTS

The authors gratefully acknowledge helpful and stimulating discussions with Professors J. Heicklen and A. Lifshitz.

ALEXANDER IVANOVITCH OPARIN AND THE ORIGIN OF LIFE: A RECOLLECTION AND APPRECIATION

C. B. van Niel

Hopkins Marine Station of Stanford University

Pacific Grove, California 93950

In 1928, shortly before leaving Holland to assume a position at the Hopkins Marine Station of Stanford University, I went to pay a visit to Professor M. W. Beijerinck, whose courses at the Technological University in Delft had introduced me to the science of microbiology and kindled my fascination with that field. Following his compulsory retirement in 1921, he had moved to a home in the country, and I had not seen him since then. Now I wanted to express my deep appreciation for the great and lasting influence his teaching had had on my life.

Toward the end of this long and to me exciting visit, the old Master inquired about my plans for research in the new environment, and confronted me with the question: Did I intend to work on the fundamental problem of spontaneous generation? My response had been in the negative because, as I was summoned to explain, it seemed too difficult an undertaking for me. His retort was characteristic: "Well, maybe you are right. But you must always remember that you won't amount to much if you don't end up studying that problem!" It was a never-to-be-forgotten challenge.

At that time I was not aware of the fact that, four years earlier, an epoch-making contribution to the subject had been published in the U.S.S.R.; nor, I suspect, was Beijerinck. I first became acquainted with Alexander Ivanovitch Oparin's ideas on the origin of life after the appearance, in 1938, of S. Morgulis' English translation of Oparin's treatise of 1936, a considerably expanded version of his earlier paper. It is not too much to say that its impact was momentous, for it immediately became clear that the Russian biochemist had discovered a scientifically plausible and wholly appealing way out of what had become a tantalizing impasse. The latter had resulted from the many experiments, notably those of Pasteur and Tyndall, which had convincingly refuted the validity of claims purporting to show that bacteria, considered to represent the simplest forms of life on Earth, could be generated spontaneously.

Oparin had resolved the dilemma by postulating that the appearance of living organisms had been preceded by, and was itself a logical perpetuation of, physical and chemical evolutions in the universe, during which systems of ever increasing complexity were generated out of simpler ones. This sweeping concept marks the ultimate extension of Darwin's principle of biological evolution. It has the further advantage of circumventing the difficulty of having unambiguously to define "life" other than as a state of matter at a particular level of complexity. Thus it can readily accommodate divergent opinions as to precisely where the line of demarcation between living and non-living should be drawn.

Now, half-a-century after Oparin first formulated his theory, a vast amount of experimental work has yielded an imposing body of information which firmly supports his basic thesis. And it is fortunate that, during his lifetime, Academician Oparin has been able to witness the far-reaching influence his brilliant concept and research have exerted on subsequent advances.

Save by expounding the significance and implications of Oparin's grand generalization and its experimental foundations on every suitable occasion, I have not contributed to the spectacular developments in this area. But Beijerinck's exhortation has had the inevitable effect that I have ever been alert to, and closely followed, them. It has been a profoundly satisfying and stimulating intellectual experience.

For having initiated the theoretical and experimental approaches to this branch of scientific endeavor, all of us owe a debt of gratitude to its originator, on a par with that due to those other great innovators who have opened up novel and wide-ranging vistas to our understanding of nature.

THE ORIGIN AND EVOLUTION OF ENZYME CATALYSIS

M. D. Williams and J. L. Fox

Department of Zoology
University of Texas
Austin, Texas 78712

INTRODUCTION

The semicentennial anniversary of Professor Oparin's first reading of a theory on the origin of life (1) honors the many opportunities and stimuli afforded for experimental correlation, such as we see active in the field today.

This article will examine the basis, origin, and evolution of enzyme function. A few years ago we discussed our related views on origins (2). How our ideas relate to Oparin's period of chemical evolution (3) is not clear. Most of the discussion of enzyme evolution will assume a strict genetic control of the amino acid sequence during protein synthesis. Whether such genetic control was a necessary prerequisite for a living cell is a debatable topic that we will not pursue here.

A discussion of this sort, as is true of most discussions of the origin of life, is largely unconstrained due to the paucity of "factual" data. As with classical thermodynamics, we can make some observations of the end-points of enzyme evolution, but we know little of the path taken between those points. Contemporary biology presents the modern end-point for this evolutionary process. Deductions from scant data help us to guess at properties of the primordial end-point. Even less is known about the paths between these points.

ORIGIN OF ENZYMES

The Oparin theory of the origin of life describes a scenario with a hot and scant reducing atmosphere on the primordial Earth (3). It is reasonably apparent that to produce a living cell, macromolecular synthesis and the packaging within a cell-like structure of these macromolecules were essential. The experiments with thermal and adenylate proteinoid offer one of the few experimentally examined systems for primordial type enzymic activities (4). The inferences drawn from that work are:

1) Proteinoid possesses a very limited set of amino acid sequences, far from the astronomical random sequence numbers calculated by numerologists.

2) Proteinoids, of a fairly "homogeneous" nature, possess a variety of catalytic activities. Those weak catalytic powers are less specific than what is found in the enzymes of today.

3) Varying the amino acid contents of the proteinoids yields different enzymic properties.

4) Michaelis-Menten type kinetics appear to be operative. The primary meaning of this is that an enzyme-complex intermediate is formed during the course of the reaction with substrate.

Thus, it is possible both experimentally and logically to infer that protobionts possessed a limited number of proteins which at some instant could be classified in a finite number of ways, and each of which possessed a not-totally specific set of catalytic potentials of a weak nature. In the absence of a strictly regulated mechanism for protein synthesis, it is difficult to discuss evolution of function. For this reason, the following discussion will assume a functioning genetic apparatus. Some discussion is possible in the absence of a genetic apparatus, but little evidence exists to support such arguments.

BASIS OF ENZYME CATALYSIS

We will spend some time at this point to develop what catalysis represents mechanistically, based on recognition of the fact that proteins are the possessors of this function. A number of discussions of enzyme catalysis exist in the literature, e.g. Jencks (5), Henderson and Wang (6), Leinhard (7), and Kirsch (8). Many of these discussions revolve around the origin of energetic terms in mathematical representations of catalysis, but fail to present mechanistic detail. They are all essentially based on binding and orientation of the substrates by the enzyme. Where one draws the line between binding and orientation determines which description is followed in detail. We contend that the arguments are basically semantic.

Several corollaries of a contemporary enzyme may be cited:

a) Enzymes accelerate reaction rates on the order of 10^4 to 10^{14} times, i.e. they lower activation energy barriers (5).

b) Enzymes control reactions, i.e. they can often be modulated to alter their catalytic rates as an adaptation to changing conditions.

c) Enzymes are released at the end of the reaction essentially unmodified from the form in which they entered the reaction.

The acceleration of reaction rates is poorly understood. It is generally agreed that the upper rates of enzyme function are diffusion-limited. However, in most cases the Michaelis-Menten formulation, or the Briggs-Haldane formalism (9) of enzyme kinetics requires that many more enzyme-substrate associations dissociate and yield no product than those which overcome the activation barrier and produce product(s). And yet this process of product formation is orders of magnitude more probable for an enzyme-catalyzed reaction than for a diffusion-limited reaction in solution.

We propose that the phase correlation of the oscillating magnetic and electric dipoles of the substrate(s) and enzyme are the important mechanistic feature of catalyzed reactions (10). The binding of substrate to enzyme is a generally favorable process yielding a lower energy state for the system. One could plot a three-dimensional topological map to demonstrate this effect, if desired. The radiations of the oscillating substrate and enzyme dipoles that enhance movement down the sides of the binding energy well are much more effective at fairly short ranges, so diffusion-limited kinetics must obtain. However, at close contact the tendency to adapt to a coherent phase relationship will be appreciable. It is likely that this will require alterations in both the substrate and the enzyme, although the alteration would be expected to be proportionally greater for the substrate. Bond distortions resulting from basically non-covalent forces will be responsible for these alterations although this model also takes into account covalently linked intermediates as well.

What we have described so far is roughly the same as E. Fischer's lock-and-key model for enzyme-substrate interactions proposed nearly eighty years ago (11). In more modern form it conforms to Koshland's induced fit (12) and orbital steering (13) models as well.

Phase correlation of the reactants is essential if reactivity is to occur. Kineticists generally describe the successful reaction between two atoms as occurring within a single coherent oscillation (14). If forces act to increase the likelihood of phase correlation, the successful rate of reaction should increase. These forces are any that enhance phase correlation at the level of electromagnetic radiation, or of electrons, atoms, and molecules. At the electron level this would include resonance stabilization by electron delocalization. Our phasing model fits these requirements. We can also then propose that the more our reactants and products conserve their levels of symmetry, the more likely it is that phase correlation will be observed, and hence, the greater the likelihood of reaction will be. In this sense, we can also point out that the Woodward-Hoffman rules of reactivity (15) and the Wyman, Monod, and Changeux rules of symmetry (16) are followed.

A consequence of the reaction occurring through binding to an enzyme is that the energy released by product formation may be dissipated to the enzyme first instead of being lost to solvent, as is the case for uncatalyzed reactions. It is tempting to speculate that, within the microenvironment of the active site of an enzyme, this dissipative energy can be used to help stabilize the enzyme structure. The rate of dissipation of energy through the protein would be a function of the levels of organization at various locations. Proteins are often considered to possess both fluid and highly rigid regions. Dissipation of energy through a fluid region would be slower. The intersystem crossover barrier could rather easily prove to be rate-limiting for this process. This leads us to conclude that catalysis could provide energy that extends the average lifetime of the native conformation of the enzyme. The validity of this statement for all enzymes is not clear, but it is reasonable for those catalyzing exothermic reactions. This would help explain in mechanistic detail the function of enzymes as dissipative structures as described by non-equilibrium thermodynamics (17). One of the properties predicted by a non-equilibrium thermodynamic analysis of enzymes is that a considerable energy throughput is expended merely to maintain the organization of these metastable states. This is how the notion of dissipative structures arises.

EVOLUTION OF ENZYME FUNCTION

Contemporary enzymes are noted for their specificity and catalytic powers. We believe that, through mutation of duplicated structural genes (several mutational mechanisms can yield duplicated genes), more specific and more highly catalytic proteins evolved from a simple, polyfunctional, and weakly catalytic set of enzymes. The level of evolution and specificity obtained by an evolving enzyme would determine the remaining level of its other activities. It is quite possible that so-called "contaminating" levels of activity observed in the laboratory within "pure" enzyme preparations may be low-level intrinsic levels of other catalytic functions due to the evolutionary history of this "pure" protein itself, rather than to small amounts of other proteins.

The natural selection process for mutant enzymes would require increased efficiency of catalysis once sufficient numbers of proteins were available for the cell's reactions. At later stages of evolution, adaptability would become increasingly important as competition and diversity increased. This would have led to the emergence of modulation of enzyme activity.

At the level of the enzyme, evolutionary selection would occur for:

1) Increasing the depth and/or breadth of the enzyme-substrate complex binding-energy well. Two aspects are important here. First, alterations in the enzyme active-site structure which would enhance the phase correlation between the substrate and the enzyme would deepen the binding-energy well, and perhaps, increase repulsion of other molecules at short distances. Secondly, structural modifications which would enhance the ability of the enzyme to alter its conformation upon close interaction with the substrate would increase binding stability.

2) Decreasing the activation energy barrier. Modifications of the protein which enhance coherence between the substrates would create larger reaction probabilities and hence faster reaction rates. This would cause a lowering of the requirements for energies necessary to achieve the proper reactive substrate conformations. An alternative statement of this is that in a coherent or resonant state there is a minimal energy input required to maintain the coherence.

3) Enhancing the relative dissociation constant for product(s) as opposed to reactant(s). A subtle set of influences is essential here. We require that the substrate and product possess similar symmetries to enhance reaction rates. Yet, we need to free the enzyme of product as quickly as possible to optimize the amount of time the enzyme is involved in substrate binding and reaction. A small conformational change following the catalytic step would benefit this process. The next corollary best explains how this could occur.

4) Selecting energy dissipative pathways. The crossing of the activation energy barrier hump is attended by energy release. With the reaction restricted to the environment of the enzyme, the released energy will dissipate to its immediate surroundings. These higher energy states within the enzyme may yield small modifications of local conformation as the energy progressively dissipates through the enzyme from the active site. It is tempting to speculate that

the dissipative energy pathway may serve to help maintain macromolecular structure. We would conclude that a "functioning" enzyme would possess longer term stability than an "unused" one.

5) Symmetry adapting modifications which would allow subunit interactions for pathways in which modulation is advantageous. All popularly known modulatory enzymes are multimeric (9). The complex structures possess various levels of lower orders of symmetry than their component protomeric units. Such a complex structure would tend to exhibit a higher level of coherence within its structure and could thus be more effective in its interactions with substrate molecules.

CONCLUSIONS

We have presented a model of enzyme function that emphasizes phase correlation between substrate and enzyme magnetic and electric dipoles. We have shown how many concepts of enzyme function proposed earlier by others can easily be assimilated into this general picture and we emphasized where evolutionary changes could enhance catalytic activity.

At the enzyme level, attempts to correlate our theory to experimental data of primary protein structure levels are known and in the cases of such proteins as ferredoxin, cytochrome c and hemoglobin, much comparative data are available (18). However, at the three-dimensional level, new insights have been revealed by M. Rossman and his group (19). Rossman has shown that nucleotide-binding proteins such as flavodoxins (single chain), malate dehydrogenase (dimeric), and lactate dehydrogenase (tetrameric) possess largely homologous tertiary structures, especially about the nucleotide binding site. There are no primary structural homologies apparent between these proteins. Thus, primary structural data are shown to be of quite limited value when looking at the evolution of different but related enzyme functions. This has been rather clear in phylogenetic studies based upon primary structural homology comparisons, since the merit of cytochrome c and hemoglobin derives from their conservative mutation rates.

This analysis may lead to new approaches to enzyme function studies and can, perhaps, permit a modeling which will in turn permit new insights into the specialization of enzymes during evolution.

ACKNOWLEDGMENTS

We wish to thank an NIH Training Grant for support (MDW) and the NSF (JLF) research support which led to these proposals.

REFERENCES

1. Oparin, A. I., "Proiskhozhdenie zhizni," Izd. Moskovskii Rabochii, Moscow, 1924.

2. Fox, J. L., in: "Molecular Evolution: Prebiological and Biological" (Rohlfing, D. L., and Oparin, A. I., eds.), Plenum Press, New York, 1972.

3. Oparin, A. I., "The Origin of Life," 2nd ed.,Dover Publications, New York, 1938.

4. Rohlfing, D. L., and Fox, S. W., Advances Catal. 20, 373 (1969); Fox, S. W., and Dose, K., "Molecular Evolution and the Origin of Life," W. H. Freeman, San Francisco, 1972.

5. Jencks, W. P., "Catalysis in Chemistry and Enzymology," McGraw-Hill, New York, 1969.

6. Henderson, R., and Wang, J. H., Ann. Rev. Biophys. Eng. 1, 1 (1972).

7. Leinhard, G. E., Science 186, 149 (1973).

8. Kirsch, J. F., Ann. Rev. Biochem. 42, 205 (1973).

9. Lehninger, A. L., "Biochemistry," Worth, New York, 1970.

10. Fröhlich, H., Nature 228, 1093 (1970).

11. Fischer, E., Chem. Ber. 27, 2985 (1894).

12. Koshland, D. E., Jr., Proc. Nat. Acad. Sci. U.S. 44, 98 (1958).

13. Storm, D. R., and Koshland, D. E., Jr., Proc. Nat. Acad. Sci. U.S. 66, 445 (1970).

14. Frost, A. A., and Pearson, R. G., "Kinetics and Mechanism," 2nd ed., ed., John Wiley, New York, 1961.

15. Woodward, R. B., and Hoffman, R., Angew. Chem. Int. Ed. 8, 781 (1969).

16. Monod, J., Wyman, J., and Changeux, J. P., J. Mol. Biol. 12, 88 (1965).

17. Prigogine, I., Nicolis, G., and Babloyantz, A., Physics Today 25 (11) 23 and 25(12), 38 (1972).

18. Dayhoff, M. O., "Atlas of Protein Sequence and Structure," Vol. 5, National Biomedical Research Foundation, Washington, D.C., 1972.

19. Rao, S. T., and Rossman, M. G., J. Mol. Biol. 76, 241 (1973).

OPARIN AND THE ORIGIN OF LIFE:

COSMOLOGICAL CONSIDERATIONS

Richard S. Young
Planetary Biology, Planetary Programs
Office of Space Science, National Aeronautics and Space Administration
Washington, D.C. 20546

A. I. Oparin has considered the problem of the origin of life since the early 1920s. In 1924 he published a small booklet entitled, "Proiskhozhdenie Zhizni," where he first put forth his views in which the origin of life appears as the gradual evolution of organic substances. In 1936 he published his second book, "Vozniknovenie Zhizni Na Zemle" (The Origin of Life on the Earth). The second book elaborated and expanded his views and attempted to collate data from astronomy, geology, and biochemistry with his views on the origin of life. The striking thing about Oparin's early views on the origin of life problem was that indeed they were cosmological in scope. He considered the origin of life to be an integral part of the early evolution of a planet; in the 1920s and 1930s this was hardly a widely held view. His study of the literature available on the nature of the stars, planets, meteorites, and moons was very thorough in view of the amount of data available at that time. From these early papers he was able to draw some remarkably cogent conclusions which, even by comparison with the tremendous amount of data available today from telescopic and spacecraft observations, holds up remarkably well.

He considers the concept of panspermia, first put forth by Richter and later by Arrhenius, but rejects it after careful consideration. He recognized very early that the panspermia hypothesis really offers no solution to the origin of life problem, although it could, in principle, contribute to the question of the origin of life on Earth.

In the early 1900s, it was generally felt that carbon on the surface of the Earth was available in its most completely oxidized form, therefore virtually incapable of further chemical transformation.

Oparin was among the first to use the growing literature of astronomical and geological data to refute this assumption. He noted in his early books, for example, that even at the comparatively high temperatures of the surface of stars, atoms of carbon commence to unite to produce C_2, or they unite with nitrogen or hydrogen to produce CN or CH, and that these were the typical forms of carbon to be seen spectroscopically in stellar spectrum, and in comets. In this context it is interesting to note that many of the data on stars, planets, and comets which we read about in papers in our most recent journals were already available in papers published in the early 1930s. Of course, it should also be pointed out that many of the data available in the 1930s were quite incorrect. In his 1936 book, Oparin points out that in 1934 it was known that Mercury had little or no atmosphere and that Venus had an extremely dense atmosphere through which the surface of the planet was not visible, that neither oxygen nor water had been detected in the atmosphere of Venus, and that large quantities of carbon dioxide were present. He also points out that water can be seen in the very thin atmosphere of Mars although very little oxygen is present. This was thought by Wildt to be due to the formation of ozone under the influence of ultraviolet rays. It was also reported at that time (incorrectly) that no carbon dioxide existed on Mars.

It was known as early as 1932 that the atmospheres of Jupiter and Saturn contained ammonia and methane. These identifications were confirmed in 1934 by Adel and Slipher. Oparin was also well aware of the presence of carbon-carbon and carbon-hydrogen compounds in meteorites, which had been demonstrated earlier, and in fact in his book in 1936 he stated that carbon is found either in its elementary form or in the form of compounds with nitrogen (CN) but more often with hydrogen in the atmosphere of large planets, comets, the Sun, and meteorites. He also pointed out that in this respect only the Earth and Venus offer exceptions in that their atmospheres contain carbon dioxide. He concluded, therefore, that the carbon dioxide in the Earth's atmosphere was of secondary origin and that carbon first appeared on the Earth's surface in a reduced form, particularly as hydrocarbons. Oparin also concluded that the molecular oxygen found in our present-day atmosphere on the Earth was formed secondarily as a result of biological activity. Thus by looking at the question of the origin of life on a cosmological scale, Oparin was able to conclude in the 1920s and early 1930s that, "In the process of formation of our planet from the original incandescent mass of gas, heavy clouds of carbon must have very quickly condensed into drops or even solid particles and entered the primitive nucleus of the Earth in the form of a carbonaceous rain or snow. There the carbon came into immediate contact with the elements of heavy metals forming the nucleus, primarily with iron which constitutes such an essential component of the central core of our present Earth. Mixed with the heavy metals, the carbon reacted chemically as the Earth cooled off, whereby carbides were produced, which are the carbon compounds most stable at high temperatures. The crust of primary igneous rocks which were formed subsequently separated the carbides from the Earth's atmosphere. The atmosphere at that period differed materially from our present atmosphere in that it contained neither oxygen nor nitrogen gas, but was filled instead with superheated aqueous vapor. The crust separating the carbides from this atmosphere still lacked rigidity to resist the gigantic tides of the interior molten liquid mass, caused by the attractive forces of Sun and Moon. The thin layer of igneous rock would rupture during these tides and through the crevices so formed the molten liquid mass from the interior depths would spread over the Earth's surface. The superheated aqueous vapor of the atmosphere coming in contact with the carbides reacted chemically, giving rise to a great variety of derivatives (alcohols, aldehydes, ketones, organic acids, etc.) through oxidation by the oxygen component of water. At the same time, these hydrocarbons also reacted with ammonia which appeared at that period on the surface of the Earth. Thus amides, amines and other nitrogenous derivatives originated."

It was no accident, then, that Oparin was able to produce a truly new insight into the origin-of-life question. It was based largely on data scattered throughout the literature, most of which was on the frontier of its discipline. It is worth noting how well the concept has withstood the test of time and the flood of experimental data in the last 20 years, and in fact how much of the early astronomical data which went into his thinking has been "re-discovered" or confirmed in recent years. The student of Oparin would do well to reread his early books.

LIST OF CONTRIBUTORS TO

THE ORIGIN OF LIFE AND EVOLUTIONARY BIOCHEMISTRY

ALLAKHVERDOV, B. L., Department of Plant Biochemistry, Biological Faculty, Moscow University, Biophysical Institute, USSR Academy of Sciences (Pushchino)

BAKARDJIEVA, N., Department of Plant Physiology, University of Sofia, Bulgaria and Institute of Plant Physiology, Bulgarian Academy of Sciences, Sofia

BALTSCHEFFSKY,H., Department of Biochemistry, Arrhenius Laboratory, University of Stockholm, S-104 05, Stockholm, Sweden

BAR-NUN, A., Department of Physical Chemistry, The Hebrew University, Jerusalem, Israel

BJORNSON, L. K., University Hospital, New York Medical Center, New York, New York 10016

BRESLER, S. E., Leningrad Institute of Nuclear Physics, Academy of Sciences of USSR, Moscow, USSR

BUVET, R., Laboratoire d'Energétique Electrochimique et Biochimique, Centre pluridisciplinaire, Avenue du Général de Gaulle 94000, Creteil (France)

CAIRNS-SMITH, A. G., Chemistry Department, University of Glasgow, Glasgow, G12 8QQ, Scotland

CALVIN, M., Laboratory of Chemical Biodynamics, Lawrence Berkeley Laboratory, University of California, Berkeley, California 94720

CAMMACK, R., Department of Plant Sciences, University of London King's College, London SE24 9JF

DEBORIN, G. A., A. N. Bakh Institute of Biochemistry, USSR Academy of Sciences, Moscow, USSR

DOSE, K., Institut für Biochemie, Universität Mainz, D-6500 Mainz BRD

EGOROV, I. A., Bakh Biochemical Institute, USSR Academy of Sciences, Moscow, USSR

EVREINOVA, T. N., Department of Plant Biochemistry, Biological Faculty, Moscow University, Biophysical Institute, USSR Academy of Sciences (Pushchino)

EVSTIGNEEV, V. B., Institute of Photosynthesis, USSR Academy of Sciences, Moscow, Putschino, USSR

EVSTIGNEEVA, Z. G., A. N. Bakh Institute of Biochemistry, USSR Academy of Sciences, Moscow, USSR

FERRIS, J. P., Department of Chemistry, Rensselaer Polytechnic Institute, Troy, New York

FOX, J. L., Department of Zoology, University of Texas, Austin, Texas 78712

FOX, S. W., Institute for Molecular and Cellular Evolution, University of Miami, Coral Gables, Florida 33134

GEORGIEV, G., Department of Plant Physiology, University of Sofia, Bulgaria and Institute of Plant Physiology, Bulgarian Academy of Sciences, Sofia

GROTH, W., Institute for Physical Chemistry, The University of Bonn

HALL, D. O., Department of Plant Sciences, University of London King's College, London SE24 9JF

HALMANN, M., Isotope Department, Weizmann Institute of Science, Rehovot, Israel

HARADA, K., Institute for Molecular and Cellular Evolution and Department of Chemistry, University of Miami, Coral Gables, Florida 33134

HASEGAWA, M., Department of Biophysics and Biochemistry, Faculty of Science, University of Tokyo, Hongo, Tokyo, Japan

KENYON, D. H., Department of Cell and Molecular Biology, California State University, San Francisco, California 94132

KEOSIAN, J., Marine Biological Laboratory, Woods Hole, Massachusetts 02543

KRASNOVSKY, A. A., A. N. Bakh Institute of Biochemistry, Academy of Sciences of the USSR, Moscow, USSR

KRETOVICH, W. L., A. N. Bakh Institute of Biochemistry, USSR Academy of Sciences, Moscow, USSR

KRITSKY, M. S., A. N. Bakh Institute of Biochemistry, USSR Academy of Sciences, Moscow, USSR

KULAEV, I. S., Institute of Biochemistry and Physiology of Microorganisms, USSR Academy of Sciences, Pushchino, USSR

KURSANOV, A. L., K. A. Timiryazev Institute of Plant Physiology of the USSR Academy of Sciences, Moscow

KVENVOLDEN, K. A., Chemical Evolution Branch-Planetary Biology Division, Ames Research Center, NASA, Moffett Field, California 94035

LACEY, J. C., JR., Laboratory of Molecular Biology, University of Alabama in Birmingham, University Station, Birmingham, Alabama 35294

LEMMON, R. M., Laboratory of Chemical Biodynamics, Lawrence Berkeley Laboratory, University of California, Berkeley, California 94720

LIPMANN, F., The Rockefeller University, New York, New York 10021

MITZ, M. A., National Aeronautics and Space Administration, Washington, D.C.

MIZUTANI, H., Department of Biophysics and Biochemistry, Faculty of Science, University of Tokyo, Hongo, Tokyo, Japan

MULLINS, D. W., JR., Laboratory of Molecular Biology, University of Alabama in Birmingham, University Station, Birmingham, Alabama 35294

NICODEM, D. E., Department of Chemistry, Rensselaer Polytechnic Institute, Troy, New York

NODA, H., Department of Biophysics and Biochemistry, Faculty of Science, University of Tokyo, Hongo, Tokyo, Japan

NOVAK, V.J.A., Czechoslovak Academy of Sciences, Prague, Czechoslovakia

OKIHANA, H., Department of Biophysics and Biochemistry, Faculty of Science, University of Tokyo, Hongo, Tokyo, Japan

ORGEL, L. E., The Salk Institute for Biological Studies, San Diego, California 92112

PAECHT-HOROWITZ, M., Polymer Department, The Weizmann Institute of Science, Rehovot, Israel

PAVLINOVA, O. A., K. A. Timiryazev Institute of Plant Physiology of the USSR Academy of Sciences, Moscow, USSR

PAVLOVSKAYA, T. E., A. N. Bakh Institute of Biochemistry, Academy of Sciences of the USSR, Moscow, USSR

PESHENKO, V. I., Department of Plant Biochemistry, Biological Faculty, Moscow University, Biophysical Institute, USSR Academy of Sciences (Pushchino).

RAO, K. K., Department of Plant Sciences, University of London King's College, London SE24 9JF

ROHLFING, D. L., Department of Biology, University of South Carolina, Columbia, South Carolina 29208

RUBIN, A. B., Moscow State University, Moscow, USSR

RUBIN, B. A., Faculty of Biology, Moscow State University, Moscow, USSR

SCHWARTZ, A. W., Department of Exobiology, University of Nijmegen, The Netherlands

SERGIYENKO, I. Z., Bakh Biochemical Institute, USSR Academy of Sciences, Moscow, USSR

SHAVIV, A., Department of Physical Chemistry, The Hebrew University, Jerusalem, Israel

SOROKINA, A. D., A. N. Bakh Institute of Biochemistry, USSR Academy of Sciences, Moscow, USSR

VAN NIEL, C. B., Hopkins Marine Station of Stanford University, Pacific Grove, California 93950

WILLIAMS, M. D., Department of Zoology, University of Texas, Austin, Texas 78712

YANO, T., Department of Biophysics and Biochemistry, Faculty of Science, University of Tokyo, Hongo, Tokyo, Japan

YOUNG, R. S., Planetary Biology, Planetary Programs, Office of Space Science, National Aeronautics and Space Administration, Washington, D.C. 20546